ADVANCES IN CHEMICAL PHYSICS

VOLUME 126

Advances in
CHEMICAL PHYSICS

Edited by

I. PRIGOGINE

Center for Studies in Statistical Mechanics and Complex Systems
The University of Texas
Austin, Texas
and
International Solvay Institutes
Université Libre de Bruxelles
Brussels, Belgium

and

STUART A. RICE

Department of Chemistry
and
The James Franck Institute
The University of Chicago
Chicago, Illinois

VOLUME 126

AN INTERSCIENCE PUBLICATION
JOHN WILEY & SONS, INC.

For general information on our other products and services please contact our Customer Care
Department within the U.S. at 877-762-2974, outside the U.S. at 317-572-3993 or fax 317-572-4002.

Wiley also publishes its books in a variety of electronic formats. Some content that appears in print,
however, may not be available in electronic format.

Library of Congress Catalog Number: 58-9935

ISBN 0-471-23582-2

Printed in the United States of America

10 9 8 7 6 5 4 3 2 1

CONTRIBUTORS TO VOLUME 126

DAVID M. BISHOP, Department of Chemistry, University of Ottawa, Ottawa, Canada

ALFREDO CÁRDENAS, Department of Computer Science, Cornell University, Ithaca, New York, U.S.A.

BENOÎT CHAMPAGNE, Department of Chemistry, University of Ottawa, Ottawa, Canada. Permanent address: Laboratoire de Chimie Théorique Appliquée, Facultés Universitaires Notre-Dame de la Paix, Namur, Belgium

WILLIAM T. COFFEY, Department of Electronic and Electrical Engineering, School of Engineering, Trinity College, Dublin, Ireland

CLIFFORD E. DYKSTRA, Department of Chemistry, Indiana University–Purdue University at Indianapolis, Indianapolis, Indiana, U.S.A.

RON ELBER, Department of Computer Science, Cornell University, Ithaca, New York, U.S.A.

AVIJIT GHOSH, Department of Computer Science, Cornell University, Ithaca, New York, U.S.A.

YURI P. KALMYKOV, Centre d'Etudes Fondamentales, Université de Perpignan, Perpignan, France

MARTIN H. MÜSER, Institut für Physik, Johannes Gutenberg-Universität, Mainz, Germay. Current address: Department of Applied Mathematics, University of Western Ontario, London, Ontario, Canada.

MARK O. ROBBINS, Department of Physics and Astronomy, The Johns Hopkins University, Baltimore, Maryland, U.S.A.

HARRY A. STERN, Department of Computer Science, Cornell University, Ithaca, New York, U.S.A.

SERGEY V. TITOV, Institute of Radio Engineering and Electronics of the Russian Academy of Sciences, Fryazino, Moscow Region, Russian Federation

MICHAEL URBAKH, School of Chemistry, Tel Aviv University, Tel Aviv, Israel

INTRODUCTION

Few of us can any longer keep up with the flood of scientific literature, even in specialized subfields. Any attempt to do more and be broadly educated with respect to a large domain of science has the appearance of tilting at windmills. Yet the synthesis of ideas drawn from different subjects into new, powerful, general concepts is as valuable as ever, and the desire to remain educated persists in all scientists. This series, *Advances in Chemical Physics*, is devoted to helping the reader obtain general information about a wide variety of topics in chemical physics, a field that we interpret very broadly. Our intent is to have experts present comprehensive analyses of subjects of interest and to encourage the expression of individual points of view. We hope that this approach to the presentation of an overview of a subject will both stimulate new research and serve as a personalized learning text for beginners in a field.

I. PRIGOGINE
STUART A. RICE

CONTENTS

INTERMOLECULAR INTERACTION: FROM PROPERTIES TO POTENTIALS AND BACK

CLIFFORD E. DYKSTRA

Department of Chemistry, Indiana University–Purdue University at Indianapolis, Indianapolis, Indiana, U.S.A.

CONTENTS

Advances in Chemical Physics, Volume 126, Edited by I. Prigogine and Stuart A. Rice.
ISBN 0-471-23582-2. © 2003 John Wiley & Sons, Inc.

I. INTRODUCTION

The subject of intermolecular interaction, and specifically the weak noncovalent attraction among molecules, is fascinating because it comprises a chemistry of its own, one in which whole, essentially intact molecules are the building blocks instead of atoms. Clusters of molecules exist via the attractive interactions; and there are, or likely will be, rules for cluster structure akin to the structural rules for the bonding of atoms. The subject is a sizable piece of molecular physics, one that has been reviewed in different ways for about a half-century, if not longer, and a number of those reviews and monographs [1–17] provide a useful context for this report. An always-present research objective in the study of inter-molecular interaction is developing interaction potentials, because these encapsulate the understanding we have of the fundamental aspects of interaction. Also, interaction potentials are the tools for the next level of study of this chemistry, namely, the dynamics of clusters and energetics of bulk systems. It is model interaction potentials that are the focus of this review, not the whole of intermolecular interaction phenomena.

Intermolecular interaction potentials have application at the junction of molecule-building-block chemistry and atom-building-block chemistry. Study-ing the influence of surrounding molecules or solvent molecules on chemical reactions or other atom-building-block phenomena requires knowledge of the weak interaction of the surrounding molecules with the reactive species. This brings forth an important issue for the "industry" of intermolecular interaction potential construction, that of level of detail. A level of detail for an interaction potential that may be tractable for the intricate dynamics of a two-molecule cluster probably won't be tractable for the study of an extended aggregation of molecules—that is, a solvated system. Detail translates into accuracy, even if not directly; hence, the question in front of anyone pursuing the construction of intermolecular interaction potentials is, "What level of accuracy is needed?" There is no single answer, of course, because the potentials are used in ways ranging from those that require the highest attainable accuracy to those that require, or seem to require, merely a coarse form of the overall effect. In the history of interaction potentials, and likely for the immediate period ahead, what is learned from working at the detailed level drives, influences, or provides the knowledge base for everything less detailed. It is a logical progression of simulation technology. For the purposes of a review, there is some organizational advantage in selecting for the focus one certain level of detail, such as one of the extremes of highest accuracy or of being simplest for computation. Rather than that, the objective here is to look at interaction potentials from the perspective of *differing* levels of accuracy and the type of understanding that aids the progression from the detailed to the practical.

In a rough sense, molecules do not change on weak intermolecular interaction; and the subject, as normally defined, excludes anything as abrupt as the breaking or forming of a chemical bond. Again, it is essentially intact molecules that are the building blocks. Hence, it is reasonable that the intrinsic properties of isolated, intact molecules offer physical insight to intermolecular interaction and also offer a means, not necessarily a complete means, for constructing interaction potentials. Also at hand as a means for surface construction but usually more laborious is direct quantum mechanical determination. Perhaps best of all is to do both by drawing on ab initio calculations to study the isolated constituent molecules and also to explore the role of properties in subtle electronic structure responses due to weak interaction.

An interesting outcome of starting with molecular properties to understand weak interaction energetics is that property changes often can be put on an equivalent footing. That is, property surfaces for clusters can often be constructed along with potential surfaces using some of the same elements. In that sense, the construction of intermolecular interaction potentials can yield back property information akin to that which may have gone into the potentials. This is important for gathering insight into the physical basis and electronic structure features of weak interaction; and in turn, it is important for the advantageous extraction of coarse potentials from the best detailed potential information available. This review is meant to focus on those property connections and hopefully to offer, thereby, a perspective that augments other recent reviews of weak interaction, two of which give emphasis to potential construction [16,17].

The subject, especially in the context of considering the progression from detailed to coarse, could be extended to include continuum representations of surrounding molecules. We won't take that extension; it is better a subject on its own. Furthermore, it is distinct from the notion of analyzing interactions using intrinsic molecular properties as much as possible. The objectives, then, are to take a fundamental view of constructing potentials, one that allows for a range of accuracy and a wide range of applications and one that exploits property information. This requires integrating some topics that are easily review subjects on their own. Consequently, certain major development areas in weak interaction do not have the in-depth discussion that usually goes with reviews of weak interaction and hydrogen bonding [1–17], instead going only as far as needed to connect with the area of potential construction. The ultimate applications of the sort of intermolecular interaction potentials that are considered here are (i) detailed quantum dynamical treatment of rotation, vibration, and tunneling of small clusters, (ii) aggregation energetics, structures, and properties for clusters with up to hundreds of molecules, and (iii) molecular dynamics simulation for solvents in biomolecular and reactive problems. Were one to take those three in order as defining a kind of axis of detail, work from my group to date would be a distribution peaked somewhere past (i) and closer to (ii). Certain points of discussion will reflect that experience or, unintentionally, that bias.

II. CONTRIBUTING ELEMENTS OF INTERMOLECULAR INTERACTION

A. Distinction from Chemical Bonding

Weak intermolecular interaction refers to hydrogen bonding, van der Waals attraction, London forces, long-range forces, and so on, through a number of different concepts that have been introduced throughout the history of chemical science. The essential unifying idea to define intermolecular interaction is that molecules attract, repel, or both, in the absence of forming ionic or covalent bonds. The formation of covalent bonds involves rather sharp changes in orbital character, and often more than one electron configuration dominates the wave function at such points. This is easy to distinguish from weak interaction where the largely intact nature of the interacting molecules means that the electronic structure does not show sharp orbital or configurational changes. Ionic bonding is a little different to distinguish in that orbital changes may take place over longer distances and therefore seem less abrupt. As well, a single configuration may dominate throughout the process. Nonetheless, orbitals change significantly as a result of forming an ionic bond, and this is a distinction from intermolecular interaction. There are changes at the atomic building block level that are a part of ionic bonding. For this discussion, interaction potentials for a partner molecule that is charged will be excluded as being at the fringe with respect to the definition of weak intermolecular interaction. It may be that certain cases of ionic bonding can be approached uniformly along with weak intermolecular interaction, and that certainly is true for ionic systems at long range; however, the adequacy of notions discussed here for ionic systems, even as simple as $H_2O \cdot H_3O^+$, is something to be considered separately as a possible extension of the ideas. With that, the broad meaning of hydrogen bonding will be limited herein to that of all the constituents being neutral.

The language of chemical bonding is sometimes used to discuss weak intermolecular interaction. Interactions of occupied and empty orbitals, mixing of orbitals, and so on, seem to be ready concepts that can be used to account for structural features of small clusters. This language puts the conceptual picture (only) in the context of quantum mechanics; however, in the event that there is simpler physics at work because of the molecules' electronic structures being largely intact, then this language may be adding unnecessary complexity. For instance, consider accounting for the fact that the equilibrium structure of N_2–HF is linear whereas that of the isoelectronic cluster HCCH–HF is T-shaped. The quadrupole moments of N_2 and acetylene are opposite in sign, which corresponds to negatively charged ends for one and positively charged ends for the other. The favorable interaction of the proton end of HF's dipole, the end that can approach the closest, with an axial quadrupole will be collinear for ends negatively charged and will be pointing into the middle of the molecule (T-shaped) for the oppositely signed quadrupole. Chemical bonding language can say the same thing, though not as directly. For the two clusters, the positively charged proton end of HF is favorably located near the $2p$ σ orbitals of N_2, and the alignment is collinear for the sake of σ–σ overlap. With acetylene as HF's partner, the proton can be said to interact with the carbon–carbon $2p$ π cloud. In this case, the classical multipole and the quantum arguments are really the same; however, as long as the orbitals are not changing too much, the introduction of orbital interactions can be extra baggage. Furthermore, it may not be useful to bring to mind concepts used for changes in electronic structure, when mostly that is something not central to weak interaction. Hence, in the discussion of contributing elements to intermolecular interaction, we often don't need to draw on the language of orbital interactions and chemical bonding.

Can long-range weak interaction evolve into chemical bonding? Yes, but apart from ionic bonding, we should anticipate a potential surface bump between a region of weak attractiveness and the region of the chemically bound system. The sharp changes for covalent chemical bonding tend to develop closer than do the attractions of intermolecular interaction, and those attractions do not smoothly evolve into the sharp changes of bond formation.

B. Types of Elements

The language of molecule building block interactions is that of several types of effects whose juxtaposition yields the net interaction potential. These effects or contributing elements include (1) the electrostatic interactions of the unrelaxed charge distributions, (2) dispersion or the interaction of instantaneous multipoles, (3) the electrostatic effects of polarization and hyperpolarization of the molecular charge distributions, (4) the effects of penetration or overlap of constituent charge distributions and intermolecular electron exchange, and (5) charge transfer if not already included in the other pieces. The first of these can be evaluated, in principle, from permanent moments of the molecular charge distributions, and hence labeling this as a "multipole" contribution is appropriate. With the subscript M for these potential terms, D for dispersion, P for polarization, X for exchange and penetration, and R for the remaining contributions of all types, a potential function, V, can be considered to have the form of a sum:

$$V = V_M + V_D + V_P + V_X + V_R \tag{1}$$

This is a phenomenological approach, suggesting the construction of potentials by finding individual pieces. The pieces might be obtained from ab initio calculation, from experimentally measured properties for some of the pieces, or from outright modeling, even guessing, of the individual pieces.

An important limitation inherent in this phenomenological approach was noted by Coulson a half century ago [1]. The pieces in Eq. (1) are both attractive and repulsive contributions; in fact, some may change from one to the other as one moves across a potential surface. Coulson did a simple evaluation of pieces for the water–water interaction and made it clear that the magnitude of V would tend to be comparable to that of the individual elements. That is, the juxtaposition of positive and negative elements yields a net interaction that tends to be small, and the magnitude of the net interaction may be comparable to the magnitudes of one or more contributing elements. Consequently, accuracy of V demands considerable accuracy in each of the individual elements. Also, it is not meaningful to argue about one or another of these terms being dominant. Instead, it is the subtle result of their combining that determines the structure and energetics of a cluster.

It is useful to recognize that certain of the elements of interaction can arise in the absence of any electronic structure change while the others develop only because of some slight change in electronic structure. For instance, the multipole term is in the first category, whereas mutual polarization is in the second. This really amounts to using the language of perturbation theory and distinguishing first-order from higher-order energy effects. Buckingham [18] did this at a very early point in the theory of intermolecular interactions. Furthermore, this view suggests that a base-level or starting form for any sort of interaction potential is that wherein effects of electronic structure changes are ignored as being a higher-order contribution. Improving detail and accuracy then comes with steps to include higher-order elements.

C. Pairwise Additivity and Cooperativity

Pairwise additivity—namely, that a potential consists of additive pieces for each pair of species—is used here mostly in the sense of the species being the largely intact molecules making up a cluster, not the atoms in the molecules. This reference to whole molecules is done even if there are multiple sites in a molecule used for expressing a model potential. Were we to refer to atoms in a molecule, that would allow for three-body or nonpairwise additive potential elements in the simple cluster Ar–HF, for example, even though the weak interaction is between only two largely intact units, Ar and HF. For any fixed H–F distance, those three-body terms get reclassified (not changed) as two-body terms if we refer to molecular units, not atomic sites, as the interacting bodies. This holds even with there being multiple sites in an individual molecular unit so long as the geometrical structure of the molecule is unchanging.

Some of the types of contributing elements combined in Eq. (1) can give rise to potential pieces that are not additive. These would involve products of property or parameter values for more than two molecules, and these are often referred to as cooperative or nonpairwise additive elements. A simple illustration is in the electrical interaction contributions. While the interaction of permanent moments is pairwise additive, involving products of moments of only two different molecules at a time, the polarization energy can have a cooperative part. For some cluster of the molecules A, B, and C, the dipole polarization energy of A will be the polarizability of A, α_A, multiplied by the square of the field experienced at A, F. That field is a sum of contributions from B and C ($F = F_B + F_C$) proportional to their multipoles, and its square has a cross term, $F_B F_C$, involving a multipole of B times a multipole of C. The net interaction element includes $\alpha_A \ F_B F_C$, thereby giving an overall A–B–C or three-body term. Mutual or back polarization can be shown to produce contributions up to N-body for a system of N species.

If we are willing to consider even the smallest of contributions for the sake of completeness, then it is important to realize that contributors in the category that requires a change in electronic structure can affect other contributors and give rise to cooperative terms not explicit in Eq. (1). For instance, if molecule A in a cluster is polarized by molecule B, then its dispersion interaction with molecule C may be affected. Perturbation theory provides a means for organizing the pieces and for realizing that there are higher-order and mixed terms, in general [10,11,14]. The task is to identify which are important for some desired level of accuracy.

D. Functional Forms and Uniqueness in Potentials

The various types of elements in Eq. (1) have a variety of functional forms. The multipole interactions among neutral species vary with separation distance, R, as R^{-3} for dipole–dipole interaction, R^{-4} for dipole–quadrupole interaction, R^{-5} for quadrupole–quadrupole interaction, and so on. Polarization by a dipole has a distance dependence of R^{-n} with $n > 5$, and polarization by higher multipoles has a more rapid falloff with R. The interaction of fluctuating multipoles, or dispersion, also has a distance dependence of R^{-n} with $n > 5$. Hence, without even considering all the types of contributions, it is clear that not only will the interaction energy at one point be a subtle consequence of juxtaposition of elements, but also the whole potential surface may involve positive and negative terms that can be grouped into sets exhibiting like distance dependence.

The similar distance dependence developing with different contributing elements raises the important issue of uniqueness in the potential terms. If one term partly offsets another, do we know that those terms aren't both in error by amounts that are getting offset? Likewise, do we even need to know such terms precisely? From a practical standpoint, a particular judgment—good or bad—of a total potential's accuracy cannot be applied with the same certainty to the accuracy of the individual elements. There could be a seemingly good potential where too much of one element might be okay because of too little of another that happens to have the same variation with geometry over the surface. In other words, the juxtaposition of small competing terms that complicates accurate determination of a surface also implies that a correct total might not have an accurate breakdown into its pieces. Should we be concerned? To answer that, we can look from the standpoint of the "industry" of generating potentials and realize that the pieces are tools. Perhaps not as basic a separation as particle kinetic and potential energies that form the fundamental Hamiltonian for the complete quantum description of the system, they are certainly ways (1) to break down the interaction for interpretation, (2) to guide the representation (functional form) of a surface, or (3) to set up model potentials by combining elements.

The very definition of some of the elements shows the difficulty of trying to go beyond taking them as tools. For example, charge transfer can be defined as net flow of charge across some chosen dividing surface between two interacting molecules. The choice of the dividing surface, as well as distinguishing charge flow due to charge transfer from that due to polarization, can be done in different ways. Different definitions can produce different pieces. While a consistent set of definitions and even partitioning of ab initio energies is possible, there is probably a limit in how far one can or should go in pinning down all the individual pieces. The case of $(HCN)_n$ clusters, where cooperative effects are important [19], has shown the type of disagreement that can ensue. Essentially, one interpretation of the electronic structure changes of molecules in HCN chains was primarily that of charge transfer [20], while another was that of polarization [21,22]. One can go through more thorough analysis [23] for the sake of separating the two types of electronic structure changes, and what typically remains as charge transfer after entirely accounting for polarization seems to be a relatively small close-in effect. From the standpoint of constructing potentials, the interpretation is less important than the fact that polarization energetics are the more straightforward to represent, being based on the simple laws of classical electrostatics. The pieces are the tools to build a potential; and the potential, not the tools, is the objective. The simplest level of physics that can account for the features of the interaction is likely to be the best blueprint for building potentials. Anything less simple that remains can be added subject to the accuracy demands of the application.

Uniqueness of potentials is an issue in a different sense. Two surfaces might yield like predictions and yet have different parameters or even different functional forms. This happens if the set of predictions, such as geometries or vibrational frequencies, sample surface regions where there is not a significant difference between the potentials. Hence, since not every twist and bump on these complicated surfaces affects every feature or phenomenon, some non-uniqueness may need to be tolerated by virtue of being unimportant.

Along this line, it is interesting to mention studies of Ar_n–HF clusters. In one of these [24], the measured rotational constants of $Ar_{1,2,3,4}HF$ clusters [25–28] and the measured HF vibrational red shifts arising from interaction with argon atoms [29,30] were used as target values. A quite simple potential was built on fixed electrical properties with very few adjustable parameters for the other terms. Repeated dynamical calculations were performed to find the sensitivity to the parameters and select a set that gave the best match with the target values. It proved possible to get quite close to the target values, and that led to evaluations for like properties in larger clusters with many more argon atoms. There are more elaborate and more accurate potentials [31] that have been applied to the vibrational dynamics of Ar_nHF clusters [32–34], and the results of the most recent of these studies [34] showed considerably better agreement with experimental values than those with the simpler model [24] (see Table I). The characterization made [34] of the simple model's pair potential as "poor," though, shouldn't obscure a point from the simple model results [24] of how much of the dynamical behavior and structural features of these clusters is not sensitive to the more intricate elements of a high-quality potential. We may argue that *unique* determination of individual pieces becomes an objective only to the extent that there are phenomena or quantitative determinations that call for a correspondingly detailed form of the overall potential. In all model potentials, though much more so in simple potentials, there can be offsetting errors, errors that are not manifested in whatever is being studied, and nonuniqueness. Rather than seeking only the highest accuracy, it is important

TABLE I
Spectroscopic Features of Ar_n–HF Clusters

		Ground Vibrational State Rotational Constants (MHz)		HF(n: 0–1) Red Shift (cm^{-1})		
		Calculated Dykstra [24]	Experiment[a]	Calculated Dykstra [24]	Calculated Hutson et al. [34]	Experiment McIlroy et al. [30]
Ar_2HF	$\langle A \rangle$	3591	3576	15.42	14.78	14.83
	$\langle B \rangle$	1720	1739			
	$\langle C \rangle$	1158	1161			
Ar_3HF	$\langle A = B \rangle$	1190	1188	20.58	19.25	19.26
Ar_4HF	$\langle A = B \rangle$	619	624	20.71	19.57	19.70
$Ar_{12}HF$				37.31	39.20	

[a] Experimental rotational constants: Ar_2HF [26], Ar_3HF [27], Ar_4HF [28].

for dealing with a range of applications of model potentials that we understand offsetting errors, nonuniqueness, and so on. This is to achieve the best descriptions when simple potential forms are required. Assessments of effects from simplifying potentials for computational advantage are therefore important.

III. AB INITIO INTERACTION ENERGIES AND SURFACES

The capability of the highest level of ab initio treatment has reached the point where ab initio calculations increasingly provide benchmark information on structures, force constants, and stabilities, at least for small clusters. While there are still only small numbers of weak interaction problems for which (1) ab initio surface calculations have been carried to the highest levels, (2) correspondingly high-level vibrational–rotational calculations have been performed, and (3) extensive comparisons have been made with rotational and vibrational spectroscopic data—$(HF)_n$ clusters being a leading example [35,36]—there is reason for good confidence in calculational capability. Calculations have been pushed to a point of very small error through extensive treatment of electron correlation and huge basis sets. Even if such calculations are not commonplace, they provide some of the most definitive information on interesting features such as tunneling barriers that are only obtained indirectly from experiment. They can also provide highly useful information on systems that have yet to be fully studied in the laboratory. Furthermore, what is determined from the very-high-level and benchmark studies can be used to work the most effectively at lower levels of treatment so as to keep the errors as small as possible. Specifically, the crucial concerns of electron correlation and the adequacy of the basis set are ultimately guided by the high-end studies. With proper attention to basis-set needs and electron correlation, along with attention specific to the physics of weak interaction, useful potential surfaces (even in the form of a collection of grid points) can be obtained from ab initio calculation.

A. Basis-Set Completeness and Superposition

Perhaps more challenging to resolve than the choice of electron correlation treatment for weak interaction problems is the basis-set selection. Partly, this reflects the fact that basis sets have traditionally been devised for describing chemical bonding, not for the subtle juxtaposition of effects in weak interaction. They have to do both for weak interaction potential evaluations. This issue in basis-set selection, the adequacy of the basis to describe the electronic structure effects that comprise the interaction, can impose more stringent requirements than for describing an isolated molecule. Consider polarization as a contributor to interaction. Since the adequate determination of molecular polarizabilities requires basis sets augmented with diffuse and higher-l basis functions [37], an adequate accounting of the polarization response of two interacting molecules will have essentially the same large basis set needs as a polarizability evaluation.

A second problem in basis-set selection in conventional supermolecule calculations is that of superposition error or BSSE (basis-set superposition error). The deficiency of a basis for a single molecule, even if it is a small deficiency, may be corrected in part by the flexibility of another molecule's basis set. There can be an "attraction" of a molecule for another molecule's basis functions. The elimination of this artificial interaction effect tends to go slowly with increasing basis set size, except in the case of explicitly correlated basis functions as in R12 methods [38–41]. Hence, there is an extensive literature on ways to exclude BSSE. The original idea, that of Boys and Bernardi [42], is to evaluate the artificial attraction of a molecule for a set of ghost functions, those being the ones that would be found on an interacting molecule. They are ghost functions because the nuclei and electrons of the approaching molecule are not included in the evaluation. This artificial attraction is subtracted from the normally evaluated interaction in the Boys–Bernardi or counterpoise correction scheme. When large, flexible basis sets are used, as they should be to carefully describe the interaction contributions, the Boys–Bernardi scheme is generally accepted as the proper correction. BSSE is a difficulty that is manifested in quite subtle ways, and considerable discussion [43–46] and developments with corrections for BSSE have continued.

An interesting idea that may prove to be related to basis-set needs is how the extra electron of an anion is bound to a molecule. From a recent review of electrostatically formed anions by Simons and Skurski [47], it is clear that there are interesting localizations of negative charge. If part of the polarization and/or charge transfer between monomers in a cluster happens to mimic this behavior, then it will be important to ensure that flexibility of the sort needed to describe molecular anions exists within a chosen basis set, too.

The two issues in basis-set selection, describing the interaction contributors and minimizing BSSE, are intertwined. For instance, we have found that small or modest basis sets augmented in regions between monomers (i.e., with bond functions) can display relatively large counterpoise corrections [48]. Yet, rather good values seem possible from these types of calculations with bond functions [49–51]. Hence, it may be that some capability to describe certain interaction contributions should be included at the expense of BSSE. In fact, an opinion we expressed [48], that atom-centered functions relative to bond functions may be the best improvement of a small basis for a given computational cost, has met with a solid counterargument relating, in part, to describing dispersion [52]. The deepening of our understanding of what needs to be described, such as electrostatic-like binding of small excess charge mimicking anions, dispersion, or something else, will likely continue to refine and improve ideas for basis set selection.

High-accuracy calculations clearly require very flexible basis sets extended in a number of ways such as with multiple polarization sets, with diffuse basis function augmentation, and with other than atom-centered functions. Use of smaller bases goes along with less reliability; however, with bases smaller than double-zeta in the valence plus one well-chosen set of polarization functions on all centers, including hydrogens, the reliability is so limited that results are not likely to be meaningful for most contemporary problems of weak interaction.

B. Electron Correlation Effects

Having taken weak interaction to refer to systems where molecular electronic structure is largely intact, there is little reason to expect nondynamical electron correlation to be of major importance. In fact, the inclusion of dynamical correlation effects via the low level of second-order perturbation theory (i.e., MP2 [53] or MBPT-2 [54]) has become fairly standard for weakly bound clusters. This is not to say that MP2 is sufficient for the highest accuracy. It is more a consequence of MP2 being relatively more suitable for weak interaction potentials than for the potentials of breaking or forming chemical bonds. More complete treatment of correlation through higher order perturbative treatment or coupled cluster (CC) approaches [55–59] assures greater reliability. A significant approach for all types of correlation treatments, that of R12 theory [38–41], uses many-electron basis functions that depend explicitly on interelectronic distances to deal with the interelectronic cusp problem. Basis-set requirements are thereby reduced, though at the expense of certain computational steps to use the R12 bases. R12 treatments have been applied to problems of weakly bound clusters [60,61] with very good results.

Density functional theory (DFT) [62] incorporates electron correlation at a very small computational cost, but its suitability for weak interaction still seems a somewhat open question. There have been comparisons of DFT and conventional methods [63–68]. Mostly, an improvement over SCF level treatment seems possible, but there is a clear dependence on the choice of functional and on basis-set size. Also, DFT may be more sensitive to BSSE in smaller basis sets than conventional treatments [64]. A single functional choice for spectroscopic accuracy in treating weakly bound clusters does not yet seem at hand, but that alone does not preclude application of DFT for lower levels of accuracy. With the computational cost advantage of DFT, the capability exists for treating large, extended clusters.

C. Perturbative Analysis

As Buckingham [18] has done, the intermolecular interaction can be formulated as a perturbing Hamiltonian and then elements of the interaction are associated with specific terms and specific orders of perturbation theory. For an ab initio calculation, this begins by assigning electrons to the specific subunits of a cluster in order to define the zero-order Hamiltonian. However, this introduces the complication of ensuring full electron antisymmetrization (i.e., for electrons assigned to different molecules). Symmetry-adapted perturbation theory (SAPT) [11,14,69–73] adapts the perturbative treatment to incorporate the antisymmetrization requirement as opposed to imposing it on the zero-order wave function.

SAPT avoids the subtraction of large energy values that is necessarily part of a supermolecule ab initio calculation. A supermolecule calculation obtains the interaction energy of monomers A and B, ΔV_{AB}, as $V_{AB} - V_A - V_B$, whereas SAPT finds ΔV_{AB} directly. The interaction evaluated in SAPT is defined so as to be free of BSSE; however, the other requirements on basis-set quality and for correlation effects still hold. SAPT has yielded highly accurate interaction data, first for rare gas atoms interacting with small molecules [72–74] and more recently with molecule–molecule clusters such as the CO_2 dimer [75]. Further examples are the very accurate results achieved for Ne–HCN [76] and a pair potential for water [77]. Another example study of perturbative treatment of the interaction potential has been a study of rare gas–HCN clusters [78] which included vibrational analysis.

D.   Representing Ab Initio Surfaces

In principle, ab initio calculational results can provide a conventional grid of surface points spanning the geometrical degrees of freedom for the molecules in a weakly interacting cluster, and conventional techniques of surface fitting might be used to find an interaction potential. The $3N - 6$ degrees of freedom for a cluster with N atoms, though, is often sizable enough to make evaluation of a full grid unapproachable. Of course, the largely intact nature of molecules in a cluster can be a help, because it means that sometimes it will be reasonable to hold the intramolecular structural parameters fixed. This amounts to a Born–Oppenheimer-like separation of the relatively fast intramolecular vibrations from the usually slower intermolecular (weak mode) vibrations. Even then, the number of degrees of freedom for M rigid, nonlinear molecules is $6M - 6$. Another possibility is to limit the grid to an equilibrium or tunneling region of the geometrical space. While that may be suitable for finding certain features of the interaction potential, dynamical treatments will usually require a full representation of the surface, not one limited to a small region. For instance, diffusion quantum Monte Carlo treatment of cluster vibrational dynamics requires repeated potential evaluations at structures that evolve through random stepping. In other words, the points can't be anticipated and in any case can't be made to match selected grid points. (Special techniques do exist for effectively using grid points in QMC [79,80].) For general purposes, it is essential that the energy can be evaluated as needed. In certain of our DQMC calculations [24], the number of energy evaluations have been on the order of 10^7 to 10^8. Hence, the degrees of freedom of an interaction surface and its ultimate use should enter the considerations for collecting grid data.

If ab initio energies have been evaluated as a grid of points, property analysis of the interaction can establish at least certain key functional forms for fitting the points or representing the surface in an explicit functional form. In an ab initio study or Ar/Ne–H_2S, we found that we could use the form of a simple model potential based on response properties of the rare gas and H_2S in order to fit a grid of ab initio points [81]. One valuable outcome of analyzing the contributors to intermolecular interaction potentials is finding the pieces that are important. This will tend to keep the number of fitting parameters to a small number and keep the potential function as concise as possible for the targeted level of accuracy. In turn, this probably will keep the number of ab initio grid points to the fewest possible.

A good idea for improving surfaces based on ab initio calculations is to empirically adjust the parameters in the potential so as to make the surfaces agree with experimental observations, or with any other feature. Klopper et al. [36] did this very effectively for the HF dimer. Morphing is the contemporary jargon for this process, and it is a good idea because it can correct for systematic problems in a surface. For instance, if the ab initio level of treatment used on a problem is known to produce a 5% error in the dipole moments, then empirically adjusting the dipole moment where it enters into the potential in order to match observations or other information is a way to morph the original surface into one that corrects that systematic deficiency of the ab initio treatment. If one fits ab initio grid data to a potential whose form is based on the interacting pieces of weak interaction, as in Eq. (1), and one achieves a concise representation (fewest parameters), the morphing process is automatically easier by having the fewest parameters.

IV. DEVELOPING MODEL INTERACTION POTENTIALS

A. Interaction-Specific Versus Transferable Potentials

Weak interaction potentials can be constructed specific to a single system, such as the dimer of two molecules of water or the trimer of hydrogen fluoride. They may contain terms that correspond to the usual contributing elements and may incorporate cooperative effects. In any case, the objective is a potential that is expressed in terms of geometrical parameters of the specific system as a whole. The alternative approach is to construct potentials with parameters tied to the molecule building blocks. This alternate approach imposes a type of constraint in that functional forms used to represent a surface must reflect the types of parameters. The contributing element of permanent moment interactions is clearly an example of how parameters intrinsic to the building blocks (i.e., each molecule's multiple moments) determine the contribution. The pieces making up multiple moment interaction each have a product of moment values from any pair of interacting molecules, and this goes along with a specific functional form.

It is possible that many of the contributing elements to weak interaction can be represented by forms that assign parameters to the building blocks. This has long been recognized for the interaction of fluctuating dipoles, or dispersion, between rare gas atoms. Once experimental information was sufficient to obtain the C_6 and C_{12} parameters for 6–12 interaction potentials $[V(r) = C_6/r^6 + C_{12}/r^{12}]$ of Ne_2, Ar_2, and Kr_2, potentials of good quality were obtained for the heterodimers (e.g., NeAr) by taking the C_6 and C_{12} parameters be products of the square roots [82] of the parameters for Ar_2 and Ne_2: $C_{6[NeAr]} = (C_{6[NeNe]} C_{6[ArAr]})^{1/2}$. This type of relation is sometimes designated as a combining rule. It amounts to assigning transferable parameters to rare gas atoms, and there are two important things to notice. First, the number of parameters is smaller with imposed transferability. For the set Ne, Ar, Kr, there are three transferable dispersion parameters to describe Ne_2, Ar_2, Kr_2, NeAr, NeKr, and ArKr, whereas there would be six interaction-specific C_6 parameters to do the same thing. Second, the transferable approach, compared to an interaction-specific approach, has less adjustability and cannot be expected to offer the same level of quality. It is fair to say that imposing transferability sacrifices a certain share of accuracy for the sake of broader applicability. How much is sacrificed depends on the whether there is a genuine basis for a transferable form. Clearly, for certain interaction elements, such as multipole moment interaction, a genuine basis is definitely the case.

In the case of dispersion, a strong basis for a transferable form via a combining rule [83,84] has been evaluated and tested by Thakkar [85]. The combining rule uses C_6 coefficients for a pair of identical species and their dipole polarizabilities, α, to find a C_6 coefficient for interactions of unlike species.

$$C_6^{A-B} \cong \frac{2\alpha_A \alpha_B C_6^{A-A} C_6^{B-B}}{\alpha_B^2 C_6^{A-A} + \alpha_A^2 C_6^{B-B}} \qquad (2)$$

Thakkar [85] found that for 210 interactions, the combining rule produced an rms error of 0.52% with a maximum error of only 2.36%. A formal expression that can be given for C_6 coefficients [16] is that of an integral over frequency of the product of the dynamic dipole polarizabilities of the two interacting species. For like species, as in C_6^{A-A}, this would be an integral over the square of the dynamic dipole polarizability. If one takes as an approximation that the ratio of C_6

coefficients will be proportional to the squares of static ($\omega = 0$) dipole polarizabilities,

$$\frac{C_6^{\text{A--A}}}{C_6^{\text{B--B}}} \approx \frac{\alpha_A^2}{\alpha_B^2} \tag{3}$$

then the two terms in the denominator on the right-hand side of Eq. (2) become identical. A cancellation is then possible with the following result:

$$C_6^{\text{A--B}} \approx \sqrt{C_6^{\text{A--A}} C_6^{\text{B--B}}} \tag{4}$$

This is the combining rule mentioned above, and it is fairly good for at least a specific set of interacting species such as the rare gas atoms. Going from use of Eq. (4) to Eq. (2) in a model potential represents a way of achieving a very high level of accuracy while maintaining transferability. Hence, it seems possible that interaction elements that do not immediately seem to have a physical basis via intrinsic molecular property values might nonetheless be cast in a transferable form.

B. Ab Initio Determination of Potential Elements

There are two ideas used for ab initio determination of potential elements. One is to partition the ab initio energy into contributions in line with, or similar to, those in Eq. (1). Another idea is to use ab initio calculations to evaluate properties that enter the interaction expression. This has a limitation if it is not possible to cast every interaction element as involving a "property" of the interacting species, but it fits well with the objective of transferability because any properties are tied to the building-block molecules.

1. Ab Initio Evaluation of Properties

The most immediate properties for ab initio evaluation are electrical response properties. The electrical response properties are all derivatives of the moleculer energy. Therefore, they can be evaluated by methods that determine energy derivatives (gradients, etc.) directly or by finite fields. For finite field evaluations, the Romberg approach of Champagne and Mosley [86] is especially good at ensuring numerical reliability and is easily implemented [87]. By finite field or by direct evaluation of energy derivatives, there are requirements for reliability. First, basis sets must be flexible enough to describe relatively slight polarization changes in electronic structure. This calls for diffuse valence sets and extensive polarization function sets, and they should augment good-quality core/valence bases to avoid an unbalanced description. The correlation consistent basis sets of Dunning and co-workers [88] are among available bases that include a number of extended sets suited to property evaluation or that are easily augmented for this purpose. Also, there are very large, multiply polarized bases that have been used in very critical evaluations [e.g., Refs. 89–91].

Electron correlation plays a role in electrical response properties; and where nondynamical correlation is important for the potential surface, it is likely to be important for electrical properties. It is also the case that correlation tends to be more important for higher-order derivatives. However, a deficient basis can exaggerate the correlation effect. For small, light molecules that are covalently bonded and near their equilibrium structure, correlation tends to have an effect of 1–5% on the first derivative properties (electrical moments) [92] and around 5–15% on the second derivative properties (polarizabilities) [93–99]. A still greater correlation effect is possible, if not typical, for third derivative properties (hyperpolarizabilities). Ionic bonding can exhibit a sizable correlation effect on hyperpolarizabilities. For instance, the dipole hyperpolarizability β of LiH at equilibrium is about half its size with the neglect of correlation effects [100]. For the many cases in which dynamical correlation is not significant, the nondynamical correlation effect on properties is fairly well determined with MP2. For example, in five small covalent molecules chosen as a test set, the mean deviation of α elements obtained with MP2 from those obtained with a coupled cluster level of treatment was 2% [101].

To give an example of correlation effects and the differences in correlation treatments, several calculations were done for *trans*-1,3-butadiene. The basis set for these calculations was a cc-pVTZ basis [88] but with only one *p*-function set (exponent $= 0.9$) on hydrogen, no *f* functions for carbons, and no *d* functions for hydrogens. The structure of butadiene used for these calculations had the carbons at ± 0.325078 Å from the longitudinal axis (x) with 1.453 Å as the C—C bond length and 1.331 Å as the C=C bond length. The dipole polarizability and second hyperpolarizability were obtained by finite field evaluations, and Fig. 1 shows the correlation energy as a function of field strength relative to the correlation energy at zero field. The results for α and γ based on these curves are given in Table II. Correlation diminishes the polarizability by about 10% but increases the second dipole hyperpolarizability substantially. MP2 does a rather good job of accounting for the correlation effect when compared with the highest level treatment used, a Brueckner orbital (BO) double substitution coupled cluster level [102]. A much more complete study of the dipole polarizability of butadiene has been reported by Maroulis et al. [107], while a prior study of this molecule [108] has produced some special attention to the role of correlation [109,110].

Part of the process of building a model potential using electrical properties of interacting molecules is representing the permanent charge field. The most direct ab initio approach is to evaluate the moments of the charge distribution to some desired order and use them as the representation. As molecule size

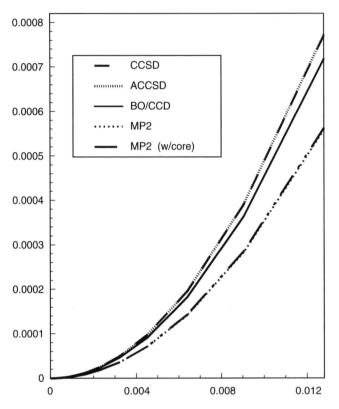

Figure 1. Electron correlation in butadiene as a function of the strength of an electric field applied along the longitudinal (x) axis relative to the correlation energy at zero field strength. The top two curves are nearly coincident, showing the similarity between CCSD and the approximate form ACCSD (see Table I). The bottom two curves are also nearly coincident. They correspond to MP2 calculations done without correlating the carbon $1s$ orbitals and with the inclusion of correlation from these core orbitals. All the other correlation treatments were done without including core correlation effects. The middle curve is the Brueckner orbital ACCD curve.

increases, the use of a single center representation is less and less suitable. The alternative is to distribute low-order moments to selected sites in a large molecule. It is possible to do this on the basis of simply reproducing the molecule-centered moments. For instance, we found that a good representation of the charge field of benzene was obtained by placing identical dipoles on the C—H bonds [111]; however, symmetry was working to advantage. Placing the identical dipoles only required choosing their position along a C—H bond (one parameter) and determining their size (second parameter). The more

TABLE II

Ab Initio Results (in a.u.) for the Longitudinal (x-direction) Components of α and γ for Butadiene

Level of /100 Calculation[a]	α_{xx}	$\gamma_{xxxx}/100$	α_{xx}	$\gamma_{xxxx}/100$	α_{xx}	$\gamma_{xxxx}/100$
SCF					80.11	88.4
	Doubles correlation		Singles correlation		Total	
MP2	−6.98	123.4	0.21	4.9	73.12	211.8
CCSD	−9.35	171.1	0.27	13.9	70.76	259.5
ACCSD	−9.37	172.3	0.27	14.5	70.74	260.6
BO/ACCD	−8.94	153.1			71.44	241.4

[a] The correlation treatments were done at the levels of second-order perturbation theory, coupled cluster theory with single and double substitutions (CCSD) [55–59], an approximate form of CCSD, and that form with Brueckner orbitals (BO) [102]. The approximate form [103,104], using herein the designation ACCSD or ACCD, has been shown to yield potential curves, potential surface slices, and properties very close to the corresponding CC results [104–106].

general approach is Stone's distributed moment analysis (DMA) [112,113], which gives a distribution of dipoles, quadrupoles, and point charges directly from the ab initio calculation.

The C_6 dispersion coefficients for dipole–dipole dispersion between pairs of interacting species, the coefficients for terms involving higher multipolar dispersion, and coefficients for three-body dispersion terms can be and have been evaluated by ab initio techniques [114–119] as well as through relations to experimental optical data based on moments of the dipole oscillator strength [120–122]. These are parameters of the interaction, not properties. However, as noted in Section IVA, values for C_6 coefficients of like pairs (e.g., A–A), and possibly for other dispersion coefficients, can be used in simple [Eq. (4)] or in more complete forms [Eq. (2)] as an intrinsic property of a molecule. The basis set and correlation requirements for adequate evaluation show, in part, the same requirements for describing polarizabilities; however, there are further needs and other than atom-centered functions are seen as being suited [49–52].

2. Perturbative Partitioning

The idea of partitioning is that in the course of an ab initio calculation, different elements can be extracted and/or isolated. At the most basic level, this serves as a means of interpreting ab initio energetics more so than a distinct means for obtaining the energetics. Morokuma [123–125] and Kollman [126] devised the key computational strategies to extract from ab initio calculations the contributions that could be associated with the different elements of noncovalent weak interaction. One immediate outcome was the confirmation that electrical effects, that is the interaction of permanent moments and polarization energetics, tend to be very important. Different partitionings have been developed [127]; and perturbation theory, especially SAPT [11,14,69–73], directly gives an extensive partitioning of the interaction energies. This may offer the type of information to construct system-specific, full potential surface models from a relatively small number of ab initio surface points.

C.   Models for Parameters Used in Interaction Potentials

The use of properties intrinsic to molecules for model interaction potentials requires obtaining those properties through ab initio calculations or in some cases through models of the properties themselves. Ideas for this have existed for a long time independent of their use in interaction potentials [128–130]. Quasi-additivity of atomic and/or bond contributions is often apparent from molecular polarizability or related data [101,131–134], and this has been invoked in a number of ways for models or other predictive schemes [134–140] that give polarizabilities. We have followed the idea of roughly additive atomic contributions [133] to accomplish transferability. That is, local α_i tensors for a given type of atom in a specific bonding environment (e.g., sp, sp^2, sp^3) are taken to represent the dipole polarization of an atom of that type in whatever molecule it occurs. A whole set of values can be built forcing this transferability (i.e., forcing all like atom-bonding sites in a sizable collection of molecules to have the same local α_i), and the results are quite good for molecular α's. It is even useful, though less accurate, for the second dipole hyperpolarizabilities, γ.

A rather novel scheme for modeling molecular polarizabilities as distributed dipole polarizabilities has recently been reported [141]. In this approach, the overall quadrupole induced in a molecule by an external field, as calculated with ab initio methods, is decomposed into induced dipoles distributed to atomic sites. In turn, this yields the dipole polarizability values at those sites. In effect, this relates the overall dipole–quadrupole polarizability to a distribution of dipole polarizabilities.

Recently, an additive scheme for bond polarizabilities has been incorporated with the MM3 force field [142] to facilitate evaluation of induced dipoles and other features associated with polarization. It is likely that approaches for modeling of polarizabilities, apart from thier role in interaction potentials, will continue to be developed and explored.

D.   Interaction Potential Models

There are a growing number of interaction schemes based on properties to generate full potentials or else potential surface information for specific regions or specific objectives. There have long been interaction potentials that are empirical or entirely system-specific. Indeed, there are hundreds of potentials that have been used for the specific interaction of a water molecule with another water molecule. However, in the spirit of the essential value of properties in interaction potentials, the view of interaction models given here is limited to those that (a) are not specific to a single pair interaction, at least in their development, and (b) utilize electrical properties.

In 1985, the Buckingham–Fowler model [143] for the geometries of van der Waals clusters was introduced. This scheme uses the distributed multipole analysis (DMA) [112,113] representation of the permanent charge fields to obtain the electrostatic interaction. The repulsive part of the potential was treated with hard spheres of assigned diameters. The approach has worked very well in giving the preferred orientations of monomers in clusters [5,8]. The hard sphere form of the repulsive part of the potential is sufficient for finding potential minima but not for representing the potential any closer. A more flexible potential for the close-in regions was that of Spackman's model [144,145] where an atom–atom form for the repulsion and dispersion was added to electrical interaction energy based on a partitioned multipole moment analysis.

Our notions of weak interaction led us to put electrical effects and polarization upfront in modeling. The result was a potential energy surface scheme designated molecular mechanics for clusters (MMC) [146] which uses ab initio information on molecular electrical properties (permanent moments, multipole polarizabilities, hyperpolarizabilities) in the evaluation of the classical electrical interaction of a cluster. This naturally combines the permanent charge field interaction with polarization energies. The other MMC potential elements were treated empirically and not as fully. They were represented collectively via atom–atom "6–12" potential terms with parameters selected so that the overall potential gave the best match with available spectroscopic information. The "6–12" parameters, like electrical properties, were assigned to monomers [as in Eq. (4)], and this inforced transferability meant that once such an MMC representation of a molecule was obtained, an interaction surface could be calculated for it with any other MMC-represented monomer. This imposed transferability of parameters from one cluster to another has shown modest to quite good success for a number of mixed trimers [146–149], for instance. MMC has provided useful quantitative information on stabilities and structural parameters, though not always to an accuracy that can answer every question, especially those related to detailed dynamical behavior. Again, there is an unavoidable trade-off between simplicity and accuracy. Sorenson, Gregory, and Clary [150] reported a study of the cluster of benzene and two water molecules which used the MMC representation for benzene [151], and Kong and Ponder have reported constructing an MMC type of model for water, with some interesting variations, as part of a force field program [152].

The effective fragment potential (EFP) is a more recent scheme that goes beyond interaction modeling. It incorporates electrical response and is targeted at reactions in solution and solvent effects on chemical properties [153,154]. With EFP, "active" molecules are treated by ab initio electronic structure methods apart from "spectator" molecules of the solvent. For reactions (for example, Ref. 155), EFP is a true hybrid of ab initio and model treatments. For the spectator segment of the approach, the permanent moment interactions are corrected by a screening function to account for charge penetration effects [156] and is a stand-alone model potential scheme for small, weakly bound clusters [154,157,158].

Xantheas and co-workers [159,160] have incorporated polarization in a model scheme and have used that to provide a clear basis for the enhancement of water's dipole in ice. A model potential with polarization has been reported for the formaldehyde dimer [161]. It is an example of a carefully crafted potential, which is system-specific because of its application to pure liquid formaldehyde, but which has terms associated with properties and interaction elements as in the above models. As well, some of the earliest rigid-body DQMC work, which was by Sandler et al. [162] on the nitrogen-water cluster, used a potential expressed in terms of interaction elements derived from ab initio calculations with adjustment (morphing). Stone and co-workers have developed interaction potentials for HF clusters [163], water [164], and the CO dimer [165], which involve monomer electrical properties and terms derived from intermolecular perturbation theory treatment. SAPT has been used for constructing potentials that have enabled simulations of molecules in supercritical carbon dioxide [166]. There are, therefore, quite a number of models being put forth wherein electrical analysis and/or properties of the constituents play an essential role, and some where electrical analysis is used to understand property changes as well as the interaction energetics.

A final point in this section relates to transferability. Seeking transferable models via intrinsic properties/parameters of the interacting species as opposed to seeking system-specific model potentials is advantageous because of the broader range of application; however, as already noted, doing so limits adjustability and hence it limits accuracy. There is, though, a useful connection. A transferable scheme may provide an initial model potential that may then be adjusted (or morphed) to improve accuracy for a specific application. We have done essentially this for pure acetylene clusters [167]. We started with the MMC representation for acetylene, but changed it slightly, using ab initio calculations, to achieve a better match of the measured rotational constant of the acetylene dimer [168] with a DQMC calculational result for the ground vibrational state. Tested on the trimer, tetramer, and deuterated forms of the dimer, the potential showed very good agreement with experimental rotational constants. We even calculated structures of larger acetylene clusters and their relation to the structures of the dimer and tetramer (Fig. 2).

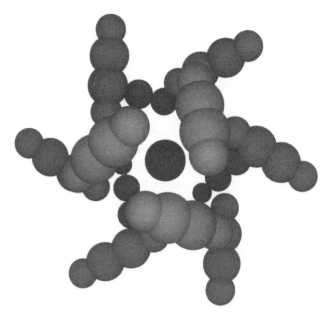

Figure 2. The equilibrium structure of $(HCCH)_{13}$ calculated [167] with a polarizable model potential. The central molecule is seen essentially end-on. The 12 surrounding molecules are in three layers. The upper and lower layers have three molecules and resemble the structure of the cyclic trimer of acetylene. The middle layer of six acetylene molecules has a pinwheel-like arrangement. Puckering of the rings in the layers yields T-shaped orientations between acetylenes in different layers, along with the essentially T-shaped arrangements for adjacent molecules within each layer. The number of favorable (T-shaped) quadrupole–quadrupole interactions among acetylenes is thereby enhanced.

V. CALCULATIONAL ASPECTS OF PROPERTY-BASED POTENTIAL MODELS

Basing the construction of intermolecular interaction potentials on properties means basing them first on electrical moments, and perhaps on polarizabilities and higher-order response. The detail to which the electrical response can be treated is something that can be improved in steps; however, the calculational organization stays essentially the same. As already mentioned, many properties may be the basis for constructing potentials; but to a good extent, the pieces needed in the evaluation for electrical analysis are the most involved to calculate. Sometimes, those pieces provide the values for using other property terms. Hence, it is the electrical analysis which deserves particular care and attention. Also, there are three key reasons for considering these calculational aspects.

1. For purposes of surface fitting ab initio grid points, as mentioned earlier, it is advantageous to know the functional forms of major terms—that is, the electrical property terms. A straightforward, formal analysis readily provides this information.

2. In terms of number of surface points and possibly gradient evaluations, the most extensive use of potential surfaces for weak interaction will be in dynamical treatments, either classical or quantum mechanical. For these, the cost of evaluation can be important; and with property-based models, one may have to consider the trade-off between cost and accuracy associated with how extensive is the treatment of electrical interaction.

3. To use electrical response to determine property surface information as discussed in the next section, or simply how to evaluate property changes associated with polarization of charge distributions, full electrical analysis is crucial.

The polytensor approach introduced by Applequist [169] is a terrific organization of the problem of electrical interaction for high-level calculation because it can be continued uniformly to any order of multiple moment, any distribution of moments, and any number of interacting species. Furthermore, it can incorporate multipole polarization and hyperpolarization [170]. As such, it provides a scheme that can be coded for computer application in an open-ended fashion while also providing the formal analysis needed to extract functional forms of different electrical interaction pieces.

For a known charge distribution $\rho(\mathbf{r})$, such as obtained from an ab initio electron density and nuclear geometry of a molecule, the elements of the Cartesian moments, M, of the distribution relative to the coordinate system origin can be expressed as

$$M_{x^I y^J z^K} = \frac{\int x^I y^J z^K \rho(\mathbf{r}) d\mathbf{r}}{(I + J + K)!} \tag{5}$$

The sum $N = I + J + K$ is the order of the moment to which a given element belongs. The moment polytensor, $\mathbf{M}$, is the single index array of the usual moment elements in canonical order from the zeroth moment (charge) to the first (dipole), second (quadrupole), and so on, to any desired level of termination:

$$\mathbf{M} = (M_0, M_x, M_y, M_z, M_{xx}, M_{xy}, \ldots) \tag{6}$$

$\mathbf{M}$ is a first-rank polytensor. The usual multipole polarizabilities can be arranged in a second-rank (two index) polytensor, $\mathbf{P}^{(2)}$, by using the ordering of element labels in $\mathbf{M}$ for the rows and columns in $\mathbf{P}^{(2)}$. For instance, $\mathbf{P}^{(2)}_{2,5}$ or $\mathbf{P}^{(2)}_{x,xx}$ is a dipole(x)–quadrupole(xx) polarizability tensor element. Hyperpolarizabilities are third- and higher-rank polytensors, $\mathbf{P}^{(3)}$, $\mathbf{P}^{(4)}$, and so on, constructed the same way. The $\mathbf{M}$ and $\mathbf{P}$ tensors are fixed arrays of values based on properties of the interacting species and/or interacting subunits. The polytensor organization casts the electrical interaction evaluations in a form that is independent of the orders of multipoles included.

The polytensor organization makes it very clear that the key computational step for electrical parts of interaction potentials is where the geometry information (separation distance and orientations) enters. This is best done with a second-rank polytensor, $\mathbf{T}^{(2)}$, and its structure and elements for a pair of interacting sites can be understood in terms of operations performed on $1/|\mathbf{r}|$:

$$\mathbf{T}^{(2)} = \begin{pmatrix} 1 \\ \nabla \\ \nabla^2 \\ \nabla^3 \\ \cdots \end{pmatrix} \begin{pmatrix} 1 & -\nabla & \nabla^2 & -\nabla^3 & \cdots \end{pmatrix} \frac{1}{|\mathbf{r}|} \tag{7}$$

with the individual operators in Eq. (7) following the order of the $\mathbf{M}$ tensor:

$$\begin{pmatrix} 1 & -\nabla & \nabla^2 & \cdots \end{pmatrix} = \begin{pmatrix} 1 & -\dfrac{\partial}{\partial x} & -\dfrac{\partial}{\partial y} & -\dfrac{\partial}{\partial z} & \dfrac{\partial^2}{\partial x^2} & \dfrac{\partial^2}{\partial x \partial y} & \dfrac{\partial^2}{\partial x \partial z} & \cdots \end{pmatrix} \tag{8}$$

Hence, if the truncation in $\mathbf{M}$ is to include through quadrupoles, then each $\mathbf{T}^{(2)}$ would have 13 rows and 13 columns. There is an easily exploited (anti)symmetry among elements in $\mathbf{T}^{(2)}$ that reduces the evaluation cost. Overall, the evaluation of the elements needs to be done efficiently, and means are available for that [170].

In polyensor form, the interaction energy between moments at site A and moments at site B is a matrix inner product:

$$E = \mathbf{M}_A \mathbf{T}^{(2)}_{AB} \mathbf{M}^T_B \tag{9}$$

The directly induced moments are also obtained in simple matrix form:

$$\mathbf{M}_{A\text{-ind}} = \mathbf{P}^{(2)}_A \mathbf{T}^{(2)}_{AB} \mathbf{M}^T_B \tag{10}$$

For a collection of surrounding molecules, the contribution in Eq. (10) can be summed:

$$\mathbf{M}_{A\text{-ind}} = \mathbf{P}^{(2)}_A \sum_B \mathbf{T}^{(2)}_{AB} \mathbf{M}^T_B \tag{11}$$

When evaluated, the summation in Eq. (11) is a rank-one polytensor that represents the potential experienced at molecule A in terms of field components, field gradient components, and so on. This can be used with response properties such as shielding polarizabilities to find property changes dues to electrical influence. The evaluation is analogous to Eq. (11). The incorporation of mutual or back polarization/hyperpolarization requires a self-consistent solution for the induced moments, and this can be done iteratively [170]; or if there are no hyperpolarizabilities, it can be done by matrix inversion.

A first point of discussing the polytensor organization for evaluating electrical interaction energies is to see how computational effort grows with multipole order. The numbers of elements in $\mathbf{M}$ associated with a multipole of order 0 (charge) 1 (dipole), 2, and 3 are 1, 3, 9, and 27, respectively. Though these numbers can be reduced by converting from Cartesian moments to irreducible spherical forms, the size of $\mathbf{T}$ grows as the square of the sum of these numbers—that is, as the square of the total number of elements in $\mathbf{M}$. Hence, from a computational standpoint, every order of multipole is a big step from the one before. For simulations of liquids that might involve hundreds of molecules, this computational complexity can pose a limitation. This leads to the second point in this section—that is, that the computational effort is largely in the $\mathbf{T}$ tensor, whether it is found explicitly or not.

Clearly, a truncation of $\mathbf{M}$ at one element, a point charge, makes for the simplest $\mathbf{T}$, the single element $1/r$. However, it is difficult to represent a molecular charge distribution with a distribution of only a few point charges. The charges have to be relatively large, and this yields abrupt changes at certain regions. A comparison has been presented for water that shows this [37]. Using many point charges instead of a few adds cost even with the simple (single-element) $\mathbf{T}$ tensors, and it seems that the step to distributing a few dipoles may be more advantageous from a computational standpoint. Indeed, for neutral species, the best overall scheme is probably to distribute dipoles and even quadrupoles to represent the charge field of a molecule. Further improvement might then come from a small number of point charges, each being very small in size, and this is in line with using DMA [112,113]. It may be premature to say that this is an optimum modeling strategy. We should anticipate that there is a lot more experience to be gained and that there are comparisons to be made. However, at the least, we can expect that limiting model potentials to forms that include point charges but not local dipoles is not necessarily computationally advantageous.

Two-body dispersion yields an interaction term that is relatively simple to calculate. The term corresponding to the interaction of a dipole induced by the fluctuating dipole (DD) of another center varies as $1/r^6$. More complete descriptions of dispersion will include higher-order terms. They vary as r^{-N} with $N = 8, 10, 12, \ldots$ corresponding to higher-order fluctuating and induced multipoles [16]. For dispersion sites that are not treated as spherical, there is an angular dependence via Legendre polynomials. Three-body dispersion is more complicated with three separation distances being involved—r_{AB}, r_{AC}, and r_{BC},—for the interaction of species A, B, and C, along with three angles $\theta_C, \theta_A, \theta_B$, corresponding respectively to the angles of the $\overline{AB}$–$\overline{BC}$–$\overline{CA}$ triangle. The first term is a dipole–dipole–dipole (DDD) [171] term:

$$V_{DDD} = \frac{C_{DDD}(1 + 3\cos\theta_A \cos\theta_B \cos\theta_C)}{r_{AB}^3 r_{AC}^3 r_{BC}^3} \tag{12}$$

which has an overall distance dependence of r^{-9}. Via substitution with the law of cosines, this can be expressed in terms of the distances:

$$\overline{r^N} \equiv r_{AB}^N + r_{BC}^N + r_{AC}^N \tag{13}$$

$$V_{DDD} = \frac{C_{DDD}}{4 r_{AB}^3 r_{AC}^3 r_{BC}^3}\left(1 + \frac{3\overline{r^2}\,\overline{r^4} - 6\overline{r^6}}{2 r_{AB}^2 r_{AC}^2 r_{BC}^2}\right) \tag{14}$$

For an equilateral triangle, the DDD term is repulsive, whereas for linear arrangements, it is attractive. For a system of only three interacting species, there should be little difference in using Eq. (12) versus Eq. (14). However, with many interacting species, Eq. (14) suggests an effective computational organization for simulations with repeated energy evaluations. To avoid redundant steps, every time a distance between two sites, r_{ij}, is updated or changed, the following are computed and stored with r_{ij}: $r_{ij}^2, r_{ij}^3, r_{ij}^4$, and r_{ij}^6.

Higher multipole terms [172–175], such as dipole–dipole–quadrupole (DDQ), and then the terms for DQQ, QQQ, and so on, have an overall dependence on distance that goes as r^{-N} with N being 2 greater than for the DDD term with every higher-order multipole step. Higher-order three-body dispersion coefficients (e.g., C_{DDQ}) have been determined from ab initio calculations [117,118]; and in at least one case, Ar_3, the effects of DDQ dispersion on vibrational transition frequencies have been found [176]. In a simulation of the vapor–liquid equilibrium of pure argon, Bukowski and Szalewicz [119] showed there are important three-body effects; and because of certain cancellations, these were primarily DDD dispersion.

Finally, it should be noted that the dispersion interaction that is at work at long range does not continue close-in. It has been recognized that damping out the dispersion close-in yields the correct behavior [177–183] and computationally simple damping functions have been devised [179,181]. These can be used for terms associated with dispersion in potential models.

VI. INTERMOLECULAR INFLUENCE
ON ELECTRONIC STRUCTURE

A. Electronic Structure Changes, Property Changes, and Property Surfaces

An underlying thesis of this review has been that weak interaction implies little change to the electronic structures of interacting species. Yet there are certainly slight changes occurring, and they have interesting and revealing manifestations. Polarization is one of the clearest, most direct types of electronic structure changes taking place in intermolecular interaction. It may be the dominant change in many cases. This is significant because we can account for polarization changes to the electronic structure via multipole polarizabilities and hyperpolarizabilities.

In principle, ab initio calculations of potential surfaces can be accompanied by ab initio evaluations of property surfaces. However, this is likely to be a cumbersome task. On the other hand, if many properties reflect polarization changes in the electronic structures of the interacting species, then property surfaces should be well-suited to modeling. Indeed, the potential surfaces and property surfaces can be put on an equal footing via evaluation of the electrical influence of surrounding species (i.e., fields, field gradients, and so on).

How well does this work? For one, we have shown that for carbon monoxide in a series of carbonmonoxyheme proteins, the experimentally observed correlation of ^{13}C chemical shifts, ^{17}O chemical shifts, ^{17}O nuclear quadrupole coupling, and CO vibrational frequency shifts arose because each property change had to due with polarization of the CO by particular distal ligands in the proteins [184]. We have shown that polarization accounts for the evolution of the cluster dipole hyperpolarizability of $(HF)_2$ as the monomers approach [185], how the ranges of structural nonequivalences in the chemical shifts of proteins are matched by the representative sizes of the shielding polarization [186] (first derivative of the nmr shielding tensor with respect to an external electric field), and we have developed a generalized picture of Sternheimer shielding [187]. In the study of a subtle property, the nuclear quadrupole coupling constants of weakly bound clusters [188], a hybrid calculation helped reveal the polarization nature of the effect. The charge field of a perturbing molecule was built into a one-electron operator that was used in the evaluation of the nuclear quadrupole coupling constant of the perturbed molecule, but the perturbing molecule was not otherwise included (no nuclei, electrons, or basis functions). The ab initio calculations showed a correlation between interaction strength and effect on nuclear quadrupole coupling through a series of clusters, and this was in very good agreement with experimentally determined values. An earlier and much more extensive series of ab initio studies by Cummins, Bacskay, and Hush [189–191] had provided a strong picture of nuclear quadrupole coupling being influenced electrically by surrounding molecules. In some systems they studied, 80% of the change in the field gradient at the nucleus could be associated with a polarization response. These studies [184–191] are a very good indication that the primary way in which the property change takes place is via the polarizing, charge field influence of a perturbing molecule. Hence, the basic of idea of obtaining property changes along with energy changes seems to hold in a scheme that takes good account of the electrical part of the interaction. That is, in evaluating the electrical contribution to an interaction potential, the external electrical potential at a molecule (or site in a molecule) has to be evaluated, and it is this information which can feed the computational process for calculating a property change.

B. Bridging Quantum Mechanical Treatment to Models of Surroundings

Interaction modeling based on properties of molecules, properties that of course are determined by quantum mechanics, amounts to a bridge from the quantum to the classical picture. But the reverse is also important, and we can see this from the example just discussed [188] involving work done in collaboration with the late H. S. Gutowsky and his group. Their spectroscopic work gave nuclear quadrupole coupling constants, eQq's, for the nitrogen in a series of HCN dimers, the partners being HCCH, HCN, and HF. These are a measure of the electric field gradient at the ^{14}N nucleus, and such evaluation usually requires careful quantum mechanical treatment because of sensitivity to subtle and close-in features of the electronic wavefunction. The objective, accounting for the trend in eQq with the interaction strength of the three partner molecules, was realized by an ab initio calculation done on HCN alone through representing the partner species by a one-electron operator. This is a classical to quantum bridge, one that precludes intermolecular quantum effects. It suggests that weak interaction effects of surrounding molecules on a species being described quantum mechanically can, to a good extent, be incorporated as an electrical influence, and this is much simpler than a full quantum mechanical treatment of surrounding species.

In complex systems, there is an advantage in partitioning physical space into regions described with different levels of treatment, starting with the highest levels for regions of greatest interest. The lowest-level descriptions may extend to non-quantum descriptions as in the variety of MM/QM (molecular mechanics/quantum mechanics) methods in use where peripheral molecules are treated at the MM level. Electrical analysis, if used in these methods, can go both directions at the interface. It can provide an operator that reflects the influence of the MM region on the detailed quantum region, and then the electrical properties evaluated for the quantum region can form the "external" influence for the MM region.

The recent effective fragment potential (EFP) scheme [153,154] mentioned earlier is clearly a successful connection between quantum and non-quantum mechanically described regions. The effective potentials used to represent the non-quantum mechanical spectator molecules in the ab initio treatment (quantum region) include Coulomb interactions, polarization, and exchange repulsion.

Permanent charge fields offer an influence bridge between quantum and non-quantum regions in complex systems. If the polarization response in the non-quantum region can be represented more fully, then it may be possible to bridge mutual polarization effects, too. This more physically complete interface might mean that the boundary between quantum and non-quantum regions could be pulled closer to the active region—a computational benefit.

VII. TOWARD MORE DETAILED RESPONSE FEATURES

Implicit in the electrical response discussed so far is that polarization is treated locally. That is, the external electrical environment arising from nearby molecules acts at a site or at distributed sites within a molecule. The basis for following this in a potential model is twofold. First, the electronic structure changes due to weak interaction are slight, which can be easily seen by examining ab initio electron densities of weakly perturbed molecules. Second, it seems clear from analyzing property changes due to weak interaction that, primarily, the source of the slight electronic structure changes is polarization. The future in model construction, particularly that which seeks to follow properties along with energies, will likely involve more detailed treatment of polarization response. Within the framework of the response being local, the detail is increased by the straightforward act of distributing more sites within a molecule. However, going beyond a local response is another approach to adding detail, though of a different sort; and possibly, there are ways to provide models with the capability to represent somewhat more substantial electronic structure changes than has been considered so far.

The interaction of long-chain molecules such as polymers is a problem area where the nature of polarization response can be a significant concern on its own. An example is from a study of parallel hexatriene molecules carried out to represent a truncated form of solid-state polyacetylene [192]. This study included both ab initio calculations and an electrostatic model using polarizability, α, and second hyperpolarizability, γ, tensors distributed to the carbon centers. The ab initio calculations on a single hexatriene molecule were used to find the distributed tensors for the electrical analysis. The objective in this study was not the interaction energy, but the effect on each molecule's polarizability and hyperpolarizability due to intermolecular interaction. The ab initio evaluations benchmarked the electrostatic model calculations both for the number of parallel hexatrienes, up to seven, and for the variation of properties with separation distance. Of course, the model treatment could be extended to much more than seven hexatrienes; hence, values were obtained for the asymptotic limit of an infinite number of parallel chains. The intermolecular interaction effects turned out to be substantial, with the per monomer dipole polarizability in the direction of the chain, α_L, reduced by about half and the per monomer hyperpolarizability, γ_L, reduced to about 5% of its value in an isolated molecule. As chain length increases, there is the possibility that the polarization response may be more than can be described by a sequence of point dipoles being induced at the atomic sites used in this local repsonse treatment. For long-chain molecules or otherwise extended species, polarization might result in a net flow of charge from one end of a molecule to the other. Applequist's relay approach for intramolecular interaction provides a scheme for a response at one part of a molecule to influence another part of a molecule [193]. Stone [194] and Munn [195] have considered this with respect to charge. The key is to have a charge–charge polarizability (or susceptibility) at sites in a molecule whereby a potential at one site acts to change the net amount of charge at another site, a nonlocal interaction.

Stone divides molecules into regions [194]; and somewhat similarly, Munn's work on regular crystals makes divisions into molecular volumes [195]. Stone connects the dipole induced in a region with charge–charge, charge–dipole, and dipole–dipole polarizabilities, the first two corresponding to a flow of charge into or out of the region in response to an external potential. Each region, i, is associated with a site, s_i, which serves as the center for the region's response properties. The analysis offers a good physical picture by breaking down a molecule into regions and allowing for charge to flow from region to region in addition to being shifted within the region. The charge–charge, charge–dipole, and dipole–dipole polarizabilities, though, are more complicated than the usual molecular dipole polarizabilities because they are defined for pairs of sites.

Right now, the evaluation of charge susceptibilities and the related values in interaction models is challenging. There have been signficant developments for practical approaches [196,197]. However, an important link emerges from the work of Hunt and co-workers [198–202], who have shown how geometrical derivatives are related to nonlocal charge susceptibilities. Derivatives of a molecular electronic energy with respect to geometrical parameters are routinely calculated with ab initio techniques; hence, this may provide a useful route to the necessary properties. With a different focus in the formal treatment than Stone's development [194], Hunt and co-workers have considered a wide range of response features, including frequency-dependent response [198–200,203], based on nonlocal susceptibilities and nonlocal polarizabilities. Particularly important is the connection of these response features with forces in molecules and geometrical derivatives as also discussed by Fowler and Buckingham [204] and Baker et al. [205]. There is another difficulty, that of achieving transferability; however, it may be that the need for this more elaborate treatment of polarization will tend to be quite specific, making transferability less important. As expressed at the outset of this review, it is important to realize that interaction analysis and model development is an industry serving several needs ranging from coarse forms to very detailed and precise forms and from system-specific to generally applicable.

VIII. CONCLUSIONS

The construction of potentials for weak interaction where molecules are largely intact has growing application. That the molecules are not changed by interaction in the abrupt ways that go along with chemical bonding implies that a simpler level of physics *should* be at work, one that can be exploited by relying, as much as possible, on intrinsic properties of molecules. The focus herein has been on constructing and building models tied to using properties, and specifically on doing so in a mostly fundamental way as opposed to working backwards to achieve known outcomes or making educated guesses for very simple types of potentials. The latter type of approach is widespread, and this is perhaps because there is such a strong need for hydrogen-bonding potentials, particularly for the theoretical study of liquids [206,207]. Working directly from an analysis of the interaction, its elements, the full functional forms for those elements, and the parameters in those elements is an increasingly common route to potentials for dynamical simulation. It is a route with much promise, and so that trend will continue.

There have been many creative directions for developing potentials, and a considerable understanding has developed in the last two decades, especially. However, a statement by Buckingham well before that in a 1967 report [18] continues to hold: "There is now general agreement that the significant forces between atoms and molecules have an electric origin." What exists now is greater means to utilize this understanding and go further. There are high-quality ab initio approaches for electrical response information, for dispersion, and for properties and parameters, and there are fully developed calculational approaches for very extensive treatment of electrical interaction. Within the last 5–10 years, there has been strong attention given to evaluating and dissecting short-range or close-in effects, anisotropy in dispersion, and other subtle aspects needed at some point in potential construction. These fundamental developments represent the high end of our understanding of weak interaction.

An argument of this report is that basing model interaction potentials on intrinsic properties of molecules as much as possible ensures conciseness in the potential function and the greatest prospect for transferability as opposed to being system-specific. Furthermore, property-based potentials are probably the easiest to tune so as to balance detail/accuracy against computational cost for the intended application. The future of this type of technology is in increasing the range of tunability to greater and greater accuracy, and a number of possible steps have been mentioned.

The slight changes that do take place in the electronic structure of weakly interacting species, excluding those being influenced only by rare gas atoms, seem to be largely polarization (induction) changes. This argument is made on the basis of the variety of property changes that can be accounted for by evaluating only a polarization response. This implies getting back other property information, or, in other words, obtaining property surfaces along with potential energy surfaces via the modeling that is done. We can speculate that polarization being the primary electronic structure change has another interesting outcome: It is likely to be a strong contributor to cooperative effects, perhaps a very dominant effect among cooperative elements in certain clusters. Our own evaluations on many clusters have shown four-body polarization effects to be very much smaller than three-body polarization effects. It is possible that with adequate treatment of intermolecular polarization response, maybe only through three-body effects, the crucial cooperativity needed to connect gas-phase potentials with condensed phase behavior may be realized. Three-body dispersion might be a further improvement for this connection.

Various contemporary efforts at constructing model potentials (e.g., Refs. 143–168) point to this as a rapidly emerging area of technology development, an "industry" in support of molecular simulations. It is probably at a stage like that of the technology of ab initio calculations when small basis sets, small molecules, and limited treatments of electron correlation were typical. That would be a time about 30 years ago or so. The comparison, though, does not hold for accuracy, which is already quite high in many types of models, and it does not mean that model potentials will be getting very much more complex and extensive. Rather, the comparison with the development of ab initio electronic structure technology is to suggest that this is an area of computational chemistry that is likely to be very significant on its own. There is growing consensus on means for constructing potentials, and there are more and more critical comparisons and tests. There are likely to be fundamental improvements in the technology, things that would compare to the successive developments of high-level electron correlation treatments of ab initio technology. There are issues that resemble the "art" that goes into basis set selection, such as how to achieve computational simplicity in a model potential while maintaining suitable accuracy. Also, there are issues unlike those of the ab initio methodologies, such as how far to go in devising potentials that are transferable. In many ways, this suggestion of an emerging computational area follows the idea that weak interaction is its own chemistry with molecules, not atoms, as the building blocks. Whereas ab initio electronic structure treats atomic building blocks, interaction potential technology treats molecule building blocks. This gives even more impetus to model construction that puts property changes on an equal or nearly equal footing with the energetics.

Acknowledgments

The support of the National Science Foundation for investigations related to weak interaction over a considerable period and recently via Grant CHE-0131932 is gratefully acknowledged.

References

1. C. A. Coulson, *Research (Lond.)* **10**, 149 (1957).

2. A. D. Buckingham, in *Intermolecular Interactions: From Diatomics to Biopolymers*, B. Pullman, ed., John Wiley & Sons, New York, 1978, p. 1.

3. A. Beyer, A. Karpfen, and P. Schuster, *Topics Curr. Chem.* **120**, 1 (1984).

4. C. E. Dykstra and J. M. Lisy, in *Comparison of Ab Initio Quantum Chemistry with Experiment for Small Molecules*, R. J. Bartlett, ed., Reidel, Dordrecht, 1985, p. 245.

5. A. D. Buckingham, P. W. Fowler, and A. J. Stone, *Int. Rev. Phys. Chem.* **5**, 107 (1986).

6. P. L. Cummins, A. P. L. Rendell, D. J. Swanton, G. B. Bacskay, and N. S. Hush, *Int. Rev. Phys. Chem.* **5**, 139 (1986).

7. J. H. van Lenthe, J. G. C. M. van Duijneveldt-van de Rijdt, and F. B. van Duijneveldt, *Adv. Chem. Phys.* **69**, 521 (1987).

8. A. D. Buckingham, P. W. Fowler, and J. M. Hutson, *Chem. Rev.* **88**, 963 (1988).

9. A. C. Legon, *Chem. Soc. Rev.* **19**, 197 (1990).

10. G. Chalasinski, *Chem. Rev.* **94**, 1723 (1994).

11. B. Jeziorski, R. Moszynski, and K. Szalewicz, *Chem. Rev.* **94**, 1887 (1994).

12. M. J. Elrod and R. J. Saykally, *Chem. Rev.* **94**, 1975 (1994).

13. M. Quack and W. Kutzelnigg, *Ber. Bunsenges. Phys. Chem.* **99**, 231 (1995).

14. A. J. Stone, *The Theory of Intermolecular Forces*, Oxford University Press, New York, 1996.

15. S. Scheiner, *Hydrogen Bonding. A Theoretical Perspective*, Oxford University Press, New York, 1997.

16. A. J. Thakkar, *Intermolecular Interactions* in Encyclopedia of Chemical Physics and Physical Chemistry, IOP Publishing, Bristol, 2000.

17. S. L. Price, *Rev. Comp. Chem.* **14**, 225 (2000).

18. A. D. Buckingham, *Adv. Chem. Phys.* **12**, 107 (1967)

19. A. Karpfen, *Chem. Phys.* **79**, 211 (1983); M. Kofranek, H. Lischka and A. Karpfen, *Mol. Phys.* **61**, 1519 (1987); M. Kofranek, A. Karpfen and H. Lischka, *Chem. Phys.* **113**, 53 (1987).

20. B. F. King and F. Weinhold, *J. Chem. Phys.* **103**, 333 (1995).

21. A. J. Stone, A. D. Buckingham and P. W. Fowler, *J. Chem. Phys.* **107**, 1030 (1997).

22. C. E. Dykstra, *J. Mol. Struct. (Theochem.)* **362**, 1 (1996).

23. A. J. Stone, *Chem. Phys. Lett.* **211**, 101 (1993).

24. C. E. Dykstra, *J. Chem. Phys.* **108**, 6619 (1998).

25. S. J. Harris, S. E. Novick, and W. Klemperer, *J. Chem. Phys.* **60**, 3208 (1974); T. A. Dixon, C. H. Joyner, F. A. Baiocchi, and W. Klemperer, *J. Chem. Phys.* **74**, 6539 (1981).

26. H. S. Gutowsky, T. D. Klots, C. Chuang, C. A. Schmuttenmaer, and T. Emilsson, *J. Chem. Phys.* **86**, 569 (1987).

27. H. S. Gutowsky, T. D. Klots, C. Chuang, J. D. Keen, C. A. Schmuttenmaer, and T. Emilsson, *J. Am. Chem. Soc.* **107**, 7174 (1985); **109**, 5633 (1987).

28. H. S. Gutowsky, C. Chuang, T. D. Klots, T. Emilsson, R. S. Ruoff, and K. R. Krause, *J. Chem. Phys.* **88**, 2919 (1988).

29. G. T. Fraser and A. S. Pine, *J. Chem. Phys.* **85**, 2502 (1986).

30. A. McIlroy, R. Lascola, C. M. Lovejoy, and D. J. Nesbitt, *J. Phys. Chem.* **95**, 2636 (1991).

31. J. M. Hutson and B. J. Howard, *Mol. Phys.* **45**, 791 (1982); J. M. Hutson, *J. Chem. Phys.* **96**, 6752 (1992); A. Ernesti and J. M. Hutson, *Phys. Rev. A* **51**, 239 (1995).

32. P. Niyaz, Z. Bacic, J. W. Moskowitz, and K. E. Schmidt, *Chem. Phys. Lett.* **252**, 23 (1996).

33. S. Liu, Z. Bacic, J. W. Moskowitz, and K. E. Schmidt, *J. Chem. Phys.* **100**, 7166 (1994); **101**, 10181 (1994); **103**, 1829 (1995).

34. J. M. Hutson, S. Liu, J. W. Moskowitz, and Z. Bacic, *J. Chem. Phys.* **111**, 8378 (1999).

35. M. Quack and M. A. Suhm, *J. Chem. Phys.* **95**, 28 (1991); W. Klopper, M. Quack, and M. A. Suhm, *Mol. Phys.* **94**, 105 (1998); M. Quack and M. A. Suhm, in *Conceptual Perspectives in Quantum Chemistry*, J.-L. Calais and E. Kryachko, eds., Kluwer Academic Publishers, Dordrecht, 1997, p. 415.

36. W. Klopper, M. Quack, and M. A. Suhm, *J. Chem. Phys.* **108**, 10096 (1998).

37. C. E. Dykstra, *Chem. Rev.* **93**, 2339 (1993).

38. W. Kutzelnigg and W. Klopper, *J. Chem. Phys.* **94**, 1985 (1991).

39. W. Klopper, *Chem. Phys. Lett.* **186**, 583 (1991).

40. J. Noga, W. Kutzelnigg, and W. Klopper, *Chem. Phys. Lett.* **199**, 497 (1992).

41. J. Noga and W. Kutzelnigg, *J. Chem. Phys.* **101**, 7738 (1994).

42. S. F. Boys and F. Bernardi, *Mol. Phys.* **19**, 553 (1970).

43. E. R. Davidson and D. Feller, *Chem. Rev.* **86**, 681 (1986).

44. F. B. van Duijneveldt, J. G. C. M. van Duijneveldt-van de Rijdt, and J. H. van Lenthe, *Chem. Rev.* **94**, 1873 (1994).

45. E. R. Davidson and S. J. Chakravorty, *Chem. Phys. Lett.* **217**, 48 (1994).

46. N. R. Kestner and J. E. Combariza, *Rev. Comp. Chem.* **13**, 99 (1999).

47. J. Simons and P. Skurski, in *Recent Research Developments in Physical Chemistry*, World Scientific, Singapore, 2002.

48. G. de Oliveira and C. E. Dykstra, *J. Mol. Struct. (Theochem.)* **337**, 1 (1995).

49. F.-M. Tao and Y.-K. Pan, *J. Chem. Phys.* **97**, 4989 (1992).

50. F.-M. Tao, *J. Chem. Phys.* **98**, 3049 (1993); **100**, 3645 (1994).

51. L. Zhi-Ru, W. Di, L. Ze-Sheng, H. Xu-Ri, F.-M. Tao, and S. Chia-Chung, *J. Phys. Chem. A* **105**, 1163 (2001).

52. H. L. Williams, E. M. Mas, K. Szalewicz, and B. Jeziorski, *J. Chem. Phys.* **103**, 7374 (1995).

53. J. A. Pople, J. S. Binkley, and R. Seeger, *Int. J. Quantum Chem.* **S10**, 1 (1976).

54. R. J. Bartlett, *Annu. Rev. Phys. Chem.* **32**, 359 (1981).

55. J. Cizek, *Adv. Chem. Phys.* **14**, 35 (1968).

56. J. Cizek and J. Paldus, *Int. J. Quantum Chem.* **5**, 359 (1971).

57. R. J. Bartlett and G. D. Purvis, *Int. J. Quantum Chem.* **14**, 561 (1978); *Phys. Scripta* **21**, 255 (1980).

58. N. Nakatsuji and K. Hirao, *J. Chem. Phys.* **68**, 2053 (1978).

59. J. A. Pople, R. Krishnan, H. B. Schlegel, and J. S. Binkley, *Int. J. Quantum Chem.* **14**, 545 (1978).

60. W. Klopper and J. Noga, *J. Chem. Phys.* **103**, 6127 (1995).

61. S. Tsuzuki, W. Klopper, and H. P. Lüthi, *J. Chem. Phys.* **111**, 3846 (1999).

62. R. G. Parr and W. Yang, *Density-Functional Theory of Atoms and Molecules*, Oxford University Press, New York, 1989.

63. J. E. Del Bene, W. B. Person, and K. Szczepaniak, *J. Phys. Chem.* **99**, 10705 (1995).

64. J. J. Novoa and C. Sosa, *J. Phys. Chem.* **99**, 15837 (1995).

65. H. C. Kang, *Chem. Phys. Lett.* **254**, 135 (1996).

66. L. Gonzalez, O. Mo, M. Yanez, *J. Mol. Struct. (Theochem.)* **371**, 1 (1996); *J. Chem. Phys.* **109**, 139 (1998).

67. C. Maerker, P. V. R. Schleyer, K. R. Liedl, T.-K. Ha. M. Quack, and M. A. Suhm, *J. Comp. Chem.* **18**, 1695 (1997).

68. V. Subrainanian, K. Chitra, D. Sivanesan, M. Renuka, S. Sankar, and T. Ramasami, *J. Mol. Struct. (Theochem.)* **431**, 181 (1998).

69. T. Cwiok, B. Jeziorski, W. Kolos, R. Moszynski, and K. Szalewicz, *J. Chem. Phys.* **97**, 7555 (1992).

70. T. Cwiok, B. Jeziorski, W. Kolos, R. Moszynski, and K. Szalewicz, *J. Mol. Struct. (Theochem.)* **307**, 135 (1994).

71. R. Moszynski, P. E. S. Wormer, B. Jeziorski, and A. van der Avoird, *J. Chem. Phys.* **103**, 8058 (1995).

72. H. L. Williams, K. Szalewicz, B. Jeziorski, R. Moszynski, and S. Rybak, *J. Chem. Phys.* **98**, 1279 (1993).

73. R. Moszynski, P. E. S. Wormer, B. Jeziorski, and A. van der Avoird, *J. Chem. Phys.* **101**, 2811 (1994).

74. R. Moszynski, P. E. S. Wormer, and A. van der Avoird, *J. Chem. Phys.* **102**, 8385 (1995).

75. R. Bukowski, J. Sadlej, B. Jeziorski, P. Jankowski, K. Szalewicz, S. A. Kucharski, H. L. Williams, and B. M. Rice, *J. Chem. Phys.* **110**, 3785 (1999).

76. G. Murdachaew, A. J. Misquita, R. Bukowski, and K. Szalewicz, *J. Chem. Phys.* **114**, 764 (2001).

77. E. M. Mas, R. Bukowksi, K. Szalewicz, G. C. Groenenboom, P. E. S. Wormer, and A. van der Avoird, *J. Chem. Phys.* **113**, 6687 (2000).

78. R. R. Toczylowski, F. Doloresco, and S. M. Cybulski, *J. Chem. Phys.* **114**, 851 (2001).

79. M. A. Suhm, *Chem. Phys. Lett.* **214**, 373 (1993); **223**, 474 (1994).

80. D. F. R. Brown, M. N. Gibbs, and D. C. Clary, *J. Chem. Phys.* **105**, 7597 (1996).

81. G. de Oliveira and C. E. Dykstra, *J. Chem. Phys.* **106**, 5316 (1997); **110**, 289 (1999).

82. F. T. Smith, *Phys. Rev. A* **5**, 1708 (1972).

83. K. T. Tang, *Phys. Rev.* **177**, 108 (1969).

84. W. Kutzelnigg and F. Maeder, *Chem. Phys.* **35**, 397 (1978).

85. A. J. Thakkar, *J. Chem. Phys.* **81**, 1919 (1984).

86. B. Champagne and D. H. Mosley, *J. Chem. Phys.* **105**, 3592 (1996).

87. A FORTRAN program may be downloaded from www.chem.iupui.edu/faculty/images/Romberg_fitting_program.html.

88. T. H. Dunning, *J. Chem. Phys.* **90**, 1007 (1989); R. A. Kendall, T. H. Dunning, Jr., and R. J. Harrison, *J. Chem. Phys.* **96**, 6796 (1992); D. E. Woon and T. H. Dunning, Jr., *J. Chem. Phys.* **98**, 1358 (1993).

89. G. Maroulis, *Chem. Phys. Lett.* **226**, 420 (1994).

90. D. M. Bishop, F. L. Gu, and S. M. Cybulski, *J. Chem. Phys.* **109**, 8407 (1998).

91. R. J. Doerksen and A. J. Thakkar, *J. Phys. Chem. A* **103**, 2141 (1999).

92. C. E. Dykstra, S.-Y. Liu and D. J. Malik, *Adv. Chem. Phys.* **75**, 37 (1989).

93. H. -J. Werner and W. Meyer, *Mol. Phys.* **31**, 855 (1976).

94. R. J. Bartlett and G. D. Purvis, *Phys. Rev. A* **20**, 1313 (1979).

95. R. D. Amos, *Chem. Phys. Lett.* **70**, 613 (1980); **88**, 89 (1982).

96. E.-A. Reinsch, *J. Chem. Phys.* **83**, 5784 (1985).

97. G. Maroulis and A. J. Thakkar, *J. Chem. Phys.* **93**, 652 (1990).

98. P. W. Fowler and G. H. F. Diercksen, *Chem. Phys. Lett.* **164**, 105 (1990).

99. H. Sekino and R. J. Bartlett, *J. Chem. Phys.* **98**, 3022 (1993).

100. B. K. Lee, J. M. Stout, and C. E. Dykstra, *J. Mol. Struct. (Theochem.)* **400**, 57 (1997).

101. J. M. Stout and C. E. Dykstra, *J. Am. Chem. Soc.* **117**, 5127 (1995); *J. Phys. Chem. A* **102**, 1576 (1998).

102. R. A. Chiles and C. E. Dykstra, *J. Chem. Phys.* **74**, 4544 (1981).

103. K. Jankowski and J. Paldus, *Int. J. Quantum Chem.* **18**, 1243 (1980).

104. R. A. Chiles and C. E. Dykstra, *Chem. Phys. Lett.* **80**, 69 (1981).

105. C. E. Dykstra, S.-Y. Liu, M. F. Daskalakis, J. P. Lucia, and M. Takahashi, *Chem. Phys. Lett.* **137**, 266 (1987)

106. C. E. Dykstra and E. R. Davidson, *Int. J. Quantum Chem.* **78**, 226 (2000).

107. G. Maroulis, C. Makris, U. Hohm, and U. Wachsmuth, *J. Phys. Chem. A* **103**, 4359 (1999).

108. P. Norman, Y. Luo, D. Jonsson, and H. Agren, *J. Chem. Phys.* **106**, 1827 (1997).

109. B. Kirtman, J. L. Toto, C. Breneman, C. P. de Melo, and D. M. Bishop, *J. Chem. Phys.* **108**, 4355 (1998).

110. P. Norman, Y. Luo, D. Jonsson, and H. Agren, *J. Chem. Phys.* **108**, 4358 (1998).

111. J. D. Augspurger and C. E. Dykstra, *Mol. Phys.* **76**, 229 (1992).

112. A. J. Stone, *Chem. Phys. Lett.* **83**, 233 (1981).

113. A. J. Stone and M. Alderton, *Mol. Phys.* **56**, 1047 (1985).

114. W. Rijks and P. E. S. Wormer, *J. Chem. Phys.* **90**, 6507 (1989); **92**, 5754E (1990).

115. P. E. S. Wormer and H. Hettema, *J. Chem. Phys.* **97**, 5592 (1992).

116. H. Hettema, P. E. S. Wormer, and A. J. Thakkar, *Mol. Phys.* **80**, 533 (1993).

117. A. J. Thakkar, *J. Chem. Phys.* **75**, 4496 (1981).

118. A. J. Thakkar, H. Hettema, and P. E. Wormer, *J. Chem. Phys.* **97**, 3252 (1992).

119. R. Bukowski and K. Szalewicz, *J. Chem. Phys.* **114**, 9518 (2001).

120. G. D. Zeiss, W. J. Meath, J. C. F. MacDonald, and D. J. Dawson, *Can. J. Phys.* **55**, 2080 (1977).

121. A. Kumar, G. R. G. Fairley, and W. J. Meath, *J. Chem. Phys.* **83**, 70 (1985).

122. A. Kumar and W. J. Meath, *Mol. Phys.* **75**, 311 (1992).

123. K. Morokuma, *J. Chem. Phys.* **55**, 1236 (1971).

124. K. Kitaura and K. Morokuma, *Int. J. Quant. Chem.* **10**, 325 (1976).

125. H. Umeyama and K. Morokuma, *J. Am. Chem. Soc.* **99**, 1316 (1977).

126. P. Kollman, *J. Am. Chem. Soc.* **99**, 4875 (1977).

127. W. J. Stevens and W. H. Fink, *Chem. Phys. Lett.* **139**, 15 (1987).

128. K. G. Denbigh, *Trans. Faraday Soc.* **36**, 936 (1940).

129. A. I. Vogel, W. T. Creswell, G. J. Jefferey, I. Leicester, *Chem. Ind.* **1950**, 258.

130. J. O. Hirschfelder, C. F. Curtiss, and R. B. Bird, *Molecular Theory of Gases and Liquids*, John Wiley & Sons, New York, 1964; Chapter 13.

131. N. E. Kassimi, R. J. Doerksen, and A. J. Thakkar, *J. Phys. Chem.* **99**, 12790 (1995); *J. Phys. Chem.* **100**, 8752 (1996).

132. N. E. Kassimi and A. J. Thakkar, *J. Mol. Structure (Theochem.)* **366**, 185 (1996).

133. T. Zhou and C. E. Dykstra, *J. Phys. Chem. A* **104**, 2204 (2000).

134. K. O. Sylvester-Hvid, P.-O. Åstrand, M. A. Ratner, and K. V. Mikkelsen, *J. Phys. Chem. A* **103**, 1818 (1999).

135. K. J. Miller and J. A. Savchik, *J. Am. Chem. Soc.* **101**, 7206 (1979).

136. K. J. Miller, *J. Am. Chem. Soc.* **112**, 8533 (1990); *J. Am. Chem. Soc.* **112**, 8543 (1990).

137. K. E. Laidig and R. F. W. Bader, *J. Chem. Phys.* **93**, 7213 (1990).

138. R. F. W. Bader, T. A. Keith, K. M. Gough, and K. E. Laidig, *Mol. Phys.* **75**, 1167 (1992).

139. B. T. Thole, *Chem. Phys.* **59**, 341 (1981).

140. P. Th. van Duijnen and M. Swart, *J. Phys. Chem. A* **102**, 2399 (1998).

141. J. R. Maple and C. S. Ewig, *J. Chem. Phys.* **115**, 4981 (2001); C. S. Ewig, M. Waldman, and J. R. Maple, *J. Phys. Chem. A* **106**, 326 (2002).

142. B. Ma, J.-H. Lii, and N. L. Allinger, *J. Comput. Chem.* **21**, 813 (2000).

143. A. D. Buckingham and P. W. Fowler, *Can. J. Chem.* **63**, 2018 (1985).

144. M. A. Spackman, *J. Chem. Phys.* **85**, 6579 (1986); **85**, 6587 (1986).

145. M. A. Spackman, *J. Phys. Chem.* **91**, 3179 (1987).

146. C. E. Dykstra, *J. Am. Chem. Soc.* **111**, 6168 (1989); *J. Phys. Chem.* **94**, 180 (1990).

147. H. S. Gutowsky, A. C. Hoey, S. L. Tschopp, J. D. Keen, and C. E. Dykstra, *J. Chem. Phys.* **102**, 3032 (1995).

148. H. S. Gutowsky, E. Arunan, T. Emilsson, S. L. Tschopp, and C. E. Dykstra, *J. Chem. Phys.* **103**, 3917 (1995).

149. E. Arunan, C. E. Dykstra, T. Emilsson, and H. S. Gutowsky, *J. Chem. Phys.* **105**, 8495 (1996).

150. J. M. Sorenson, J. K. Gregory, and D. C. Clary, *J. Chem. Phys.* **106**, 849 (1997).

151. J. D. Augspurger, C. E. Dykstra, and T. S. Zwier, *J. Phys. Chem.* **97**, 980 (1993).

152. Y. Kong and J. W. Ponder, Abstract in Conference Program and Abstracts of Presentations, 29th Midwest Theoretical Chemistry Conference, May 1996.

153. P. N. Day, J. H. Jensen, M. S. Gordon, S. P. Webb, W. J. Stevens, M. Krauss, D. Garmer, H. Basch, and D. Cohen, *J. Chem. Phys.* **105**, 1968 (1996).

154. M. S. Gordon, M. A. Freitag, P. Bandyopadhyay, J. H. Jensen, V. Kairys, and W. J. Stevens, *J. Phys. Chem. A* **105**, 293 (2001).

155. S. P. Webb and M. S. Gordon, *J. Phys. Chem. A* **103**, 1265 (1999).

156. M. A. Freitag, M. S. Gordon, J. H. Jensen, and W. J. Stevens, *J. Chem. Phys.* **112**, 7300 (2000).

157. G. N. Merrill and M. S. Gordon, *J. Phys. Chem. A* **102**, 2650 (1998).

158. P. N. Day, R. Pachter, M. S. Gordon and G. N. Merrill, *J. Chem. Phys.* **112**, (2000) 2063.

159. E. R. Batista, S. S. Xantheas, and H. Jonsson, *J. Chem. Phys.* **112**, 3285 (2000).

160. E. R. Batista, H. Jonsson and S. S. Xantheas, William R. Wiley EMSL Annual Report 1999, p. 6-17; C. J. Burnham, J. C. Li, M. Leslie, and S. S. Xantheas, William R. Wiley EMSL Annual Report 1999, p. 6-18.

161. J. M. Hermida-Ramon and M. A. Rios, *J. Phys. Chem. A* **102**, 10818 (1998).

162. P. Sandler, J. Oh Jung, M. M. Szczesniak, and V. Buch, *J. Chem. Phys.* **101**, 1378 (1994).

163. M. P. Hodges, A. J. Stone, and E. C. Lago, *J. Phys. Chem. A* **102**, 2455 (1998).

164. C. Millot, J.-C. Soetens, M. T. C. Martins Costa, M. P. Hodges, and A. J. Stone, *J. Phys. Chem. A* **102**, 754 (1998).

165. A. W. Meredith and A. J. Stone, *J. Phys. Chem. A* **102**, 434 (1998).

166. R. Bukowksi, K. Szalewicz, and C. F. Chabalowski, *J. Phys. Chem. A* **103**, 7322 (1999).

167. K. Shuler and C. E. Dykstra, *J. Phys. Chem. A* **104**, 4562 (2000); **104**, 11522 (2000).

168. G. T. Fraser, R. D. Suenram, F. J. Lovas, A. S. Pine, J. T. Hougen, W. J. Lafferty, and J. S. Muenter, *J. Chem. Phys.* **89**, 6028 (1988).

169. J. Applequist, *J. Math. Phys.* **24**, 736 (1983); *Chem. Phys.* **85**, 279 (1984).

170. C. E. Dykstra, *J. Comput. Chem.* **9**, 476 (1988).

171. B. M. Axilrod and E. Teller, *J. Chem. Phys.* **11**, 299 (1943).

172. W. J. Meath and R. A. Aziz, *Mol. Phys.* **52**, 225 (1984).

173. W. J. Meath and M. Koulis, *J. Mol. Struct. (Theochem.)* **226**, 1 (1991).

174. R. J. Bell, *J. Phys. B* **3**, 571 (1970).

175. M. B. Doran and I. J. Zucker, *J. Phys. C Solid State Phys.* **4**, 307 (1971).

176. T. R. Horn, R. B. Gerber, J. J. Valentini, and M. A. Ratner, *J. Chem. Phys.* **94**, 6728 (1991).

177. J. Hepburn, G. Scoles, and R. Penco, *Chem. Phys. Lett.* **36**, 451 (1975).

178. R. Ahlrichs, R. Penco, and G. Scoles, *Chem. Phys. Lett.* **19**, 119 (1977).

179. C. Douketis, G. Scoles, S. Marchetti, M. Zen, and A. J. Thakkar, *J. Chem. Phys.* **76**, 3057 (1982).

180. K. T. Tang and J. P. Toennies, *J. Chem. Phys.* **66**, 1496 (1977); **68**, 5501 (1978).

181. C. Douketis, J. M. Hutson, B. J. Orr, and G. Scoles, *Mol. Phys.* **52**, 763 (1984).

182. P. J. Knowles and W. J. Meath, *Chem. Phys. Lett.* **124**, 164 (1986); *Mol. Phys.* **59**, 965 (1986); *Mol. Phys.* **60**, 1143 (1987).

183. R. J. Wheatley and W. J. Meath, *Mol. Phys.* **80**, 25 (1993); *Chem. Phys.* **179**, 341 (1994).

184. E. Oldfield, K. Guo, J. D. Augspurger, and C. E. Dykstra, *J. Am. Chem. Soc.* **113**, 7537 (1991).

185. C. E. Dykstra, *Acc. Chem. Res.* **21**, 355 (1988).

186. J. D. Augspurger and C. E. Dykstra, *Annu. Rept. NMR Spectrosc.* **30**, 1 (1995).

187. J. D. Augspurger and C. E. Dykstra, *J. Chem. Phys.* **99**, 1828 (1993).

188. A. I. Jaman, T. C. Germann, H. S. Gutowsky, J. D. Augspurger, and C. E. Dykstra, *Chem. Phys.* **154**, 281 (1991).

189. P. L. Cummins, G. B. Bacskay, and N. S. Hush, *J. Phys. Chem.* **89**, 2151 (1985).

190. P. L. Cummins, G. B. Bacskay, and N. S. Hush, *Chem. Phys.* **115**, 325 (1987).

191. P. L. Cummins, G. B. Bacskay, and N. S. Hush, *Mol. Phys.* **61**, 795 (1987); **62**, 193 (1987).

192. B. Kirtman, C. E. Dykstra, and B. Champagne, *Chem. Phys. Lett.* **305**, 132 (1999).

193. J. Applequist, J. R. Carl, and K.-K. Fung, *J. Am. Chem. Soc.* **94**, 2952 (1972); J. Applequist, *J. Chem. Phys.* **83**, 809 (1985).

194. A. J. Stone, *Mol. Phys.* **56**, 1065 (1985).

195. R. W. Munn, *Mol. Phys.* **64**, 1 (1988).

196. C. R. Le Seur and A. J. Stone, *Mol. Phys.* **78**, 1267 (1993).

197. F. Dehez, J.-C. Soetens, C. Chipot, J. G. Angyan, and C. Millot, *J. Phys. Chem. A* **104**, 1293 (2000).

198. K. L. C. Hunt, Y. Q. Liang, R. Nimalakirthi, and R. A. Harris, *J. Chem. Phys.* **91**, 5251 (1989).

199. K. L. C. Hunt and Y. Q. Liang, *J. Chem. Phys.* **95**, 2549 (1991).

200. P.-H. Liu and K. L. C. Hunt, *J. Chem. Phys.* **100**, 2800 (1994).

201. K. L. C. Hunt, *J. Chem. Phys.* **103**, 3552 (1995).

202. E. L. Tisko, X. Li, and K. L. C. Hunt, *J. Chem. Phys.* **103**, 6873 (1995).

203. K. L. C. Hunt, *J. Chem. Phys.* **90**, 4909 (1989).

204. P. W. Fowler and A. D. Buckingham, *Chem. Phys.* **98**, 167 (1985).

205. J. Baker, A. D. Buckingham, P. W. Fowler, E. Steiner, P. Lazzeretti, and R. Zanasi, *J. Chem. Soc. Farady Trans.* 2, **85**, 901 (1989).

206. B. M. Ladanyi and M. S. Skaf, *Annu. Rev. Phys. Chem.* **44**, 335 (1993).

207. A. Wallqvist and R. D. Mountain, *Rev. Comp. Chem.* **13**, 183 (1999).

CALCULATIONS OF NONLINEAR OPTICAL PROPERTIES FOR THE SOLID STATE

BENOÎT CHAMPAGNE[*] and DAVID M. BISHOP

Department of Chemistry, University of Ottawa, Ottawa, Canada

CONTENTS

I. INTRODUCTION

The calculation of the electric properties of individual molecules as found in an infinitely dilute gas has for long been of great interest to quantum chemists. This curiosity has been spurred in recent decades by the increasing importance of the communications industry in the world and the parallel need for materials having specific properties for electronic, optical, and other devices. In particular, the nonlinear-optical quantities, defined at the microscopic level as hyperpolarizabilities and at the macroscopic level as nonlinear susceptibilities, have played a

Permanent address: Laboratoire de Chimie Théorique Appliquée, Facultés Universitaires Notre-Dame de la Paix, Namur, Belgium

Advances in Chemical Physics, Volume 126, Edited by I. Prigogine and Stuart A. Rice.
ISBN 0-471-23582-2. © 2003 John Wiley & Sons, Inc.

key role in determining the suitability of substances for practical use—for example, in electro-optical switching and frequency mixing.

With few exceptions, a useful nonlinear optical material will be in the solid phase—for example, a single crystal or a poled polymer embedded in a film. Ironically, the quantum chemical calculations of nonlinear optical properties have for the most part been concerned with a single microscopic species. Much has been learned in this way about appropriate molecular construction, but the ultimate goal must be to investigate the nonlinear optical (NLO) properties in the solid phase.

In the physics arena the theoretical determination of NLO properties of solids has been more advanced, though not to the degree that has been achieved for simple gas-phase molecules using modern quantum chemical practices. For example, density functional theory in its crudest form has been frequently adopted to find some NLO properties for semiconductors. A glaring example of lack of progress is the third-order susceptibility of quartz. There is, as yet, no rigorous calculation of this quantity, even though it is the reference point for nearly all NLO measurements. It is our opinion that in the next few decades this situation is going to change—that is, the field of single molecule calculations will be saturated and attention will turn to the more practically relevant solid phase. This makes it an opportune time to review what computational strategies have been already developed. This, in turn, will indicate the more profitable lines of research to be pursued in the coming years.

As the situation currently stands, there are two extreme approaches: (i) the oriented-gas model and (ii) the supermolecule model—and everything in between. For (i), in its simplest disguise, single molecules lie side-by-side and the properties of the solid are just an appropriate combination of the molecular ones. For (ii), the whole solid is considered as a giant molecule and the computational approach is that of standard quantum chemical calculations. An important variation of (ii) is to make use of the translational symmetry that exists in a crystal or a periodic polymer, and this leads to so-called crystal orbital methods. Here is an important break from the conventional molecular calculations and one that is likely to see a great deal more use in the NLO field in the near future. Refinement of the oriented-gas model, including more and more precise accounting for interspecies interactions, is another likely avenue for progress.

The macroscopic optical responses of a medium are given by its linear and nonlinear susceptibilities, which are the expansion coefficients of the material polarization, P, in terms of the Maxwell fields, E [1–3]. For a dielectric or ferroelectric medium under the influence of an applied electric field, the defining equation reads

$$P = P_0 + \chi^{(1)}E + \chi^{(2)}E^2 + \chi^{(3)}E^3 + \cdots \tag{1}$$

where P_0 is the electric dipole moment per unit volume in zero electric field, $\chi^{(1)}$ is the (electric dipole) linear susceptibility, and $\chi^{(2)}$ and $\chi^{(3)}$ are the second- and third-order nonlinear susceptibilities. For the sake of clarity, the tensor nature and frequency-dependence of the susceptibilities have been omitted. In Eq. (1), the polarization and Maxwell fields are macroscopic quantities in the sense that they are derived from their microscopic counterpart by a suitable averaging procedure over a region which is large with respect to the atomic dimensions but small with respect to the wavelengths of the optical waves. For an isolated molecule, the electric dipole moment, μ, is expanded in a Taylor series of the external field:

$$\mu = \mu_0 + \alpha E + \frac{1}{2}\beta E^2 + \frac{1}{6}\gamma E^3 + \cdots \tag{2}$$

where μ_0, α, β, and γ are the permanent dipole moment, the polarizability, and the first and second hyperpolarizabilities, respectively. As we will see, these microscopic responses are, in many schemes, used as a starting point to evaluate the macroscopic responses, and the effects of the surroundings are introduced subsequently. This is the case for ionic and molecular solids where disjoint volumes can easily be associated with the constituting entities, and it also applies to bond/atom charge models. In such cases, the effects of the surroundings can be decomposed into two contributions. First, the surroundings modify the molecule/ion wave function and its properties through permanent electrostatic interactions as well as by polarization, dispersion, and exchange-overlap contributions. The corresponding (microscopic) NLO properties are generally referred to as "dressed" or effective hyperpolarizabilities. The second effect is the difference between the (local) electric field that (hyper)polarizes the entities and the macroscopic field that defines the bulk susceptibility. The inhomogeneity and the nonlocality of the responses to the applied fields are consequently crucial aspects to be included in these determinations.

In tune with the above introductory remarks, we have arranged this review in the following way: Section II deals with the oriented gas model that employs simple local field factors to relate the microscopic to the macroscopic nonlinear optical responses. The supermolecule and cluster methods are presented in Section III as a means of incorporating the various types of specific interactions between the entities forming the crystals. The field-induced and permanent mutual (hyper)polarization of the different entities then account for the differences between the macroscopic and local fields as well as for part of the effects of the surroundings. Other methods for their inclusion into the nonlinear susceptibility calculations are reviewed in Section IV. In Section V, the specifics of successive generations of crystal orbital approaches for determining the nonlinear responses of periodic infinite systems are presented. Finally, in Section VI, we consider both semiempirical methods, which typically incorporate the experimental susceptibility into the evaluation of the nonlinear responses, and techniques for determining the in-crystal ion hyperpolarizabilities as intermediates for evaluation of the susceptibilities of ionic crystals. In each section, the pros and cons of the different approaches are noted and illustrated by giving selected references, which focus on particular aspects of the methodology, or the calculations, or the characterization of materials.

II. THE ORIENTED GAS MODEL

The oriented gas model was first employed by Chemla et al. [4] to extract molecular second-order nonlinear optical (NLO) properties from crystal data and was based on earlier work by Bloembergen [5]. In this model, molecular hyperpolarizabilities are assumed to be additive and the macroscopic crystal susceptibilities are obtained by performing a tensor sum of the microscopic hyperpolarizabilities of the molecules that constitute the unit cell. The effects of the surroundings are approximated by using simple local field factors. The second-order nonlinear response, for example, is given by

$$\chi^{(2)}_{IJK}(-\omega_\sigma; \omega_1, \omega_2) = \frac{1}{2V} f_I^{\omega_\sigma} f_J^{\omega_1} f_K^{\omega_2} \sum_s \sum_{i,j,k} C^s_{Ii} C^s_{Jj} C^s_{Kk} \beta^s_{ijk}(-\omega_\sigma; \omega_1, \omega_2) \qquad (3)$$

where V is the unit cell volume, $f_L^{\omega_n}$ is the local field factor appropriate for the crystal axis L, and the circular frequency ω_n. The first sum runs over the unit cell molecules, and the second sum runs over the Cartesian coordinates of the molecular frame; the $C^s_{L\ell}$ coefficients are the scalar products of unit vectors along the crystal axis $\vec{L}$ and the molecular axis $\vec{\ell}$. Equation (3) assumes that the crystal symmetry is high enough that the $f_L^{\omega_n}$ tensors are diagonal in the crystallographic frame.

The validity of this model requires that the intermolecular forces (most often van der Waals) be much weaker than their intramolecular analogs, which are associated with the chemical bonds. To be more precise, provided the exact local field factors are used, the microscopic quantities deduced from $\chi^{(2)}$ or $\chi^{(3)}$ do not correspond to the β (first hyperpolarizability) and γ (second hyperpolarizability) values of the gas phase species but rather to those of the molecule "dressed" by its surroundings. Conversely, using the hyperpolarizabilities of the isolated molecules, Eq. (3) provides the susceptibilities of an "artificial" crystal because the geometry and/or the electronic properties of the constituent units may not be consistent with the true crystal. Of course, the gap between the model and the true values depends upon how large the van der Waals forces, the intermolecular charge transfer effects, the hydrogen bonds, and so on, are. Its appeal is to be understood not only by the age-old search in science for rationalizing phenomena in terms of simple additive parameters, but also by its success (sometimes only relative) in decomposing the molecular hyperpolariz-abilities into atomic, bond, or group contributions.

In Eq. (3), two of the local field factors (associated with the incident waves of circular frequency ω_1 and ω_2) account for the difference between the macroscopic (Maxwell) field and the local field actually experienced by the molecule/unit cell. The third (related to the generated wave at ω_σ) originates from the fact that the nonlinear contribution to the displacement vector is the product of the nonlinear polarization and the local field factor [5]. The local fields, and therefore the local field factors, should account, in principle, for the microscopic nature of the unit cell; but, in practice, this is far too complicated [6] and such an approach would negate the simplicity of the model. It has therefore commonly been preferred to replace them by "effective" or "average" local fields and local field factors that assume the fields to be uniform over the unit cell. Beyond the crudest approximation that consists in neglecting these local field factors altogether ($f_L^{\omega_n} = 1$), several expressions have been used to account for the anisotropic nature of medium. The simplest is the anisotropic Lorentz spherical cavity expression, $f_L^{\omega_n} = \frac{1}{3}[n_L^2(\omega_n) + 2]$, where $n_L(\omega_n)$ is the refractive index along the L axis and where the only electrostatic interactions taken into account are dipolar in nature. It is important to recall that by using this expression for $f_J^{\omega_1}$ and $f_K^{\omega_2}$, one assumes that the nonlinear contribution to the local field is negligible. This relationship can be improved by considering an ellipsoidal cavity for which analytical expressions can be found from solving the Maxwell equations with boundary conditions [7]. Further improvements to these local field factor expressions take account of the nondiagonality of the local-field tensor in the crystallographic frame [7], the self-polarization correction, which is a type of dressing of the isolated gas species [6], and retardation effects.

Many authors often consider that the local field factor corrections are small, but this is far from being true because they come in the form of a third (in $\chi^{(2)}$) or a fourth (in $\chi^{(3)}$) power of local field factors. An error by a few percent that may originate, for example, from the frequency-dispersion of the dielectric constant or from the size and shape of the cavity is therefore magnified and can easily become 50% or more. Just consider the change from a sphere to an ellipsoid where the large axis is twice as large as the other two. Using a typical $n_L(\omega_n)$ value of 2.0, this changes the local field factor of 1.33 to 1.17 and 1.41 for the long and small axes, respectively, thereby resulting in global corrections to third-order NLO processes, which are larger than -40% and 25%, respectively.

Chemla et al. [4] used the oriented gas model in combination with knowledge of the crystal symmetry and orientation of the molecules in the unit cell to resolve macroscopic crystal second-order susceptibility tensors into the contributions from microscopic units. The aim was to assess the validity of an additivity scheme for substituted benzenes. A similar approach was also later adopted by Oudar and Zyss [8] to study methyl-(2,4-dinitrophenyl)-aminopropanoate (MAP) crystals. The combination of crystal SHG data, treated using the oriented gas approximation, with EFISH (electric-field-induced second harmonic generation) measurements on MAP in solution enabled them to demonstrate the two-dimensional character of its first hyperpolarizability tensor. Finite-field semiempirical INDO (intermediate neglect of differential overlap) calculations were able to reproduce this two-dimensional character as well as the larger value of the diagonal component associated with the D/A (donor/acceptor) pair in the para position. The differences between the theoretically and experimentally deduced β-tensor components were as much as one order of magnitude. Subsequently, this model was used to define the MAP crystal structure, which exhibits the highest phase-matched crystal nonlinearity.

In their next paper, these authors [9] addressed the effect of the crystal point group on the relation between the microscopic and macroscopic nonlinear responses for both second harmonic generation (SHG) and the electro-optic dc-Pockels (dc-P) effect. Adopting the oriented gas approximation and assuming a one-dimensional character for the chromophores, they considered the influence of phase-matching conditions (for SHG) and determined the optimal orientation of the molecular conjugation axis within the crystalline frame. They also pointed out the limits of this additivity-based scheme for the electro-optic dc-Pockels effect due to the vibrational contributions associated with low-frequency intermolecular vibrations. In the same period, Zyss and Berthier [10] applied the additivity scheme in a slightly different way in order to determine the macroscopic second-order nonlinear response of urea crystals. A Coulomb point-charge potential was located around the urea molecules to account for the hydrogen bonds and the electrostatic interactions. This potential modified the electronic properties of the molecule in the crystal with respect to its isolated state and was found in a self-consistent way using the molecular wave functions.

During the next 20 years, the oriented gas approximation was frequently applied to connect microscopic and solid-state macroscopic NLO responses of organic systems. Table I gives, for both second- and third-order processes, information on typical investigations that are discussed in more detail in the following paragraphs.

Allen et al. [11] obtained for $\chi^{(2)}$ of 3-(1,1-dicyanonethenyl)-1-phenyl-4,5-dihydro-1H-pyrazole very good agreement between the measured quantity and the theoretical value calculated by combining this mean-field approximation with semiempirical β results. Typical β or $\chi^{(2)}$ determinations also include the ab initio CPHF (coupled-perturbed Hartree–Fock) study of Hamada on 3-aminoxanthone (AX) and 2-methyl-4-nitroaniline (MNA) [12–13] in which

TABLE I

Selected References Employing the Oriented Gas Approximation to Relate the Microscopic and Macroscopic Nonlinear Optical Responses in Crystals

Author, Year	Compound	Property	Ref.
Chemla et al. 1975	m-Dinitrobenzene, m-nitroaniline, m-aminophenol, and m-resorcinol	$\chi^{(2)}$	4
Oudar and Zyss, 1982	Methyl-(2,4-dinitrophenyl)-aminopropanoate	$\chi^{(2)}$	8
Allen et al., 1988	3-(1,1-Dicyanonethenyl) -1-phenyl-4,5-dihydro-1H-pyrazole	$\chi^{(2)}$	11
Hamada, 1996	3-Aminoxanthone	$\chi^{(2)}$	12
Hamada, 1996	Polyacetylene, polydiacetylene, polyyne	$\chi^{(3)}$	20
Yakimanski et al., 1997	Bis[4-(dimethylamino)benzylidene]acetone, 2,6-Bis[4-(dimethylamino)benzylidene] cyclohexanone	$\chi^{(2)}$	14
Voigt-Martin et al., 1997	1,3,5-Triamino-2,4,6-trinitrobenzene	$\chi^{(2)}$	15
Perpète et al., 1997	Polydiacetylene, polybutatriene	$\chi^{(3)}$	22
Yamada et al., 1997	p-Nitrophenyl-nitronyl-nitroxide	$\chi^{(3)}$	25
Cheng et al., 1998	β-BaB$_2$O$_4$	$\chi^{(2)}$	27
Hamada, 1998	Polysilane	$\chi^{(3)}$	21
Perpète et al., 1998	Polysilane	$\chi^{(3)}$	23
Luo et al., 2000	Fullerenes	$\chi^{(3)}$	26
Lin and Wu, 2000	3-Methyl-4-hydroxy-benzaldehyde	$\chi^{(2)}$	16
Zhu et al., 2000	Urea, m-dinitrobenzene, (N)-(4-nitrophenyl)- (L)-prolinol, m-aminophenol, 3-methyl-4-nitropyridine-1-oxide, nitro-4-pyridino-2-(L)-prolinol	$\chi^{(2)}$	17
Gu et al., 2001	Polyacetylene	$\chi^{(3)}$	24

he used Lorentz-type local field factors. It turns out that using this approach, $\chi^{(2)}$ of MNA is underestimated (even after correcting the experimental data to account for the most recent quartz reference data). The reasons for this, and mentioned by Hamada, are (i) the absence of electron correlation and frequency-dispersion, (ii) the underestimation of the local field corrections related to the non-self-consistent evaluation of $\chi^{(1)}$, and (iii) the limitations of the additivity model as a result of strong electrostatic intermolecular interactions [13].

Yakimanski et al. [14] followed the same scheme and evaluated the appropriate equations with static PM3 (Parametrized Model 3)—molecular—polarizability and first hyperpolarizability values in order to find the nonlinear optical potential of crystals built from two-dimensional intramolecular charge-transfer molecules (also called Λ-shape molecules). They found that in these Λ-shape molecules, contrary to the one-dimensional chromophores, the largest part of the molecular nonlinearity can be transformed into the macroscopic nonlinearity. Moreover, for the one-dimensional 2,5-bis(benzilidene)cyclo-pentanone (BBCP), they found good agreement between the theoretical

(10.3 pm/V) and experimental $(7 \pm 2$ pm/V) $d_{XYZ}(-2\omega; \omega, \omega)$ coefficients. In this way, they demonstrated the efficiency of combining the quantum chemical calculations of the (hyper)polarizabilities with electron crystallography in order to characterize compounds for which it is difficult to grow sufficiently large crystals for the purpose of measuring their NLO properties. Subsequently, Voigt-Martin et al. [15] showed that the strong SHG signal of 1,3,5-triamino-2,4,6-trinitrobenzene (TATB) microcrystal powder originates from both dipolar and octupolar contributions to the second-order susceptibility which itself arises from the noncentrosymmetry of one of the TATB crystal phases.

An improvement in estimating $\chi^{(2)}$ was made in a study by Lin and Wu [16], where the static β-tensor components of 3-methyl-4-hydroxy-benzaldehyde (MHBA) were calculated at the MP2/6-311 + + G** level of approximation. For some of the SHG NLO components (at $\lambda = 1064$ nm), theory overestimated experiment but frequency-dispersion had not yet been taken into account. Because in the crystals the MHBA molecules form pairs linked by hydrogen bonds, the first hyperpolarizability of MHBA dimers was evaluated and combined with the usual Lorenz–Lorentz local field factor for spherical entities. This, however, led to little improvement, probably because the dimer is less spherical than the monomer. Within the oriented gas approximation, another direction for improving the theoretical estimations of $\chi^{(2)}$ consists in using the characteristics of small clusters of chromophores in Eq. (3). Typically crystal unit cells may contain $2, 4, \ldots$ nonequivalent molecules. This approach is related to the earlier work of Zyss and Berthier [10] as well as to that of Lin and Wu [16] and can be seen as a hybrid between the oriented gas model and the supermolecule approach (see Section III). It enables one to take into account the interactions between the molecules and, particularly, the effects that originate from their large dipole moment. It has been used by Zhu et al. [17] in combination with first hyperpolarizability values evaluated with the summation-over-states (SOS) method and a ZINDO (Zerner intermediate neglect of differential overlap) Hamiltonian. Using local field factors for ellipsoids, $f_L^{\omega_n} = 1 + [n_L^2(\omega_n) - 1]A_L$ where A_L accounts for the relative lengths of the major and minor axes $(A_L = 1/3$ for spheres), as well as the two-state approximation to get the dynamic first hyperpolarizability values, they showed, in the case of six typical organic crystals, the superiority of the method based on the unit cell over the one that considers only a single molecule.

Due to their large third-order nonlinear responses, conjugated materials have received the attention of several groups, the most studied systems being the prototypical π-conjugated polyacetylene (PA), polydiacetylene (PDA) and its mesomeric polybutatriene (PBT) form, and the σ-conjugated polysilane (PSi). Due to π-electron delocalization, the longitudinal second hyperpolarizability per unit cell of the π-conjugated systems increases with chain length until a linear region is reached where the number of unit cells is between 20 and 100. As a consequence, besides the infinite-system approaches (see Section V), large oligomers are needed to assess the polymeric bulk response. The theoretical estimation of the $\chi^{(3)}$ value of these polymers can typically involve a first step where γ of larger and larger oligomers are evaluated and then extrapolated to the infinite chain length [18]. Consequently, the surrounding effects have only to be included for the two directions perpendicular to the chain axis. However, as a result of experimental difficulties (three-dimensional disorder, chain defects, etc.) as well as computational limitations (lack of electron correlation effects on both the geometry and electronic properties, lack of frequency dispersion, and lack of vibrational contributions), the agreement between experimentally deduced third-order responses and their theoretical counterparts has always been quite poor when adopting the additivity scheme: theory underestimating experiment by one to several orders of magnitude [19]. In this regard, using local field factor corrections has always reduced the gap between theory and experiment. The local field factors evaluated by Hamada [20] for PA chains employed the calculated asymptotic longitudinal polarizability per unit cell which, like the second hyperpolarizability, was calculated at the CPHF/6-31G level of approximation. The same approach was later employed for substituted PSi chains of different conformations [21]. Recently, Perpète et al. [22,23] inserted experimental refractive index data into the Lorentz expression to deduce the second hyperpolarizability per unit cell of substituted PDA and PSi. For these studies, with inclusion of the local field factor corrections, the theoretical $\chi^{(3)}$ estimates still remained one order of magnitude smaller than in experiment. Gu et al. [24] have also considered the Lorentz local field factor to discuss the differences between the experimental $\chi^{(3)}$ data for PA and their estimates based on CPHF/CO (coupled-perturbed Hartree–Fock/crystal orbital) calculations. It should be emphasized that all these $\chi^{(3)}$ calculations have employed expressions corresponding to spherical cavities whereas the polymers are in fact one-dimensional and the associated ellipsoidal cavity local field factor is unity for the longitudinal axis direction.

Among the few determinations of $\chi^{(3)}$ of molecular crystals, the CPHF/ INDO study of Yamada et al. [25] is unique because, on the one hand, it concerns an open-shell molecule, the p-nitrophenyl-nitronyl-nitroxide radical (p-NPNN) and, on the other hand, it combines in a hybrid way the oriented gas model and the supermolecule approach. Another study is due to Luo et al. [26], who calculated the third-order nonlinear susceptibility of amorphous thin-multilayered films of fullerenes by combining the self-consistent reaction field (SCRF) theory with cavity field factors. The amorphous nature of the system justifies the choice of the SCRF method, the removal of the sums in Eq. (3), and the use of the average second hyperpolarizability. They emphasized the differences between the Lorentz–Lorenz local field factors and the more general Onsager–Böttcher ones. For C_{60} the results differ by 25% but are in similar agreement with experiment, whereas for C_{70} the difference is larger but no comparison with experiment has been given.

The oriented gas model can also, in principle, be used to evaluate the nonlinear susceptibilities of inorganic crystals with or without accounting for the Coulomb effects in the hyperpolarizability determinations (see Section VI). Without these Coulomb effects, which can be large, Cheng et al. [27] have followed the oriented gas model for determining $\chi^{(2)}$ of β-BaB_2O_4 at the SOS/INDO/S level of investigation and found that oxygen-to-barium charge transfers dominate the second-order response. The microscopic-to-macroscopic transformation and the agreement with experiment have, however, been criticized [28]. There is no doubt that such approaches deserve to be pursued further.

III. SUPERMOLECULE AND CLUSTER METHODS

In supermolecule and cluster methods, the hyperpolarizabilities of entities containing several interacting molecules, atoms, or units are evaluated as a whole, just as in standard molecular calculations, to estimate the bulk nonlinear susceptibilities. The results can be analyzed by association with the different types of bonds present, and these range from the weak van der Waals and hydrogen bonds of molecular crystals to the strong bonds of ionic and covalent crystals. The quality of the calculated macroscopic NLO responses depends on the level of the theoretical treatment as well as on the size, shape, and boundaries of the composite species. Both aspects are equally important. The first is mostly responsible for describing the modifications of the electronic and geometrical properties of the target molecules/units when present in a crystal, while the difference between the applied fields and the fields experienced by the units under study is, to a large part, determined by the second.

When dealing with molecular crystals, Langmuir–Blodgett films, or chromophores dispersed in polymer matrices, besides possible strong ion-pair intermolecular interactions, there are two types of bond: strong intramolecular and weak intermolecular bonds. Therefore the distinction between the field aspect and the molecular aspect is clear. In this case, one can refer to the method as either the supermolecule or the cluster approach, but, for the sake of clarity, we will only use the former. On the other hand, in the case of covalent or ionic crystals, the two types of effect are intertwined and one refers to the cluster approach only because the system is no longer a supermolecule. Many theoretical investigations have addressed the impact of intermolecular interactions upon molecular properties by adopting the supermolecule approach, although the number of interacting entities included is often small. There are fewer studies on covalent and ionic clusters. The specifics of these two kinds of investigation as well as a brief survey of the most representative ones are addressed in Sections III.A and III.B.

There is at least one aspect of common concern for both types of systems: the convergence of the microscopic to the macroscopic bulk NLO properties. This convergence is related to the evolution of the response as a function of the size and the shape of the aggregate and to the extrapolation procedures (if used) to determine the properties of the infinite system. This is analogous to the situation encountered in the "oligomeric approach," which has been frequently and successfully employed to evaluate the properties of stereoregular polymers. There, since the chain length dependence is very strong, several extrapolation schemes have been employed [29–32]. However, in contradistinction to polymers, the systems under study here spread in all spatial directions and the number of atoms in the system can be very large before reaching convergence or even before being able to extrapolate. This is the reason why most supermolecule/cluster studies have been carried out at semiempirical levels or have concerned few-atom clusters or molecular dimers and trimers. It is interesting to note that for extrapolating the oligomeric values, it is natural to estimate the hyperpolarizability (for example, γ) of the infinite polymer per unit cell by taking the difference $\gamma(N) - \gamma(N-1) = \Delta\gamma(N)$ instead of dividing the hyperpolarizability by the number of unit cells $[\gamma(N)/N]$. Indeed the former converges faster because the end-chain effects are mostly removed [33]. Although it has not been used so far, the advantage of the $\Delta\gamma$ (or $\Delta\beta$) approach should be even greater for three-dimensional entities because the size of the boundary region, here a surface, grows rapidly with the size of the cluster. In this framework the search for a simple scheme that could reduce the three-dimensional problem to a problem of lower dimensionality is of great value. For instance, it has been found recently for MNA clusters that multiplying together the two crystal packing ratios associated with one-dimensional MNA arrays can reproduce within a few percent the first hyperpolarizability of two-dimensional aggregates [34,35] and therefore substantially reduce the computational effort. Although this scheme has been applied to determine the bulk nonlinear second-order susceptibility of the MNA crystal [34], its validity for three-dimensional arrays has not yet been proven, nor has it been generalized to other crystals, or even justified on physical grounds.

The shape issue is certainly as crucial as the size problem and the two are connected. When the size of the cluster is very large, the shape governs the relation between the external electric fields inserted into the Hamiltonian and the macroscopic or Maxwell fields that enter into the field-dependent expression for the macroscopic polarization [Eq. (1)]. For small clusters the shape is very important. The variations of β and γ resulting from the relative positions between the different units can often be largely accounted for by mutual polarization effects—that is, induced dipole moments on neighboring units modifying the field on the targeted species and vice versa. This can be illustrated by using two interacting nonpolar molecules and writing their second hyperpolarizability as a function of the dimer configuration. Because there are no dipolar interactions (which would have an important effect), in addition to the mutual polarization, the only remaining surrounding effects are the weak higher-order multipolar, London dispersion, and Pauli-type interactions. In one such case, the molecule chosen was the π-conjugated molecule of all-*trans* hexatriene, and the calculations were made at the CPHF/6-31G level of approximation (a method unable to account for most of the weak interactions). When the two molecules are placed side by side and in a parallel configuration like the in-phase chains in the crystal of polyacetylene [36], the dominant longitudinal static electronic second hyperpolarizability, $\gamma_L^e(0;0,0,0)$, of the dimer amounts to 1.28 times the value of the monomer, whereas for aligned molecules (with an intermolecular distance of 2.0 Å), the $\gamma_L^e(0;0,0,0)$ value becomes 3.04 times the monomer value. On the other hand, the simple additivity rule is obviously characterized by a multiplication factor of 2.0. As expected from simple electrostatics, the effect of mutual polarization is thus (a) a reduction of the hyperpolarizability per molecule when the molecules are stacked perpendicular to the field direction and (b) an increase when they are parallel to it. Of course the true bulk situation cannot be represented by these two extremes but rather by their correct balance, and it is necessary to average over various orientations to determine the bulk nonlinear response.

For isotropic materials, the aggregates should have either a spherical shape like the diamond T_d clusters of Bishop and Gu [37] or a cubic shape like the $Si_{54}H_{56}$ cluster characterized by Jansik et al. [38]. Expanding the aggregate in only one or two dimensions can lead to properties that are quite different from the bulk ones. For anisotropic systems like long conjugated chains (as well as polar donor/acceptor, D/A, compounds), the aggregates may have to have different shapes to ensure rapid convergence toward the infinite system properties. When the units are polar, the changes to the ground electronic state and geometric structures upon packing can be large and are associated with variations in the dipole moment and with charge transfer. However, in addition to these effects, one should not forget the ever-present mutual polarization that arises when external electric fields are switched on.

The electrostatic effects associated with the relative orientation of the chromophores show a specific connection with the corresponding electronic excited states and dipole transition moments that can, in turn, be related to changes in the hyperpolarizabilities via summation-over-states (SOS) expressions [1–3]. When the molecules interact, the electronic excited states split. In the case of a collinear arrangement, the intensities (oscillator strengths) are shifted to the red, that is, to the states of lower energies, whereas for a side by side and parallel configuration, the intensities are shifted to the blue. Furthermore, since the excited state polarizabilities are usually larger than the ground-state ones, the electrostatic effects upon the excited-state dipole moments are enhanced with respect to the ground state [39].

A. Molecular Crystals

The supermolecule approach is the most straightforward method to account for nonadditive and specific effects. Using perturbation theory, the interaction energy—and therefore the corresponding field-dependencies which give the successive hyperpolarizabilities—between molecules can be decomposed into electrostatic (or Coulomb), induction (or polarization), dispersion (or London forces), and exchange-overlap contributions [40,41]. However, the chemist analyzes the results in terms of hydrogen bonds and donor–acceptor and van der Waals interactions, which are combinations of the various "perturbation theory" terms. The dominant interaction contributions to the hyperpolarizabilities originate from the electrostatic and induction terms, whereas the dispersion and exchange-overlap terms are smaller and much more difficult to evaluate because they require electron-correlation treatments. In the case of pairs of atoms or molecules with high symmetry, detailed expressions have been derived for the induction and dispersion contributions to the hyperpolarizabilities [42] and applied to representative cases [43]. It is, however, not necessary to employ these methodologies to account for most of the impact of the surroundings on the NLO properties in large molecular aggregates. Indeed, the electrostatic and induction interactions can generally be reproduced by much simpler levels of approximation including ab initio Hartree–Fock theory and its various semiempirical offspring (e.g., AM1, PM3, INDO, ZINDO).

When using basis sets, one aspect of concern is the basis set superposition error (BSSE) that originates from an unbalanced treatment of the aggregate and its fragments. By employing the counterpoise correction scheme, Augspurger and Dykstra [44] have found that as soon as the basis set is large enough (inclusion of diffuse and polarization functions) to determine accurately the second hyperpolarizability of the acetylene dimer, the BSSE is small. Similarly, in the case of the CPHF/6-31G longitudinal second hyperpolarizability of hexatriene clusters [45,46] as well as pairs of parallel or perpendicular polyacetylene chains [63], the BSSE correction is small and can be neglected.

Table II provides a survey of representative calculations of β and γ of molecular aggregates with an eye toward determining the bulk susceptibilities or, at least, toward addressing the effect that crystal packing has on the NLO responses. Initial studies focused on the first hyperpolarizability, and this was a consequence of the earlier work by Chemla, Zyss, and collaborators and of the necessity to improve the oriented gas model (see Section II). Dirk et al. [47] studied dimers of urea and of (N)-(5-nitropyridyl)-2(L)-prolinol (PNP) by adopting a Pariser–Parr–Pople (PPP) Hamiltonian. They showed the limits of

TABLE II
Selected References Using the Supermolecule Approach to Determine the
Macroscopic Nonlinear Optical Responses in Molecular Crystals

Author, Year	Compound	Packing Type	Ref.
	β		
Dirk et al., 1986	(N)-(5-Nitropyridyl)-2(L)-prolinol, urea	Dimer	47
Waite and Papadopoulos, 1988	Water	Oligomers	49
Waite and Papadopoulos, 1990	Hydrogen fluoride	Oligomers	48
Yasukawa et al., 1990	p-Nitroaniline	Dimers	50
Perez and Dupuis, 1991	Urea	Dimer and trimers	54
Di Bella et al., 1992	p-Nitroaniline, 3-methyl-4-nitropyridine 1-oxide	Dimers and trimers	51
Di Bella et al., 1997	p-Nitroaniline	Charged dimer	52
Moliner et al., 1998	p-Nitroaniline and hydrogen fluoride	Linear dimers and oligomers	55
Castet and Champagne, 2001	2-Methyl-4-nitroaniline	1D and 2D clusters (up to 16 molecules)	34
Okuno et al., 2001	4-Dimethylamino-4'-carboxyazobenzene 4-dimethylamino-2'-nitro-4'-carboxyazobenzene	Dimer	53
Botek and Champagne, 2002	2-Methyl-4-nitroaniline	1D and 2D clusters (up to 16 molecules)	35
Guillaume et al., 2002	2-Methyl-4-nitroaniline	1D, 2D	56
	γ		
McWilliams and Soos, 1991	Polyacetylene oligomers	Dimer and heptamer	58
Perez and Dupuis, 1991	Urea	Dimer and trimer	54
Augspurger and Dykstra, 1992	Acetylene	1D	44
Kirtman, 1995	Polyacetylene oligomers	Dimer, trimer	45
Chen and Kurtz, 1996	Polyacetylene oligomers	Dimer	61
Champagne and Kirtman, 1998	Polyacetylene oligomers	Dimer to pentamer	62
Xie and Dirk, 1998	Stacked benzene	1D	64
Kirtman et al., 1999	All-trans hexatriene	Dimer to heptamer	46
Champagne and Kirtman, 2001	Polyacetylene	1D and 2D clusters	60
Kirtman et al., 2002	Polyacetylene oligomers (up to $C_{40}H_{42}$) and polymers	Dimers	63

the simple additivity scheme for β without considering mutual polarization effects, whereas the latter are employed to explain the change in dipole moment upon packing. Moreover, Dirk et al. pointed out that these induction effects are associated with the splitting of the excited state levels and the redistribution of the dipolar transition intensities. Waite and Papadopoulos determined the first and second hyperpolarizabilities of hydrogen fluoride [48] and water [49] clusters as prototypes of hydrogen-bonded systems and reported great sensitivity of the response to the configuration and size of the clusters.

The supermolecule approach has been used to elucidate the relationship between the relative orientation of molecular packing of small clusters of *p*-nitroaniline (PNA) [50,51] and 3-methyl-4-nitropyridine 1-oxide (POM) [51] and their second-order NLO response. The AM1 (Austin Model 1) investigation by Yasukawa et al. [50] pointed out the enhancement of β in the collinear dimer and its decrease for parallel stacking. The ZINDO study of Di Bella et al. [51] demonstrated that the dominant component of β is maximized for a stacked pair of PNA molecules when they are in a slipped co-facial arrangement such that the donor of one molecule is directly above the acceptor of the other. Other packing structures (stacked dimers including the situation where their dipolar axes are rotated with respect to one another) lead to a decrease of the first hyperpolarizability with respect to the value for the isolated molecule. For dimers and trimers in eclipsed conformations, β of both POM and PNA decreases with the intermolecular distance but the decrease is smaller for POM, which possesses a small dipole moment. For distances less than 3.5 Å, β of the PNA dimer starts to increase due to very strong dipole–dipole interactions, whereas for POM it simply continues to go down. Nevertheless, the effects on β, presented in Refs. 50 and 51, mostly result from simple electrostatic and polarization effects between at the most three molecules, and therefore they cannot really assess the bulk susceptibility. Later, the study of Di Bella et al. [51] was extended by considering charged PNA dimers [52].

Similar conclusions concerning the relation between the relative molecular orientation and the first hyperpolarizability have been drawn by Hamada [13] in the case of the MNA dimer as well as by Okuno et al. [53] in a study on linear and parallel dimers of 4-dimethylamino-4'-carboxyazobenzene and 4-dimethyl-amino-2'-nitro-4'-carboxyazobenzene. The impetus was the desire for an explanation for the large second-order nonlinear optical responses of cone-shaped azobenzene dendrimers. Okono et al. also assessed the adequacy of three classical electrostatic models (Section IV).

Perez and Dupuis [54] have performed ab initio static hyperpolarizability calculations on the urea monomer and dimer as well as linear and transverse trimers using extended basis sets. They showed that the additivity relationship holds well for the average γ but not for β (influenced by the intermolecular hydrogen bonds) nor for the individual components of the γ tensor. They also concluded that the intermolecular interactions in urea crystals may not be as significant as previously thought. Moliner et al. [55] performed ab initio HF investigations on the first hyperpolarizability of the collinear PNA dimer as well as on linear and zigzag $(HF)_N$ chains. Their results for PNA are in agreement with earlier studies [50,51]. For linear hydrogen fluoride chains, they found that the norm of β evolves sublinearly with the number of HF molecules whereas for zigzag chains, β has a maximum value for $N = 2$ and then decreases.

Aside from these investigations which have mostly considered only a few molecules and their relative orientation, recent CPHF/AM1 [34] and TDHF/AM1 [35,36] studies on MNA aggregates have concerned clusters containing up to 16 molecules. It was confirmed that in noncentrosymmetric crystals of polar D/A chromophores the effect of the crystal packing on the second-order nonlinear susceptibility can be highly anisotropic and that this effect is enhanced in the case of frequency-dependent phenomena. In particular, in the static limit, packing MNA dimers in the direction parallel to the dipole moment (and to the first hyperpolarizability) leads to an enhancement of the first hyperpolarizability by a factor larger than five, whereas in the perpendicular directions packing reduces the first hyperpolarizability by about 30% to 50%. By averaging over the three directions with a simple multiplicative scheme [34], it turns out that crystal packing leads to an enhancement of the dimer's first hyperpolarizability by almost a factor of two. These effects of the surrounding have been related to the hypsochromatic and bathochromatic shifts of the absorption bands [56] and are in agreement with the model of Wagnière and Hutter [39].

Reference 34 also showed that the CPHF/AM1 scheme reproduces to within 3% the crystal packing effects determined at the MP2/6-31G level of approximation. Nevertheless, it remains of interest to know how these effects are affected by using larger basis sets—that is, containing diffuse and polarization functions—as well as by including higher-order electron correlation. In addition, in none of these supermolecule studies on second-order NLO materials has the corresponding vibrational (pure vibrational and zero-point vibrational average) [57] response been considered.

In the case of third-order NLO materials, most supermolecule studies on the crystal packing effects have concerned the prototypical polyacetylene chains, and more precisely, small bundles of oligomers where the packing is perpendicular to the longitudinal axis. By adopting a PPP (Pariser–Parr–Pople) valence-bond configuration interaction method, McWilliams and Soos [58] have found that the interchain effects resulting from the first shell of nearest neighbors increase the third harmonic generation (THG) response of an idealized hexatriene chain. This conclusion was later criticized for being generalized to infinite polyacetylene chains [59]. In 1995, Kirtman [45] conducted an ab initio investigation of these packing effects for small aggregates of *trans*-butadiene and all-*trans* hexatriene. He showed that for either parallel or perpendicular interactions, the static second hyperpolarizability per chain, γ_L/chain, decreases substantially regardless of the chain length or the level of approximation. It was, however, already known that the packing effect defined by the ratio between the dimer and twice the monomer values, $R = \gamma[2]/(2 \times \gamma[1])$, is larger ($R$ is smaller) at the HF level than at the MP2 (second-order Møller–Plesset theory) level, while the polarizability of the chain decreases when going from HF to MP2. It is interesting to note that larger basis sets containing diffuse functions significantly alter the longitudinal second hyperpolarizability but not the relative effect of the interchain interaction [60]. Later, CPHF/6-31G calculations were performed at the experimental stretched-fiber-geometry [36] on an all-*trans* hexatriene bundle containing a complete shell of six nearest neighbors plus the central chain, and it was found that the effect of a single pair is enhanced so that the (γ_L/chain) value is reduced by a factor of 4.2 [46]. Similar reductions in the γ_L/chain value have also been determined from CPHF/AM1 calculations on idealized pairs and trios of polyacetylene chains placed one on top of the other [61]. In 1998, Champagne and Kirtman [62] continued the study by assessing, in the double harmonic oscillator approximation, the packing effects upon the vibrational second hyperpolarizability, which is dominated by the Raman term $[\alpha^2]^0$, and, subsequently, upon the ratio between the vibrational and electronic components. They distinguished between contributions from the high-frequency "bond-length-alternation" intrachain and low-frequency interchain vibrations and found that the former contribution behaves with cluster size in the same way as its electronic counterpart. On the other hand, the latter contribution leads, in small oligomers, to large effects that decrease rapidly with chain length.

The most recent studies of molecular crystals [60,63] have considered two other related aspects of interchain interactions: (i) the extension of the aggregates in the third direction (which corresponds to the growing axis of the polyacetylene fibers) and (ii) the increase of the PA chain length. In collinear arrays of all-*trans* hexatriene, the γ_L/chain value increases markedly with number of chains so that for two, three, and four chains the γ_L/chain ratio is 152%, 181%, and 198% of the isolated monomer value, respectively. These studies have not yet been extended to longer polyacetylene chains. When considering pairs of longer and longer oligomers in a parallel conformation, the substantial interchain effect on γ_L/chain reaches a maximum value for $C_{10}H_{12}$ and then decreases, so that for $C_{40}H_{42}$, the γ_L/chain reduction has become 19.9%, which is about half of the value at the maximum. These results are consistent with the modest effect of interchain interactions found for pairs of infinite chains in both the parallel and quasi-perpendicular conformations [63] as well as with simple electrostatics that would predict a local field factor of unity for an infinite chain (Sections II and IV). Since the effect of a single pair is amplified by completing the first shell of nearest neighbors and, to a lesser extent, the second shell of next-nearest neighbors [46], it is anticipated that larger bundles will be needed for a more definitive answer, but the effect is expected to be much smaller than for small bundles/arrays of short oligomers. The effect of the finite chain length distribution remains an open question. These results for γ of PA aggregates are consistent with the hypothesis that the interactions between the chains arise from mutual electrostatic polarization (induction) of the electron density and suggest that electrostatic models should be used (Section IV).

Xie and Dirk [64] have shown that, provided the system possesses a $+2$ charge, stacking benzene rings one on top of each other induces a very large negative second hyperpolarizability that grows supralinearly with the number of rings. This enhancement is, however, not only related to the extended nature of the system but mostly to the highly delocalizable bipolaronic defect. This work also suggests that crystal packing effects, and therefore their modeling, could be totally different in charged systems or systems displaying through-space conjugation [65].

B. Ionic and Covalent Crystals

Within the supermolecule or cluster methods, ionic and covalent crystals can be studied in similar way to molecular crystals with the exception that, for covalent and ionic crystals, when forming finite clusters, strong bonds are broken. In addition to the size and shape of the cluster, the convergence of cluster properties toward the bulk properties depends also upon the boundaries or, in other words, upon the nature of the chemical groups/atoms used to saturate the dangling bonds. These surface effects can influence the NLO responses of the clusters as well as their rate of convergence toward the bulk limit. In this sense, the choice of Bishop and Gu [37] to use diamond clusters possessing the T_d symmetry reduces the number of capping hydrogens and is efficient. At the same time, it is important to mention that the ab initio study by Tomonari et al. [66], where the Si(111) surface is modeled by silicon clusters, shows a drastic increase of the first hyperpolarizability in the presence of a dangling bond. Furthermore, in the literature there are numerous studies on the properties of clusters that differ significantly from those of the solid state in bulk or at surface. Among these, there is the study of Rantala et al. [67] on the second- and third-order NLO properties of silicon clusters containing between 7 and 13 silicon atoms. It is shown that the symmetry of the cluster, rather than its size, is the key factor that influences the hyperpolarizabilities. These concerns are, however, beyond the scope of this survey, but they draw attention to this crucial size/symmetry aspect while they are also related to some directionality aspects encountered in the supermolecule treatment.

A list of works concerned with the bulk NLO responses is given in Table III. Korambath and Karna [68] investigated, within the TDHF scheme, the second-order NLO properties of small clusters of group III–V elements (GaN, GaP, and GaAs) with the aim of characterizing the properties of the crystalline state and thin films. They showed that the first hyperpolarizability of such systems can be large. Nevertheless, since the results depend on the cluster size and symmetry, it turns out that larger aggregates are needed before drawing any quantitative conclusions for these systems.

TABLE III
Selected References for Nonlinear Optical Responses of Clusters

Author, Year	Compound	Packing type	Ref.
		β	
Rantala et al., 1990	Silicon	3D, Si_5–Si_{13}	67
Tomonari et al., 1997	Silicon	Interface	66
Korambath and Karna, 2000	GaN, GaP, and GaAs	3D	68
		γ	
Rantala et al., 1990	Silicon	3D, Si_5–Si_{13}	67
Bishop and Gu, 2000	Diamond	3D	37
Banerjee and Harbola, 2000	Metal	3D	69
Jansik et al., 20002	Silicon	1D, 2D, 3D	38

Banerjee and Harbola [69] have worked out a variation–perturbation method within the hydrodynamic approach to the time-dependent density functional theory (TDDFT) in order to evaluate the linear and nonlinear responses of alkali metal clusters. They employed the spherical jellium background model to determine the static and degenerate four-wave mixing (DFWM) γ and showed that γ evolves almost linearly with the number of atoms in the cluster.

Bishop and Gu [37] have tackled the evaluation of the bulk third-order susceptibility of diamond by building larger and larger clusters. The average $\chi^{(3)}$ value deduced from ab initio static CPHF calculations amounts to 2.68×10^{-14} esu in comparison with the experimental $\chi^{(3)}(-\omega; \omega, \omega, -\omega)$ value of 4.92×10^{-14} esu, whereas their CPHF/AM1 result is 5.79×10^{-14} esu. This study points out that the convergence with cluster size of the second hyperpolarizability per carbon atom is rather fast in comparison with the situation encountered in π-conjugated oligomers or one-dimensional arrays of D/A molecules. Indeed, it is not an exaggeration to say that in one-dimensional conjugated systems, 10 unit cells, at least, are necessary to get close to saturation, which, after translation to three-dimensional systems, would correspond to 1000 units. An analogous study has been carried out by Jansik et al. [38] on silicon clusters. After validating the use of semilocal effective core potentials, which allowed them to study larger systems, they found that the second hyperpolarizability per silicon atom did not follow a linear dependence in terms of the cluster size. This is probably related to the symmetry of the clusters, which varies as a function of the number of atoms. It is important to realize that, contrary to molecular crystals, these kinds of clusters with covalent (or partly covalent) bonds cannot be straightforwardly characterized by employing the oriented gas approximation because of the strong bonds between the different units.

IV. INTERACTION SCHEMES

Interaction schemes are equivalent to the oriented gas model in several ways: The molecules occupy distinct molecular volumes, the nonlinear susceptibilities are considered additive, and the differences between the macroscopic, E_{mac}, and local (or internal), E_{loc}, fields are accounted for by using local field factors. The local field given by the sum of the macroscopic field and the field due to the surrounding polarized entities is an average over a region containing several entities, typically the unit cell. Since most of the higher-multipole effects often vanish by symmetry, the treatment is usually limited to the dominant dipolar terms. In particular, Lorentz [70] assumed a cubic arrangement of identical particles; therefore, $E_{\text{loc}}/E_{\text{mac}} = f = \frac{1}{3}[n^2 + 2]$, where f is the local field factor and n is the refractive index. For anisotropic crystals, different refractive indices are used to represent the local field factors associated with the different crystal axes. This is the anisotropic Lorentz approximation employed within the oriented gas model (Section II). These expressions can, however, be improved, in order to account for the shape, anisotropy, and spatial distribution of the molecules in the crystals, by considering the electric dipole interactions between the molecules [71]. It is useful to express the local field factors with respect to the macroscopic or Maxwell field not just when considering infinite crystals and the long wavelength (zero frequency) limit [72], but because the macroscopic field enters in Eq. (1).   .

A list of papers employing interaction schemes to calculate NLO responses of clusters or crystals is given in Table IV. In Ref. 72, Munn summarized the essential steps for determining the local field factors and their role in relating the microscopic and macroscopic nonlinear optical responses. The local fields are determined in a self-consistent manner since the fields due to the surrounding molecules depend on their own local fields. They are also a function of the crystal geometry (via the Lorentz factors) and of the effective or dressed molecular polarizability. They can be obtained either for finite clusters or for infinite crystals by working in reciprocal space and by using the Ewald procedure to perform the lattice summations. In this case, the use of translational symmetry allows the response of the whole crystal to be expressed in terms of that of a single unit cell. In the simplest scheme, the nonlocal character of the in-crystal polarizability is neglected and each molecule is associated with a polarizable point dipole and the molecular dipole is given by the product of its effective polarizability with the local field. This is the polarizable point-dipole model which assumes that the molecules are distinct and that the polarizability response is local. In cases where the molecular dimensions are larger than the distances between the point dipoles, this leads to unrealistic results [73]. To alleviate this problem, the molecules can be divided into fragments, each associated with a polarizable dipole. Then, each fragment can interact with all the other fragments except those belonging to the same molecule. Consequently, for polymeric chains, there is no dielectric screening along the chain and the local field factor is unity. Indeed, the screening effects are implicitly included in evaluating the chain polarizability and thus in its contribution to the unit cell. In several cases, the polarizability distribution can be found by chemical intuition. For instance, in the case of naphthalene, which is made up of two identical fragments, the polarizability can be decomposed into two equivalent parts. Also, group or atom contributions can be deduced from a variety of schemes such as Stone's approach [74], the theory of atoms in molecules [75], the localization of molecular orbitals into "chemical functions" [76], atom/bond additivity [77], the use of the acceleration gauge for the electric dipole operator [78], quantum mechanically determined induction energies [79], or calculated molecular quadrupole polarizabilities and their derivatives with respect to molecular deformations [80]. Several of these models consider charge

TABLE IV
Selected References Using Interaction Schemes to Determine the
Macroscopic Nonlinear Optical Responses

Author, Year	Compound	Property	Ref.
Hurst and Munn, 1986	m-Nitroaniline Crystal	$\chi^{(2)}$	89
Hurst and Munn, 1987	Urea crystal	$\chi^{(2)}$, $\chi^{(3)}$	85
Dykstra et al., 1986	Hydrogen fluoride dimer	$\chi^{(2)}$	95
Hurst et al., 1990	Hexamine crystal	$\chi^{(2)}$	90
Munn, 1992	4-(N,N-dimethylamino)-2-acetamido-nitrobenzene crystal	$\chi^{(2)}$	91
Augspurger and Dykstra, 1992	Acetylene oligomers	$\chi^{(3)}$	44
Hamada, 1996	2-Methyl-4-nitroaniline dimer	$\chi^{(2)}$	13
Malagoli and Munn, 1997	Diethylamino-nitro-styrene crystal	$\chi^{(3)}$	99
Nobutoki and Koezuka, 1997	Urea and m-nitroaniline clusters	$\chi^{(2)}$	111
Munn et al., 1998	Buckminsterfullerene crystal	$\chi^{(2)}$	97
Reis et al., 1998	Benzene crystal	$\chi^{(3)}$	100
Reis et al., 1998	Urea crystal	$\chi^{(2)}$, $\chi^{(3)}$	101
Kirtman et al., 1999	Hexatriene bundles	$\chi^{(3)}$	46
Reis et al., 2000	Urea crystal	$\chi^{(2)}$	102
in het Panhuis and Munn, 2000	Model Langmuir–Blodgett multilayers	$\chi^{(2)}$	103
Malagoli and Munn, 2000	p-Nitroaniline surfaces	$\chi^{(2)}$	104
Reis et al., 2000	Naphthalene, anthracene, and m-nitroaniline crystals	$\chi^{(2)}$, $\chi^{(3)}$	105
in het Panhuis and Munn, 2000	Model Langmuir–Blodgett multilayers on substrate	$\chi^{(2)}$	106
in het Panhuis and Munn, 2000	Mixed films	$\chi^{(2)}$	107
Okuno et al., 2001	Dimers of donor-acceptor azobenzene	$\chi^{(2)}$	53
Reis et al., 2001	Ih, II, VIII, and IX polymorphs of ice	$\chi^{(3)}$	108
Munn, 2001	Hexamine crystal	$\chi^{(2)}$	109

flow as a source of polarizability in small [74,79] or elongated [81] systems. This is an important way to include nonlocal characteristics of the polarizability response function.

These "molecule interaction" schemes that exclude the intramolecular interactions have, however, to be distinguished from the "atom interaction" or "relay-type" models that originally were aimed at determining the molecular polarizabilities from the dipolar interactions between effective atomic polarizabilities and that require additional terms to account for the overlap between the atomic densities and for charge transfer effects [82]. Nevertheless, this approach was recently extended to describe both intra- and intermolecular interactions; and, consequently, the corresponding atomic parameters could be used to evaluate the crystal local field factors [83].

Hurst and Munn [72,84] described an extension of this treatment to the non-linear responses and showed that the molecular hyperpolarizabilities make the local fields nonlinear functions of the Maxwell field. In practice, these higher-order contributions to the local fields are generally neglected. However, they do give rise to "cascading" effects. In addition to the direct term associated with the molecular second hyperpolarizability, the third-order susceptibility contains a term arising from the second-order response to the linear and quadratic parts of the local field. The latter term is therefore quadratic in the first hyper-polarizability. Hurst and Munn [85] showed that for urea the cascading contribution to certain $\chi^{(3)}$ tensor components was not negligible. In the same fashion as for the polarizabilities, the hyperpolarizabilities can be distributed according to several models based on chemical intuition, additivity/transfer-ability schemes [86] or on more formal developments [87]. Hurst and Munn [88] emphasized the fact that the "molecular" (hyper)polarizabilities in crystals are different from the isolated case. There are various reasons for this, one is the electrostatic field of the permanent charge distribution of the surrounding molecules, others include short-range effects like electron confinement and the nonlocal contributions to the hyperpolarizabilities (see also Section III). Since these dressing effects are generally difficult to determine, they proposed a simple scheme to account for the permanent dipole field effects that are expected to be dominant for polar molecules.

By combining experimental and theoretical quantities, Munn and collaborators [85,89–91] determined the nonlinear susceptibilities of certain organic molecular crystals. Typically, their approach consisted of first calculating the Lorentz factors from the crystal structure and then by combining them with the experimental refractive indices to determine the in-crystal dressed molecular polarizabilities. Finally, the local field factors were calculated and inserted together with the calculated hyperpolarizabilities into the nonlinear suscept-ibility expressions. By using experimental refractive indices, they explicitly ac-counted for specific crystal packing effects such as nuclear and electronic rearrangements; these modify the molecular wavefunctions and properties. On the other hand, this process of determining the molecular polarizabilities from the refractive indices yields a unique answer only in particular cases such as crystals with one molecule per primitive unit cell. In other cases, different polarizabilities can give the same susceptibility, and (symmetry) constraints have to be imposed to eliminate the arbitrary parameters [92]. However, by using theoretical free-molecule β values, they omitted hydrogen bonding and other dipolar interactions that would be expected to modify the molecular first hyperpolarizability. They showed the limitations of the anisotropic Lorentz approximation which, in the case of 4-(N, N-dimethylamino)-2-acetamido-nitrobenzene, predicted the wrong local field anisotropy [91]. The agreement between the calculated values and the experimental SHG data was poor, and this was explained by the limitation of the semiempirical first hyperpolarizability calculation and the lack of specific intermolecular interactions.

By using simple scaling laws for the polarizability and first hyperpolarizability as a function of the molecular size, Hurst and Munn [93] addressed, within the point dipole approximation, the relationship between $\chi^{(2)}$ and molecular elongation and found that very large $\chi^{(2)}$ can be obtained for compact molecules, provided that the ratio between the polarizability and the molecular volume is large.

The approach recommended by Dykstra [94] for determining the hyperpolarizabilities of clusters is based on the evaluation of the interaction energy containing terms associated with the interactions between their unperturbed charge fields and with their relaxation due to mutual polarization. The interaction energy between the charge distributions was calculated with a multipole expansion. The different quantities in these expansions (permanent and induced multipoles) were evaluated by adopting general ab initio quantum chemical procedures. The hyperpolarizabilities of a given system were then obtained by differentiating the field-dependent interaction energy that had been obtained in a self-consistent manner. Depending upon the truncation of the multipole expansion and the presence of additional terms to account for nonclassical elements, the mutual electrical (hyper)polarization approach of Dykstra is similar to the technique developed by Munn. However, since in the latter case the central quantities are the induced polarization or local fields, the frequency dependence of the nonlinear susceptibilities can also be treated, whereas time-dependent phenomena have yet to be considered in the interaction energy expressions. As in Refs. 72 and 84, the Dykstra scheme requires the distribution of the (hyper)polarizability tensors over the interacting molecules and precludes interactions between the fragments of the same molecule. Nonetheless, the approach was found suitable to account for sharp changes in the axial component of the first hyperpolarizability of the hydrogen fluoride dimer [95]. A similar conclusion was drawn for the second hyperpolarizability of the acetylene dimer [44] and for the dimer of 2-methyl-4-nitroaniline [13] in the point-dipole approximation. In the case of linear and parallel dimers of donor–acceptor azobenzene, Okuno et al. [53] showed that neither the permanent field effects represented by point charges nor the local field correction modeled by the electric dipole interactions within the point-dipole approximation can completely account for the effects of the surroundings on the first hyperpolarizability.

Dykstra's approach was applied to the study of the second hyperpolarizability of all-*trans* hexatriene bundles, and it was shown that classical electrostatic polarization was the dominant effect and that the electrical properties of individual hexatriene chains could be adequately described by a set of additive point polarizability and hyperpolarizability tensors [46]. Here, the ab initio polarizability tensor for the hexatriene chain was distributed over the six carbon centers, which were taken to be equivalent except at the chain ends where local C_2 symmetry was invoked. The hydrogen centers were not utilized because they contribute only a small part to the polarization response. For the second hyperpolarizability tensor, the dominant longitudinal component was divided unequally between the carbon atoms at the chain ends and the interior carbon atoms. In order to simulate the delocalization effects that enhance (supralinearly) γ as the chain grows, the chain-end values were chosen to fit the γ of butadiene while the interior values were chosen to close the difference between the γ of butadiene and hexatriene. All remaining elements of the hyperpolarizability tensor were located at the center of the chain, and it was verified that their distribution was unimportant. For different clusters containing up to seven chains, as well as for clusters where the intermolecular distances were uniformly scaled, this partitioning yielded very good agreement (within 1.5%) between the values obtained with the ab initio and classical treatments. It is important to note that this agreement was destroyed when the distribution of γ was done in a different manner, showing that in such highly polarizable systems, high-order effects on the local field cannot be neglected. By constructing bundles containing up to 61 unit cells, it was possible to reproduce the large reduction in the longitudinal second hyperpolarizability of all-*trans* hexatriene when the chains were packed perpendicularly to the longitudinal axis.

In 1996, Munn extended the microscopic theory of bulk second-harmonic generation from molecular crystals to encompass magnetic dipole and electric quadrupole effects [96] and included all contributions up to second order in the electric field or bilinear in the electric field and the electric field gradient or the magnetic field. This was accomplished by replacing the usual polarization of Refs. 72 and 84 by an effective polarization as well as by defining an effective quadrupole moment. Consequently, the self-consistently evaluated local electric field and electric field gradient were expressed in terms of various molecular response coefficients and lattice multipole tensor sums (up to octupole). In this

approach, the quadratic electric SHG response was given in terms of transformed and extended hyperpolarizabilities which introduced the spatial
nonlocality of the response. The electric-quadrupole and magnetic-dipole macroscopic contributions to the second harmonic generation were analyzed for
crystalline C_{60} [97] as a function of frequency by adopting a valence effective
Hamiltonian. Using an empirical energy-broadening factor, resonances were
investigated and related to the different mixed contributions.

In a subsequent paper, Munn [98] showed that the frequency-dependent
local-field tensors accounted for the shift of the poles of the linear and nonlinear
susceptibilities from the isolated molecular excitation frequencies to the exciton
frequencies. The treatment also described the Davydov splitting of the exciton
frequencies for situations where there is more than one molecule per unit cell as
well as the band character or wave-vector dependence of these collective excitations. In particular, the direct and cascading contributions to $\chi^{(3)}$ contained
terms with poles at the molecular excitation energies, but they canceled exactly.
Combining both terms is therefore a prerequisite to obtaining the correct pole
structure of the macroscopic third-order susceptibility. Munn also demonstrated
that this "local field approach" can be combined with the properties of the
effective or dressed molecule and can be extended to electric quadrupole and
magnetic dipole nonlinear responses [96].

In 1997, Malagoli and Munn [99] derived expressions for $\chi^{(3)}$ that consider
all possible frequency combinations and subsequently carried out the first
detailed investigation of $\chi^{(3)}$ for a large π-conjugated molecule: diethylamino-
nitro-styrene (DEANST). In order to take into account the environmental effects
to which the molecule is exposed in the crystal, local field factors determined
from effective polarizabilities derived from experimentally measured refractive
indices were chosen rather than the theoretical values. It was assumed that each
molecule in the unit cell and each of the four fragments gave the same contribution to the susceptibility. To obtain $\chi^{(3)}$, these were combined with second
hyperpolarizability values either calculated with a MNDO Hamiltonian or at the
ab initio Hartree–Fock level with a small basis set. Whereas the ab initio
approach performed poorly, the MNDO data gave good agreement with the
experimental data in the case of the dominant tensor component, but this was
not true for the anisotropy.

One step better for determining accurate nonlinear susceptibilities was made
by using high-level ab initio (hyper)polarizabilities [100,101], and this made
possible the calculation of nonlinear susceptibilities without recourse to experimental information except for the crystal structure. For the nonpolar benzene
crystal, the differences in the $\chi^{(3)}$ tensor between the submolecule treatment and
the point-dipole approximation were small [100]. However, the differences were
larger for some components of the $\chi^{(3)}$ tensor of the urea crystal [101], which is
polar. The effects of the surroundings, due mostly to the permanent molecular

dipole moment, were built in through a self-consistent reaction field (SCRF)-like procedure and restricted to the dipolar term. Small differences were found between the effective α and γ and the properties of the isolated molecules, whereas for β and, to a lesser extent, μ, the effects of the surroundings were big. Using effective (hyper)polarizability data, the second harmonic generation and electro-optic $\chi^{(2)}$ were calculated and, as a result of crystal symmetry, the resulting $\chi^{(2)}$ coefficient was shown to be nearly the same with and without the permanent local electric dipolar field. Such vanishing effects are, however, not expected for all polar systems. Moreover, the presumably more accurate correlated values showed higher deviations from experiment than the Hartree–Fock quantities. Vibrational contributions and specific hydrogen bonding inter-actions were given as an explanation for these differences, although a preliminary investigation of the latter displayed only a small effect. In 2000, Reis et al. [102] showed that part of the remaining discrepancy between theory and experiment for the urea crystal comes from the approximate scheme used to distribute the (hyper)polarizabilities. In a further refinement of the theory, Reis et al. [102] combined Stone's treatment of nonlocal polarizability [74] with the partitioning scheme of Bader et al. [75] to calculate the distributed (hyper)-polarizabilities. In order to account for the effects of the intramolecular charge flow between atomic basins, their self-consistent treatment also incorporated the induced charge effects and therefore the effects of changes in the local potentials. At the CPHF level, this realistic description of charge distributions led to a decrease in the static $\chi^{(2)}$ with respect to the point-dipole and sub-molecule treatments [101] and suggested that inclusion of electron correlation, frequency dispersion, and (strongly inhomogeneous) permanent local field effects would lead to matching theory with experiment.

Adopting a simple distribution scheme, the method was also applied to (i) show that in model Langmuir–Blodgett multilayers [103], the significant dipolar interactions extend only to adjacent layers, (ii) simulate the surface-induced SHG for crystals of p-nitroaniline which vanishes, within the dipole approxi-mation, in the bulk due to centrosymmetry [104], (iii) characterize the second-order nonlinear response of m-nitroaniline crystal and the third-order nonlinear response of naphthalene and anthracene crystals [105] using both density functional theory in its local approximation and second-order Møller–Plesset perturbation theory techniques, (iv) address the effect of the substrate, treated as a continuous dielectric in the image-dipole approximation, on the quadratic optical response of model Langmuir–Blodgett multilayers [106], as well as analyze the SHG of mixed films formed between stearic acid and $4''$-n-pentyl-4-cyano-p-terphenyl [107].

A very detailed study of the THG nonlinear susceptibilities of four poly-morphs of ice (Ih, II, VIII, and IX) was carried out by Reis et al. [108] at the second-order Møller–Plesset perturbation theory level of approximation. The effects of the permanent electric crystal field were shown, in the dipole approximation, to be very large (with sign inversion for three of the polymorphs) on the first hyperpolarizability, and consequently on the cascading contribution to $\chi^{(3)}$. The influence of the nondipolar interactions including hydrogen bonding was assessed by considering a cluster of four molecules and appeared to be small. After neglecting the intermolecular vibrations and by considering the zero-point vibrational average (ZPVA) contribution in its static limit, the global vibrational effects on $\chi^{(3)}$ turned out to be well below 10%. Lattice vibrations effects were also considered by Munn [109] in conjunction with translations and librations in order to derive expressions for the quadratic elastic incoherent scattering from crystals (usually referred to as hyper-Rayleigh scattering). Initial estimates for the hexamine crystal indicated that the incoherent scattering effect amounted to 7% of the coherent second harmonic generation response. Another extension of the method concerned isotropic polymer materials where the structure of the chains as well as the distribution of the local and nonlocal linear and nonlinear responses are of crucial importance [110]. Electric field poling was also addressed in this paper.

The variational-perturbation approach elaborated by Nobutoki and Koezuka [111] consisted in constructing an effective Hamiltonian for each molecule of a supermolecule by renormalizing the intermolecular interactions associated with its neighbors. This method, which introduced both local field and permanent field effects, was adopted to evaluate the dipole moment of the molecules in the cluster and, from their field-dependence, the first hyperpolarizability. Provided that the cluster is large enough, the molecular property corresponds to the crystal second-order nonlinear susceptibility. This method allowed analysis of the variations in β in terms of the ground-state polarization, atomic charge distribution, and frontier orbital characteristics. Improvements to this model include frequency dispersion, intermolecular charge transfer effects, vibrational contributions, and extrapolation to the infinite crystal size.

Another scheme to calculate and interpret macroscopic nonlinear optical responses was formulated by Mukamel and co-workers [112–114] and incorporated intermolecular interactions as well as correlation between matter and the radiation field in a consistent way by using a multipolar Hamiltonian. Contrary to the local field approximation, the macroscopic susceptibilities cannot be expressed as simple functionals of the single-molecule polarizabilities, but retarded intermolecular interactions (polariton effects) can be included.

V. CRYSTAL ORBITAL METHODS

Crystal orbital or band structure methods consist in solving the problem of the infinite system directly by exploiting the translational symmetry in order to reduce the computational effort. Such approaches have been adopted mostly for characterizing ionic and covalent solids as well as for stereoregular polymers that, in one dimension, are periodic and infinite. These methods avoid the problem of truncating the crystal structures and, subsequently, extrapolating the results of finite clusters to those for infinite ones. Provided that the level of treatment is good enough, it also implicitly incorporates effects associated with the screening of the external electric field or with the modification of the electronic and geometrical structures induced by the surroundings via van der Waals forces, hydrogen bonds, and ionic interactions. However, specific techniques are needed to properly incorporate Coulomb, exchange, and electron correlation terms such that there is convergence with respect to the lattice summations. Also, integration in the first Brillouin zone has to be handled carefully.

For field-induced properties, one can adopt the usual scalar potential representation, and in the clamped lattice approximation the dipolar perturbation potential is given by

$$V = e\vec{E}(t) \cdot \sum_i \vec{r}_i \qquad (4)$$

where e is the magnitude of the electronic charge, $\vec{r}_i$ the position operator associated with electron i, and $\vec{E}(t)$ is the external or applied time-dependent electric field. Because the potential is unbound from below, it destroys the periodicity of the crystal. This is incompatible with the Born–von Kármán cyclic boundary conditions, and, in principle, no ground state exists for the electrons in the presence of a uniform electric field.

A. General

Several approaches have been proposed to overcome these problems. The most direct one is the "supercell" approach due to Kunc and Resta [115], where the constant electric field is replaced by a saw-like potential having the translational symmetry of the superlattice. In analogy to surfaces in real crystals, these multiple capacitors constitute barriers to the flow of electrons toward low potentials. Provided that the supercell is large enough, the electric field is constant in space over several cells. Hence, this region exhibits the linear and nonlinear responses of an elementary cell in an infinite crystal with respect to the constant field. In this way, the technique avoids problems related to the truncation of clusters or the need for saturation of dangling bonds, but it is not suitable for studying species where delocalization effects dominate the nonlinear responses. Indeed, it suffers from problems related to the too-large size of the supercell and the associated enormous computational requirements. Although it has been used within density functional theory (DFT)—which raises some questions in itself— this approach can, in fact, be used at any level of theory. In many aspects, the supercell method of Kunc and Resta is similar to the long-wavelength approach where the position operator is represented by

$$\vec{r} \leftrightarrow \lim_{\vec{q} \to 0} \frac{\sin \vec{q} \cdot \vec{r}}{|\vec{q}|} = \lim_{\vec{q} \to 0} \frac{1}{2} \left(\frac{e^{i\vec{q} \cdot \vec{r}}}{i|\vec{q}|} - \frac{e^{-i\vec{q} \cdot \vec{r}}}{i|\vec{q}|} \right) \tag{5}$$

Nevertheless, such an approach suffers from a further problem: namely, the need to extrapolate the results to the $\vec{q} = 0$ limit.

Crystal orbital nonlinear susceptibilities have also been computed by adopting perturbation theory, where the susceptibilities are expressed in the form of summations-over-excited states (SOS) which, in the independent particle approximation or non-self-consistent scheme, are usually written as summations over the occupied and unoccupied bands. Provided that the dipole/velocity/momentum matrix elements and band structure are known, the SOS expressions have the advantage of being easy to evaluate for a large range of frequencies encompassing both resonant and off-resonant phenomena. The bottleneck comes in the summations and their eventual truncation, as well as integration in the first Brillouin zone, which requires more care than in simple band structure calculations. The vector potential representation (velocity gauge) has been generally preferred over the scalar potential (length gauge) representation in the nonlinear susceptibility determinations carried out from the 1970's to the early 1990's, although they are equivalent and related by a gauge transformation [116]. Indeed, for cubic crystals, the ill-defined terms, $\langle \varphi_n(\vec{k}, \vec{r}) | \vec{p} | \varphi_n(\vec{k}, \vec{r}) \rangle$, with $\vec{p}$ the momentum operator and $\varphi_n(\vec{k}, \vec{r})$ a crystal orbital for band n, vanish in the former representation. Within the velocity-gauge representation, the perturbation potential in the minimal coupling approximation is

$$V = \frac{e}{mc} \vec{A}(t) \cdot \sum_i \vec{p}_i \tag{6}$$

where e is the magnitude of the electronic charge, m is the electronic mass, c is the velocity of light, $\vec{A}(t)$ is the vector potential, and $\vec{p}_i$ is the momentum operator of electron i. However, the resulting standard expressions for the NLO properties diverge as the zero-frequency limit is approached unless special care is taken. Aspnes [116] has demonstrated that, in the case of the SHG of crystals with cubic symmetry, these divergences are only apparent and can be removed by an appropriate treatment using time-reversal invariance and the symmetry of the crystals. Aspnes [116] further showed that, in the vector potential representation, only terms involving three different bands contribute to $\chi^{(2)}$ and that the virtual-electron term is larger than the virtual-hole term. He also pointed out that in the scalar potential representation, only two- and three-band terms contribute. Following this approach, Sipe and collaborators [117] have carried out several investigations where the imaginary part of $\chi^{(2)}$—which, as they pointed out, is simpler to determine owing to the presence of Dirac δ functions—is evaluated first and then, using the Kramers–Kronig relationship, the real part is obtained. They also generalized the proof that the divergences in the SOS expressions are only apparent for third-harmonic generation (THG) phenomena in cubic crystals [118] and for the second-harmonic generation in crystals of any symmetry, with the help of a newly derived sum rule [119]. In Ref. 118, a general scheme was also developed to analyze the third-order NLO responses in terms of contributions from single-particle processes. Other approaches that do not require the use of separate sum rules for each type of susceptibility have also been derived. Following the seminal work of Genkin and Mednis [120], Sipe and Ghahramani [121] separated, within the minimal coupling Hamiltonian, the intraband and interband motions of electrons for evaluating the evolution of the density operator and for calculating the induced current density. This was achieved by choosing wave functions to be instantaneous eigenstates of the time-dependent Hamiltonian and to include Berry's phase. In this way, the unphysical divergences at zero frequency were avoided. Another step forward was made by Aversa and co-workers [122,123], who returned to the scalar potential representation of the perturbation in order to obtain general expressions for $\chi^{(2)}$ and $\chi^{(3)}$ for arbitrary frequency mixings. In a recent investigation, Sipe and Shkrebtii [124] addressed, within the independent particle approximation, the determination of the injection current and shift current in the context of second-order nonlinear susceptibilities.

In the following paragraphs we describe representative calculations of the nonlinear bulk susceptibilities which adopt the SOS scheme; starting with early applications employing simplified models and moving toward semi-ab initio and ab initio calculations that have progressively appeared in parallel to the development of ab initio band structure calculations. A summary is given in Table V.

The first attempt was to calculate the dispersion of the SHG $\chi^{(2)}$ of zinc-blende crystals by Bell [125] in 1971, who simply considered the three-band term of the scalar potential expression and phenomenologically modeled the bands around a few critical points associated with the main features of the linear absorption spectrum. The study of Aspnes [116] was performed using the constant-energy-gap model extended to include explicitly the contributions of high-symmetry regions by means of parabolic approximations to the actual band structure. Like Bell [125], the momentum matrix elements were assumed to be constant over the entire Brillouin zone. His results underestimated experimental values, but by including band nonparabolicity and Coulomb effects the discrepancy was reduced. Fong and Shen [126] adopted the empirical-pseudo-potential band method including spin–orbit interactions to determine the $\vec{k}$-dependent band structures and momentum matrix elements necessary for the

TABLE V

Selected References Employing Crystal Orbital Methods to Evaluate the
Nonlinear Optical Susceptibilities of Periodic Crystals

Author, Year	Compound	Property[a]	Ref.
Bell, 1971	GaAs, InAs, InSb	$\chi^{(2)}(-2\omega;\omega,\omega)$	125
Aspnes, 1972	GaAs, GaP, GaSb, InAs, InSb, ZnS, ZnSe, ZnTe, CdTe	$\chi^{(2)}(-2\omega;\omega,\omega)$	116
Fong and Shen, 1975	GaAs, InAs, InSb	$\chi^{(2)}(-2\omega;\omega,\omega)$	126
Moss et al., 1987	GaP, GaAs, GaSb, InAs, InSb	$\chi^{(2)}(-2\omega;\omega,\omega)$	117
Ghahramani et al., 1990	$(Si)_n/(Ge)_n$ superlattices	$\chi^{(2)}(-2\omega;\omega,\omega)$	127
Moss et al., 1990	Si, Ge, GaAs	$\chi^{(3)}(-3\omega;\omega,\omega,\omega)$	118
Ghahramani et al., 1991	$(Si)_n/(Ge)_n$ superlattices	$\chi^{(2)}(-2\omega;\omega,\omega)$	119
Ghahramani et al., 1991	ZnSe, ZnTe, CdTe	$\chi^{(2)}(-2\omega;\omega,\omega)$, $\chi^{(3)}(-3\omega;\omega,\omega,\omega)$	128
Levine and Allan, 1991	AlP, AlAs, GaP, GaAs	$\chi^{(2)}(0;0,0)$	140
Zhong et al., 1992	Se	$\chi^{(2)}(0;0,0)$	141
Ghahramani and Sipe, 1992	$(GaAs)_n/(GaP)_n$ superlattices	$\chi^{(2)}(-2\omega;\omega,\omega)$	127
Huang and Ching, 1992	GaAs	$\chi^{(2)}(-2\omega;\omega,\omega)$	129
Huang and Ching, 1993	AlP, AlAs, AlSb, GaP, GaAs, GaSb, InP, InSb, ZnS, ZnSe, ZnTe CdS, CdSe, CdTe	$\chi^{(2)}(-2\omega;\omega,\omega)$	130
Ching and Huang, 1993	C, Si, Ge, AlP, AlAs, AlSb, GaP, GaAs, GaSb, InP, InAs, InSb, ZnS, ZnSe, ZnTe, CdS, CdSe, CdTe	$\chi^{(3)}(-3\omega;\omega,\omega,\omega)$	131
Levine and Allan, 1993	Urea	$\chi^{(2)}(0;0,0)$	142
Sipe and Ghahramani, 1993	CdTe	$\chi^{(2)}(-2\omega;\omega,\omega)$	121
Levine, 1994	GaP, GaAs	$\chi^{(2)}(-2\omega;\omega,\omega)$	146
Dal Corso and Mauri, 1994	One-dimensional model system	$\chi^{(2)}(-2\omega;\omega,\omega)$	149
Chen et al., 1994	SiC	$\chi^{(2)}(0;0,0)$	143
Aversa et al., 1994	Two-band model	$\chi^{(3)}(-\omega;\omega,\Omega,-\Omega)$	122
Aulbur et al., 1995	Si_1Ge_1	$\chi^{(2)}(0;0,0)$	144
Chen et al., 1995	BN, AlN, GaN	$\chi^{(2)}(0;0,0)$	145
Hugues and Sipe, 1996	GaAs, GaP	$\chi^{(2)}(-2\omega;\omega,\omega)$, $\chi^{(2)}(-\omega;\omega,0)$	132
Dal Corso et al., 1996	AlP, AlAs, AlSb, GaP, GaAs, GaSb, InP, InAs, InSb	$\chi^{(2)}(-2\omega;\omega,\omega)$	151
Hugues et al., 1997	GaN, AlN	$\chi^{(2)}(-2\omega;\omega,\omega)$, $\chi^{(2)}(-\omega;\omega,0)$	133
Hugues and Sipe, 1998	ZnS, ZnSe, ZnTe, CdS, CdSe in cubic and hexagonal phases	$\chi^{(2)}(-2\omega;\omega,\omega)$	134
Rashkeev et al., 1998	GaAs, GaP, GaN, AlN, SiC	$\chi^{(2)}(-2\omega;\omega,\omega)$	159
Adolph and Bechstedt, 1998	GaAs, GaP, InAs, InP, SiC	$\chi^{(2)}(-2\omega;\omega,\omega)$	157
Rashkeev et al., 1998	SiC	$\chi^{(2)}(-2\omega;\omega,\omega)$	160
Rashkeev et al., 1999	II–IV–V_2 (II = Zn, Cd; IV = Si, Ge; V = As, P)	$\chi^{(2)}(-2\omega;\omega,\omega)$	161
Adolph and Bechstedt, 2000	SiC	$\chi^{(2)}(-2\omega;\omega,\omega)$	158

TABLE V   (*Continued*)

Author, Year	Compound	Property[a]	Ref.
Zhu et al., 2000	*m*-Aminophenol, *urea*, m-nitrobenzene, 3-methyl-4-nitro-pyrydine-1-oxide, *m*-nitroaniline	$\chi^{(2)}(-2\omega;\omega,\omega)$	164
Rashkeev and Lambrecht, 2001	I–III–VI$_2$ (I = Ag, Cu; III = Ga, In; VI = S, Se, Te)	$\chi^{(2)}(-2\omega;\omega,\omega)$	162
Chang et al., 2001	AlP, AlAs, GaP, GaAs	$\chi^{(2)}(-2\omega;\omega,\omega)$	147
Rérat et al., 2001	GaN	$\chi^{(2)}(-2\omega;\omega,\omega)$	163

[a]ω and Ω stand for both finite and zero frequencies.

evaluation of $\chi^{(2)}$ of several zinc-blende semiconductors. Their results were obtained by considering four occupied and four unoccupied bands and were better than those of Bell but still significantly underestimated the experimental values.

Moss et al. [117] employed an empirical tight-binding method to evaluate the dispersion and zero-frequency limit of the second-order nonlinear susceptibility of zinc-blende crystals. Provided that the conduction–conduction momentum matrix elements were deduced from the observed effective masses, good agreement with experiment was obtained. They also showed that the momentum matrix elements were underestimated and thus explained the limitations of the previous investigations [125,126]. Adopting a minimal basis of orthogonalized linear combinations of Gaussian orbitals (LCGO) in conjunction with the Xα method for constructing the potential of the constituent bulk materials, Ghahramani et al. [119,127] studied the second-order nonlinear optical susceptibilities of various Si/Ge and GaAs/GaP superlattices. Here also empirism is not absent because α is chosen in such a way as to reproduce the band gap of the bulk materials. Both the empirical tight-binding and LCGO-Xα methods were used for the calculation of dispersion and anisotropy in $\chi^{(3)}(-3\omega;\omega,\omega,\omega)$ for Si, Ge, and GaAs [118]. It was found that in the zero-frequency limit, the LCGO-Xα method overestimated experiment whereas the tight-binding method underestimated it. Assessment of the efficiency of the methods for dynamic phenomena was, however, hampered by the low precision of the experiments. The LCGO-Xα method was also applied to the second- and third-order NLO responses of II–VI semiconductors [128] without accounting for the spin–orbit coupling effects that are important for getting the linear response.

Huang and Ching [129–131] followed the scheme elaborated by Sipe and co-workers while using the self-consistent orthogonalized-LCAO method in the local density approximation (LDA) to determine the band structures and momentum matrix elements. Here, the adjustment of the band gap from the LDA value to the experimental value was carried out by applying the scissors operator. Due to the large magnitude of some conduction-band/conduction-band momentum matrix elements, well-converged results required the inclusion of high-energy conduction bands in the SOS expressions. The estimates of $\chi^{(2)}(-2\omega; \omega, \omega)$ and $\chi^{(3)}(-3\omega; \omega, \omega, \omega)$ turned out to be in good agreement with experiment. The large number of systems Huang and Ching studied enabled them to address relationships between, on the one hand, the NLO responses and, on the other hand, the direct band gap, the dielectric constant, and Miller's constant.

Hugues and co-workers [132–134] employed the velocity gauge approach of Sipe and Ghahramani [121] to determine the SHG $[\chi^{(2)}(-2\omega; \omega, \omega)]$ and electronic electro-optic susceptibility $[\chi^{(2)}(-\omega; \omega, 0)]$ in semiconductors using the full-potential linearized augmented plane wave (FLAPW) method within the local density scheme. The underestimation of the LDA band gap was corrected by using the scissors approximation, and its repercussions on the momentum matrix elements were fully accounted for; they had been neglected in other studies [129–131]. In particular, they compared the SHG susceptibilities of II–VI semiconductors in both the cubic and hexagonal phases [134] and showed that most of the differences appeared for frequencies above the first transition.

Although the general schemes elaborated by Sipe and co-workers [117,121, 123] are not limited to the independent particle approximation and can, in principle, include self-energy corrections, applications (as shown later) have been so far limited. In this context, electron–electron interactions are only accounted for in the description of the ground state, and the schemes neglect electron–hole interactions for the description of excitons as well as other quasi-particle effects. The latter are, however, often introduced indirectly by using the scissors approximation, which turns out to be a viable approximation for the self-energy [135]. These excitonic effects are in fact crucial in molecular crystals of conjugated organic molecules.

Self-consistency has been introduced in different ways either by adopting time-dependent (TD) DFT [136] and its LDA [137] or by using density functional perturbation theory (DFPT) [138]. Within the TDLDA formalism, Levine [139] derived self-consistent expressions to evaluate the crystal linear and nonlinear responses to longitudinal electric fields and hence for evaluation of certain linear combinations of elements of the second harmonic generation tensor. This method uses a one-electron Green's operator, adopts the long-wavelength formalism, includes field-induced effects on the electron–electron interactions (referred to as local-field corrections), and exploits sum rules to eliminate apparent divergences. It has been implemented by Levine and Allan [140] by using a plane-wave basis with ab initio pseudopotentials. Self-energy corrections were included in the form of a scissors operator, and the consequent renormalization of the velocity operator was performed. Agreement within 15% was obtained between the calculated $\chi^{(2)}(0;0,0)$ of GaP and GaAs and a particular set of the experimental values, provided that the experimental lattice constants were used. It was also found that a change of 1% in the lattice constant induced variations in $\chi^{(2)}$ by about 25%. This method was later used to characterize $\chi^{(2)}(0; 0, 0)$ for trigonal (helical) selenium [141], urea [142], prototypes of SiC [143], and Si_1Ge_1 [144], as well as for III nitrides in both zinc-blende and wurtzite structures [145] and the off-resonant $\chi^{(2)}(-2\omega; \omega, \omega)$ of GaP and GaAs [146]. Reference 142 was unique in that it dealt with organic crystals and because the scissors correction was found unnecessary to correct the band gap. The use of the LDA potential turned out to provide a $\chi^{(2)}(0;0,0)$ value for the urea crystal in very good agreement with experiment. However, the interactions between the different molecules are both specific (hydrogen bonds) and large as shown by the importance of local field factors (corresponding to field-induced electron-electron relaxation effects). On the other hand, Ref. 144 emphasized the importance of the choice of exchange-correlation potentials and of nonlinear core charge corrections to reproduce the sign and magnitude of the second-order nonlinear susceptibility of Si/Ge systems, where the two atoms have very similar electronegativities. Although based on TDLDA, because of the mean-field approximation to the exchange-correlation kernel, this approach [139] does not consider the electron–hole interactions and cannot account for excitonic effects. A recent TDLDA study by Chang et al. [147], which combined the use of a scissors shift to include the self-energy with the use of single excitations to describe the electron–hole interactions, demonstrated that for the usual III–V semiconductors, excitonic effects increase $\chi^{(2)}$ by about 20%.

Within DFPT, self-consistency is built in, as in any variation–perturbation approach, by the iterative determination of the successive field-induced changes of the wave functions. Also, external perturbations of arbitrary wavelength can be considered. Furthermore, the $2n + 1$ theorem has been generalized to DFT [138,148], and it can be used advantageously for reducing the computational requirements. Nevertheless, due to the ill-defined character of the position operator, the first-order response of the wave function combined with the $2n + 1$ theorem is not sufficient to obtain, in a straightforward fashion, the second-order NLO responses [149]. By switching to a Wannier representation of the electronic orbitals that are constrained to be localized in finite regions of real space, a total energy functional can be written for a periodic insulating solid inside a uniform electric field [150], and its derivatives with respect to the field can be determined analytically. For computational purposes, the approach of Dal Corso and Mauri [149] also includes a back transformation to Bloch functions. In fact, the Nunes and Vanderbilt energy functional [150] can also be used within a finite field procedure to determine numerically the derivatives with respect to the field. In such a case, the ground state corresponds to a metastable state for the real solid and, as shown in Ref. 149, it provides the same $\chi^{(2)}$ results as the analytical procedure. In 1994, this method was applied to a model one-dimensional Hamiltonian. Two years later, it was generalized to time-dependent phenomena and applied to the $\chi^{(2)}$ of an ensemble of III–V cubic semiconductors [151] by using the local-density approximation (LDA) with norm-conserving pseudopotentials, a plane wave kinetic energy cutoff of 24 Ry, and nonlinear core corrections. Although the direct band gap at the origin of the Brillouin zone is underestimated by about 30–50 %, the theoretical results are in good agreement with experiment for both static and SHG (second harmonic generation) processes. However, the suitability of the Wannier representation approach for evaluating NLO responses of molecular crystals has yet to be established. As far as the exchange-correlation potentials and kernels allow, the excitonic contributions to the hyperpolarizabilities are implicitly included in the self-consistent DFPT procedures. Moreover, Ref. 152 showed that, after including higher-order local field effects in Ref. 139, the approaches of Dal Corso et al. [151] and of Levine [139] for determining $\chi^{(2)}$ are equivalent, although the computational schemes are different.

Nunes and Gonze [153] have recently extended DFPT to static responses of insulating crystals for any order of perturbation theory by combining the variation–perturbation approach with the "modern theory" of polarization [154]. There are evident similarities between this formalism and (a) the developments of Sipe and collaborators [117,121,123] within the independent particle approximation and (b) the recent work of Bishop, Gu and Kirtman [24, 155,156] at the time-dependent Hartree–Fock level for one-dimensional periodic systems.

It is important to note that these treatments, both at the DFPT and SOS level, pertain to filled bands—that is, for insulators or cold semiconductors; the treatment of partially filled bands requires the characterization of scattering effects as well. Moreover, since incident rather than macroscopic electric fields (or vector potentials) generally enter into the perturbation Hamiltonian [Eqs. (4) and (6)], the various linear and nonlinear susceptibilities have to be corrected if there are no implicit or explicit local field corrections.

In the LDA, Adolph and Bechstedt [157,158] adopted the approach of Aspnes [116] with a plane-wave-pseudopotential method to determine the dynamic $\chi^{(2)}$ of the usual III–V semiconductors as well as of SiC polytypes. They emphasized (i) the difficulty to obtain converged Brillouin zone integration and (ii) the relatively good quality of the scissors operator for including quasiparticle effects (from a comparison with the GW approximation, which takes into account wave-vector- and band-dependent shifts). Another implementation of the SOS $\chi^{(2)}(-2\omega; \omega, \omega)$ expressions at the independent-particle level was carried out by Raskheev et al. [159] by using the linearized muffin-tin orbital (LMTO) method in the atomic sphere approximation. They considered explicitly the convergence of the nonlinear susceptibility with respect to the maximal value of the angular momentum included in the LMTO Hamiltonian, the number of bands in the SOS expressions, and the integration procedure. They noticed that in the static limit a direct evaluation of $\chi^{(2)}$ is more efficient than using the Kramers–Kronig transformation. The gap problem of the LDA was corrected by using the scissors operator with renormalization of the momentum matrix elements. Alternatively, the scissors operator was directly included in the Hamiltonian to provide consistency between the eigenstates and the Hamiltonian. In both cases, the band gap was correctly reproduced but the second approach was found to better match the experimental $\chi^{(2)}$ data. They drew the attention to possible error cancellations between the simple scissors approximation and the lack of local field corrections. This approach was subsequently used with the scissors approximation to analyze the influence of the crystal structure and composition on the SHG response [160–162]. In particular, it was found that the high value of $\chi^{(2)}$ in CdGeAs$_2$ [161] comes from the very small interband term that does not compensate for the large intraband contribution, whereas a study of I–III–VI$_2$ chalcopyrite semiconductors showed that the dominating factor influencing $\chi^{(2)}$ is the anion rather than the two groups of cations [162]. Rérat et al. [163] have employed the SOS scheme with the velocity-gauge representation and a scissors shift to determine $\chi^{(2)}$ $(-2\omega; \omega, \omega)$ of GaN at both the Hartree–Fock and GGA–DFT levels of approximation. To our knowledge, this study is the first ab initio crystal-orbital Hartree–Fock and GGA (generalized gradient approximation) calculation of $\chi^{(2)}$. It was found that correcting the band gap overestimation in Hartree–Fock or underestimation in GGA–DFT by using the scissors operator led to $\chi^{(2)}(-2\omega; \omega, \omega)$ values similar to the investigations carried out with the LDA and a scissors correction. By using the extended Hückel method, Zhu et al. [164] performed crystal orbital calculations on a series of organic crystals within the SOS formalism. The agreement with experiment was good, but the validity of the method is difficult to assess due to the lack of detailed information.

B. Density Functional Theory

It is necessary to point out that although ab initio treatments have been employed, they remain unsatisfactory for several reasons: the use of a scissors operator to adjust quantities with respect to experiment, the disagreement between the different theoretical treatments, and differences with experiment. This illustrates the limitations of the associated DFT treatments and of their exchange-correlation functionals. It should be noted that for evaluation of $\chi^{(2)}$ and $\chi^{(3)}$, several problems linked to the use of DFT have been high-lighted. Gonze, Ghosez, and Godby [165] have shown that in a correct treatment of an infinite system in an electric field, a linear term will arise in the exact exchange-correlation potential, and this term cannot be simulated by a functional dependence on the periodic bulk density alone. This has led to the development of polarization-dependent density functional theory (PDDFT) [166], which is a generalization of DFT and takes the polarization of the system into account. As shown in Ref. 167, PDDFT is connected to the scissors approach which, a posteriori, justifies its use in many SOS and DFPT calculations. For practical purposes, Aulbur et al. [167] showed that one should use the scissors shift that reproduces the linear susceptibility of the system rather than the shift that matches the LDA and quasiparticle band gap. For finite systems, the polarization-dependence of the exchange-correlation functional is no longer required but the short-sightedness of the conventional exchange-correlation functionals [including LDA (local density approximation), and GGA as well as those ensuring a correct asymptotic behavior of the potential] manifests itself by leading to errors in the calculated hyperpolarizabilities of extended molecules which are so large that the deduction of structure–property relationships is impossible [168]. In large π-conjugated molecules, this leads to excessive field-induced charge transfers or, in other words, inadequate screening and, subsequently, to the overestimation of the first and second hyperpolarizabilities by orders of magnitude, whereas for ionic and covalent crystals the present survey shows that the errors are smaller. These deficiencies in molecular calculations have been traced to an incorrect electric field dependence of the response part of the exchange-correlation potential which, contrary to the exact potential as well as Hartree–Fock and post-Hartree–Fock treatments, lacks a linear term counteracting the external field and reducing the induced polarization [169,170]. Such a counteracting field is, however, present when adopting the orbital-dependent Krieger–Li–Iafrate (KLI) exchange-only potential, an approximation to the optimized-effective potential, that constitutes a first-order remedy to the problem, although it is computationally very demanding [170]. Therefore, for molecular crystals, one expects that both aspects should be addressed carefully if one wants to obtain accurate nonlinear susceptibilities.

Since the current density in the bulk measures the surface charges, the time-dependent current-density functional theory (CDFT) appears to be a way to investigate this problem. At least the results presented by de Boeij et al. [171] for the bulk susceptibility and by van Faassen et al. [172] for the polarizability of linear chains are encouraging, although this may not be the case for second- and third-order effects.

C. Polymers

Crystal orbital methods have also been used and developed in order to obtain the nonlinear optical responses of stereoregular polymers (Table VI). The rationale is related to the large nonlinear susceptibilities that π-conjugated polymers can have. The initial study, carried out by Agrawal et al. [173], employed the Genkin and Mednis [120] formalism within the tight-binding (Hückel) model and

TABLE VI
Selected References Employing Crystal Orbital Methods to Evaluate the
Nonlinear Optical Susceptibilities of Infinite Periodic Polymers

Author, Year	Compound	Property	Ref.
Agrawal et al., 1978	Polyacetylene, polydiacetylene, pairs of polyacetylene	Static and dynamic β/N and γ/N	173
Champagne et al., 1995	Polymethineimine	Static β/N	174
Otto et al., 1999	$(H_2O)_\infty$ and polymethineimine	Dynamic β/N	177
Bishop et al., 2001	$(LiH)_\infty$, $(HF)_\infty$, $(H_2O)_\infty$, polyacetylene, polymethineimine	Static β/N and γ/N	156
Gu et al., 2001	Polyyne, polyacetylene, polymethineimine	Static and dynamic β/N and γ/N	24
Kirtman et al., 2002	Polydiacetylene, polybutatriene, pairs of polyacetylene	Static γ/N	63
Gu et al., 2002	Polydiacetylene, polybutatriene	Dynamic γ/N	179

studied the effects of bond and atom alternations and of chain pairing on the NLO properties of one-dimensional periodic semiconductors, with special emphasis on polydiacetylenes (PDA). The properties were referred to as $\chi^{(2)}$ and $\chi^{(3)}$, but we prefer—and this is more correct—to write $\beta(N)/N$ and $\gamma(N)/N$ with N, the number of units, tending toward infinity. Another Hückel investigation [174] concentrated on (a) the second-order NLO responses of asymmetric unit cell polymers that modeled polymethineimine (PMI) and (b) the relations between bond alternation, atom alternation, and the sign and magnitude of $\beta(N)/N$. At this level of theory the equations reduced to the non-self-consistent scheme detailed above and can be called uncoupled.

Three self-consistent schemes for computing the first and second hyperpolarizabilities per unit cell of stereoregular polymers have been proposed at the Hartree–Fock level of theory. One is the approach taken by the Erlangen group to deal with the unbound position operator and is summarized in Refs. 175, 176, and 177. It consists of expressing the scalar potential as a sum of two terms,

$$e\vec{E}(t) \cdot \vec{r} = ie\vec{E}(t) \cdot e^{i\vec{k}\cdot\vec{r}} \nabla_{\vec{k}} e^{-i\vec{k}\cdot\vec{r}} - ie\vec{E}(t) \cdot \vec{\nabla}_{\vec{k}} \qquad (7)$$

where the first term describes the polarization and the second term describes the polarization current. The first term does not destroy periodicity. The second is incommensurate with the periodicity and is dropped. Using Frenkel's variational principle, they obtained a set of coupled equations of which the solution enabled the evaluation of the total field-induced dipole moment. By numerically differentiating twice the dipole moment with respect to the electric field amplitudes, the different components of the static and dynamic first hyperpolarizability per unit cell tensor were determined. The application of this method [177], however, raised several questions regarding the satisfaction of tensor symmetry, the equivalence between the polymeric values and the large-oligomer limit, and the inclusion of the field-dependence of the HF orbital energies (see, for instance, the discussion in Ref. 155).

In addition to the investigation of the components of the hyperpolarizability tensors associated with directions perpendicular to the chain axis, Schmidt and Springborg [178] proposed a method based on Wannier functions to determine the longitudinal components. Their method is related to the one of Kunc and Resta [115] since the number of $\vec{k}$ points is finite and the dipole moment operator (longitudinal component) is rewritten to display the same periodicity as the Born–von Kármán zone. Consequently, the accuracy of the longitudinal hyperpolarizabilities per unit cell will increase with the number of $\vec{k}$ points.

Kirtman, Gu, and Bishop [155] have derived a fully self-consistent procedure by using a noncanonical form of perturbation theory. Following Genkin and Mednis [120], they employed the vector potential and replaced the quasi-momentum $\vec{k}$ by

$$\vec{\kappa} = \vec{k} + \frac{e}{c}\vec{A}(t) \tag{8}$$

which leaves the one-electron Hamiltonian translationally invariant. They solved order by order the time-dependent Hartree–Fock equations and obtained fully analytical expressions for the static and dynamic hyperpolarizabilities of one-dimensional infinite periodic systems. Noniterative (less iterative) formulas were also obtained by taking advantage of the $2n + 1$ rule [156].

The method was tested by application to prototypical polymers [24,156] including the π-conjugated polyacetylene, polyyne, and polymethineimine systems and the hydrogen-bonded one-dimensional chains, namely, $(HF)_\infty$, $(HLi)_\infty$, and $(H_2O)_\infty$, for both static and dynamic phenomena. The first application [156] utilized a minimal basis set of Gaussian lobe orbitals while the second [24] used conventional Gaussian-type atomic orbitals. In particular, good convergence of the polymeric values with respect to the number of $\vec{k}$ points for integration in the first Brillouin zone and to the number of interacting unit cells was obtained, although long-range Coulombic contributions were not accounted for. From comparison with finite oligomer values it was shown that extrapolation errors in those earlier calculations were sometimes greater than had been reported. Moreover, the method was able to reproduce the equivalence between the first coefficients in the dispersion relations which characterize different NLO phenomena [24]. The method was also applied to the determination of the static [63] and dynamic [179] second hyperpolarizabilities of polydiacetylene and polybutatriene chains as well as to pairs of polyacetylene chains [63]. The γ/N results were analyzed in terms of bond-length alternation and types of carbon–carbon bonds.

So far, this method has only been applied to one-dimensional periodic systems, but, except for the involved coding, there is no perceived difficulty to extend it to three-dimensional crystals. As shown by studies on oligomers, these TDHF results suffer from the lack of electron correlation which can modify estimates by as much as one order of magnitude. Generalizing this approach by using traditional electron correlation methods (Møller–Plesset and coupled-cluster) or ad hoc DFT treatments is therefore a step to be pursued.

VI. OTHER APPROACHES

Since the concepts of atoms and bonds are central to chemical understanding, approaches based on atom-additivity and bond-additivity are very appealing. Due to their simplicity, they were used in the early days for actual calculations, but nowadays they continue to be employed for interpretative purposes. Needless to say, their accuracy can be surpassed by methods based on quantum mechanics. As with field-free isolated molecules, early models used to estimate second- and third-order macroscopic nonlinear responses considered such simple schemes. In the following, we describe methods that treat either chemical bonds or atoms as the central quantities for evaluating the bulk NLO responses. The philosophy consists in incorporating in the description of these central constructs the effects of the surroundings. In this way the connection with more elaborate methods, such as the oriented gas model that focuses on one molecule with local field factor corrections, or with the crystalline orbital approach that reduces the system to its unit cell, is more obvious. In what follows, a selection of such schemes is analyzed and listed in Table VII.

In 1969, Flytzanis and Ducuing [180] produced a variational perturbation method to evaluate the second-order nonlinear optical susceptibility of sp^3-bonded III–V semiconductors, where $\chi^{(2)}$ is given in terms of bond hyperpolarizabilities. In their method, the ground state was given by an antisymmetrized product of two-electron bond functions of which the degree of ionicity was determined self-consistently. The model effectively assumed localized non-interacting bonds and determined $\chi^{(2)}$ from the response of the bond-charge density to the field by assuming axial symmetry. After adjusting parameters in such a way that the calculated linear susceptibility matched experiment, they reproduced most of the experimental trends in $\chi^{(2)}$. They further showed that adopting an uncoupled approximation such as in [181] destroyed the agreement with experiment.

The same year, Levine [182] proposed a model that related the anharmonic motion of the bond charge, located approximately halfway between two neighboring atoms, to the second- and third-order nonlinear optical susceptibilities of

TABLE VII

Selected References Employing Other Approaches to Evaluate the
Nonlinear Optical Susceptibilities of Crystals

Author, Year	Compound	Property	Ref.
Flytzanis and Ducuing, 1969	InSb, InAs, InP, GaSb, GaAs, GaP, AlSb, BP	$\chi^{(2)}$	180
Levine, 1969	Crystals with diamond, rocksalt, zinc-blende, and wurtzite structures	$\chi^{(2)}, \chi^{(3)}$	182
Phillips and Van Vechten, 1969	Zinc-blende and wurtzite crystals	$\chi^{(2)}$	185
Flytzanis, 1969	III–V semiconductors	$\chi^{(2)}$	188
Kleinman, 1970	Zinc-blende crystals	$\chi^{(2)}$	186
Levine, 1970	Zinc-blende, wurtzite, and quartz crystals	$\chi^{(2)}$	187
Flytzanis, 1971	II–VI, I–VII semiconductors	$\chi^{(2)}$	190
Tang and Flytzanis, 1971	III–V and II–VI semiconductors	$\chi^{(2)}$	192
Levine, 1973	Many families of crystals	$\chi^{(2)}$	193
Shih and Yariv, 1980	Zinc-blende, wurtzite, and $LiNbO_3$ crystals	$\chi^{(2)}$	191
Tsirelson et al., 1984	$LiHCO_2 \cdot D_2O$ and related crystals LiF, NaF, NaCl, MgO, MgS	$\chi^{(2)}, \chi^{(3)}$	194
Lines, 1990	Pre-transition metal halides and chalcogenides	$\chi^{(3)}$	195
Xue and Zhang, 1996	$NdAl_3(BO_3)_4$	$\chi^{(2)}$	196
Xue and Zhang, 1997	Urea	$\chi^{(2)}$	198
Xue and Zhang, 1998	$NaClO_3$, $NaBrO_3$	$\chi^{(2)}$	200
Xue et al., 2000	Mg-doped $LiNbO_3$	$\chi^{(2)}$	201
Xue and Bishop, unpublished	Diamond and quartz-type crystals	$\chi^{(3)}$	202
Fowler and Madden, 1984	F^-, Cl^-	$\chi^{(3)}$	203
Johnson et al., 1987	Alkali halides	$\chi^{(3)}$	205
Johnson and Subbaswamy, 1989	Alkali halides	$\chi^{(3)}$	206
Mestechkin, 1997	Urea, 1,3,5-triamino-2,4,6-trinitrobenzene	$\chi^{(2)}$	208

monoatomic or diatomic crystals for which the entire range of ionicity was covered. It was based on the dielectric theory of Phillips [183] and Van Vechten [184], who had proposed an isotropic two-band (one-gap) model corresponding to the bonding and antibonding states for the electronic band structure. In their model that used sp^3-hybridized orbitals, the band gap was composed of a covalent part and an ionic part modeled by a function of the number of valence electrons and the covalent radii of the atoms. These different contributions to the gap were determined from the measured low-frequency linear susceptibility and the bond distance by assuming that the bond polarizability is isotropic. By applying an external electric field, the bond charge was displaced (the ionic radii changed) and produced a polarization for which the successive orders in the field amplitude provided the second- and third-order NLO responses. Since these were given in terms of the macroscopic linear susceptibility, this method implicitly accounted for local field effects. Dispersion effects could also be partly included by using the frequency-dependent linear susceptibility; however, the method was inadequate when near the resonance region. Although based on a formalism different from that of Flytzanis and Ducuing [180], the final expressions are similar and the results are also in good agreement with experiment.

Phillips and Van Vechten [185] themselves adopted the one-gap model within the SOS framework to evaluate the $\chi^{(2)}$ of crystals of zinc-blende and wurtzite types. Approximating the difference of the dipole moment between the ground and excited state, they found that an *extra* correction, proportional to the square of the ionicity, was required to obtain better agreement with the experimental data. Such a correction was, however, unnecessary in the reformulation by Kleinman [186], who evaluated more carefully the dipole moment difference and used the f-sum rule. It is important to mention that though the final $\chi^{(2)}$ expressions of Flytzanis and Ducuing [180], Levine [182] and Kleinman [186] were different, they all provided reasonable agreement with experiment. Levine's model was later improved to account for the acentric position of the bond charge which induced a $\chi^{(2)}$ contribution associated with a field-dependent change of the covalent gap [187].

Flytzanis [188] generalized the model of Levine [182] to evaluate the ionic contribution to the linear electro-optic (dc-Pockels) effect of III–V semiconductors that results from the coupling between the low-frequency field and the ionic displacements. This model reproduced the experimental Faust–Henry coefficients [189], which characterize the relative ionic and electronic contributions to the linear electro-optic effect. This method was extended to II–VI and I–VII semiconductors [190]. An analogous approach was later followed by Shih and Yariv [191] and also appeared suitable for zinc-blende and wurtzite crystals. Later, Tang and Flytzanis [192] proposed that charge transfer between the atoms, as opposed to the displacement of the bond charge [182], was the origin of $\chi^{(2)}$. This approach reproduced agreement with the experimental data and also was able to rationalize certain discrepancies between the sp^3 theories and experiment in terms of the effects of d electrons. However, they pointed out that a treatment of these d-electron effects required the use of molecular orbital techniques. They also determined the ionicity that gives the maximum $\chi^{(2)}$.

In 1973, Levine [193] extended the range of applicability of his original model by considering in a simple way the bond ionicity, the difference in atomic radii, and the d electrons as contributions to second-order nonlinear optical susceptibilities. The electronic nonlinear optical susceptibility arose therefore from the acentric field-induced displacement of the bond charge in a potential which had symmetric (covalent) and antisymmetric (ionic) components. The d-electron effect was introduced by substituting core charges with effective core charges. The scheme was applied to a wide variety of different types of compounds including III–V, II–VI, I–VII, IV–IV, and multibond crystals as well as highly anisotropic crystals and ferroelectrics. Considering the limited parameterization, the overall agreement with experiment was impressive and was

related to the implicit inclusion of the troublesome local field effects by the use of the measured macroscopic linear susceptibility. Frequency dispersion and electronic delocalization effects appear, however, as features that are difficult to handle.

Tsirelson et al. [194] extended Levine's model to account for the different types of bonds in second-order nonlinear susceptibility calculations. Applications to lithium formate deuterate ($LiCOOH \cdot D_2O$) and related crystals with and without water molecules showed that the water can play a significant role in $\chi^{(2)}$. They also proposed an expression for $\chi^{(3)}$ which gave qualitative agreement with experiment for five ionic crystals.

The bond-orbital theory proposed by Lines [195] to determine the third-order nonlinear susceptibility of pre-transition-metal halides and chalcogenides can be viewed as an extension to $\chi^{(3)}$ of the variational–perturbational method of Flytzanis and Ducuing [180], although in the latter model the orbitals are orthogonal. For each pair of atoms a bond was associated with its bonding and antibonding molecular orbitals, and $\chi^{(3)}$ was determined as the third-order response of the bond dipole moment to a longitudinal local electric field. The final expression for $\chi^{(3)}$ was given in terms of several parameters, which, with the exception of the local-field factor, could be deduced from the experimental geometric (bond length and bond center) and electronic (bonding–antibonding energy gap and linear response) values. The comparison between the theoretical and experimental nonlinear refractive indices, n_2, for halides showed that the local field factor was essentially independent of the linear refractive index for all the species studied. Lines [195] therefore set the local field factor to be unity. Using this parameterized method, the root-mean-square accuracy for the relative n_2 values of the halides was about 9%.

The 1973 Levine bond charge method [193] was extensively employed by Xue and collaborators to evaluate the second-order nonlinear susceptibilities of inorganic crystals and one organic crystal (urea). In Ref. 196, the SHG response of $NdAl_3(BO_3)_4$ crystals was determined and shown to be mostly associated with the boron–oxygen bonds. Combining Levine's approach for determining the electronic component of $\chi^{(2)}$ with the Shih and Yariv [191] treatment for its vibrational (ionic) counterpart, Xue and Zhang [197] evaluated the electro-optic coefficient, $r_{ijk}(-\omega;\omega,0) = -(2\epsilon_0/\epsilon_i(\omega)\epsilon_j(\omega)) \times \chi_{ijk}^{(2)}(-\omega;\omega,0)$, for diatomic crystals. This method turned out to be good for characterizing the second-order nonlinear susceptibilities of organic crystals such as the urea crystal, though they pointed out its limitations in dealing with systems with delocalized bond charges [198]. In the case of the urea crystal, $\chi^{(2)}$ was dominated by the carbon–nitrogen bond contributions, and hydrogen bond contributions were negligible. On the other hand, adopting the same strategy, the hydrogen bonds were demonstrated to be a key factor maximizing the second-order NLO responses of crystals in the family of dihydrogenophosphate ammonium [199]. Xue and

Zhang [200] incorporated the cations into their treatment and determined the SHG response of the isomorphous $NaClO_3$ and $NaBrO_3$ crystals. As in Ref. 193, they reproduced the experimental fact that, although of the same chirality, these two crystals possess $\chi^{(2)}$ values of opposite sign. In another study, Xue et al. [201] investigated the effects of magnesium-doping on the SHG responses of lithium niobate. Recently, Xue and Bishop [202] have extended the treatment of Refs. 182, 187, and 193 to evaluate the crystal macroscopic THG second hyperpolarizability and applied it to diamond- and quartz-type crystals.

Since the third-order nonlinear susceptibility contains one- and two-photon allowed terms and the linear susceptibility does not, further extensions of the models of Refs. 193 and 202 are needed to account for the specifics of these contributions. Moreover, to our knowledge, no expressions have yet been provided for the vibrational (ionic) counterparts of $\chi^{(3)}$.

In 1984, Fowler and Madden [203] proposed a three-step ab initio procedure to determine the second hyperpolarizabilities of F^- and Cl^- in lithium halide crystals. This method, which analyzed the differences, with respect to the free-anion values, in terms of crystal field effects and overlap compression, is based on the fact that the electronic density is localized in ionic crystals. The crystal field effects correspond to the pure Coulombic interactions between the reference ion and the surrounding lattice of point charges. These were incorporated by calculating the second hyperpolarizabilities of the ion in the presence of a finite fragment of the lattice of point charges. For the anions, for which the electron density is rather extended, the crystal field effects resulted in a compression of the electronic cloud and a big reduction in γ. γ of the cation is much smaller and, because its density is more localized, the crystal field effects here were neglected. The overlap compression originates from the interactions with the electronic clouds of neighboring species and leads to a further decrease in γ for the anion. Fowler and Madden [203] modeled this by adding one shell of six cations around the central anion. The second hyperpolarizability of this small cluster surrounded by the rest of the lattice point charges led to the in-crystal anion second hyperpolarizability. In principle, one should have removed from the cluster the second hyperpolarizabilities of the $(Li_6)^+$ shell and the dipole–induced-dipole contributions. The former contribution was negligible and the latter was assumed to be negligible on the basis of a related investigation on the polarizability [204]. For the fluorine anion, the average static second hyperpolarizability was reduced by a factor of 50 at the Hartree–Fock level. They [203] also showed that the crystal environment destroyed the isotropy of the γ tensor as given by the values of $\Delta\gamma = (\gamma_{xxxx} - 3\gamma_{xxyy})/\gamma_{xxxx}$: -0.71 and -0.17 for F^- and Cl^-, respectively. The small size of the cluster prevented the incorporation of possible long-range charge-transfer effects which can be masked by basis-set superposition effects. The in-crystal hyperpolarizabilities correspond in fact to the dressed hyperpolarizabilities of the ions by the surrounding

crystal; thus, for comparison with the experimental $\chi^{(3)}$, local field factors have to be included. Adopting the usual Lorentz local field factors to describe the local field corrections, they found good agreement between their theoretical results and experiment.

The approach followed by Johnson and co-workers [205,206] within the LDA of DFT was similar to that of Ref. 203. These authors also concluded that the anion γ is sensitive to the crystalline environment whereas the cation is not. Several shells of ions were considered in combination with a lattice of point charges for which the Madelung potential was computed using the Ewald sum. Using a partially self-consistent pseudopotential approach [205], which was found preferable over a fully self-consistent scheme, the determination of the ground-state electron density was made by imposing several constraints: (i) All the anions/cations had the same density and (ii) the density was spherically averaged as were its related potentials. When the field was switched on, the electron density on the neighboring atoms was kept frozen so that the field-induced quantities corresponded to average in-crystal, dressed, second hyper-polarizabilities. They found that adding a second shell of atoms left the in-crystal hyperpolarizabilities unchanged and that, after accounting for local field effects, the computed values were in good agreement with the experimental data extrapolated to zero frequency. In another study [206], they removed the constraint of sphericity by treating the cubic potential of the alkali halides as a perturbation and then they evaluated the second hyperpolarizability anisotropies. The poor agreement with IDRI (intensity-dependent refractive index) experi-mental data was attributed to the lack of anisotropic local field corrections rather than to the vibrational Raman-like component that was thought to be negligible.

Experimental investigations by Adair et al. [207] later confirmed the ultrasensitivity of the anion second hyperpolarizability to the nature of the cation. In particular, they found a good correlation between the nearest-neighbor interionic distance, R (a function of the cation), and the in-crystal anionic second hyperpolarizability: $\gamma(\text{in-crystal}) = \gamma(\text{isolated}) \times \exp(-D/R^2)$, where D is a measure of the anion sensitivity. Nevertheless, further investigation on the second hyperpolarizability of other crystals will decide whether this is a viable method. It would also be interesting to know how good these approaches are for estimating the second-order susceptibilities. Furthermore, it will be useful to know whether the dominant contribution always comes from the anion as in the above γ investigations or whether the cation can also play a role through, for instance, anion-to-cation charge transfer contributions.

Yet another Coulomb approach was proposed by Mestechkin [208] for molecular crystals and illustrated in the case of the determination of the dressed first hyperpolarizability of urea and 1,3,5-triamino-2,4,6-trinitrobenzene (TATB). The full crystal lattice potential was expressed in the form of Madelung sums and evaluated by using the calculated atomic charges within the Mulliken scheme. This potential was incorporated into the Fock matrix of the target molecules, the molecules were considered to be identical, and the problem was solved iteratively until self-consistency was achieved in a manner similar to that of Zyss and Berthier [10]. Contrary to supermolecule calculations, this approach has the advantage of leading to a hyperpolarizability tensor that has the crystal symmetry. The "Coulomb dressing" of the urea molecule increased its hyper-polarizability by about 30%, whereas for TATB the increase was much larger and the crystal Coulomb field was found to break the symmetry of the β tensor of the isolated molecule.

Aside from these approaches, empirical relations have been proposed to relate the second- and third-order nonlinear susceptibilities to their linear analogs. Among them are Miller's rule [209] for the second-order nonlinear susceptibility of ionic crystals as well as the expressions of Wang [210] and of Boling, Glass, and Owyoung [211] for the third-order nonlinear susceptibility in the low-frequency limit. The latter have been shown, by comparison with experiment, to possess their own range of validity.

Acknowledgments

The authors thank the Natural Sciences and Engineering Research Council of Canada for funding. B.C. thanks the Belgian National Fund for Scientific Research for his Senior Research Associate position.

References

1. P. N. Butcher and D. Cotter, *The Elements of Nonlinear Optics*, Cambridge University Press, Cambridge, 1990.
2. R. W. Boyd, *Nonlinear Optics*, Academic Press, San Diego, 1992.
3. D. L. Mills, *Nonlinear Optics, Basic Concepts*, Springer, Berlin, 1998.
4. D. S. Chemla, J. L. Oudar, and J. Jerphagnon, *Phys. Rev. B* **12**, 4534 (1975).
5. N. Bloembergen, *Nonlinear Optics*, Benjamin, New York, 1965.
6. S. L. Adler, *Phys. Rev.* **126**, 413 (1962).
7. C. J. Böttcher, *Theory of Electric Polarization*, Vol. 1, *Dielectrics in Static Fields*, Elsevier, Amsterdam, 1973.
8. J. L. Oudar and J. Zyss, *Phys. Rev. A* **26**, 2016 (1982).
9. J. Zyss and J. L. Oudar, *Phys. Rev. A* **26**, 2028 (1982).
10. J. Zyss and G. Berthier, *J. Chem. Phys.* **77**, 3635 (1982).
11. S. Allen, T. D. McLean, P. F. Gordon, B. D. Bothwell, M. D. Hursthouse, and S. A. Karaulov, *J. Appl. Phys.* **64**, 2583 (1988).
12. T. Hamada, *J. Phys. Chem.* **100**, 19344 (1996).
13. T. Hamada, *J. Phys. Chem.* **100**, 8777 (1996).
14. A. V. Yakimanski, U. Kolb, G. N. Matveeva, I. G. Voigt-Martin, and A. Tenkovtsev, *Acta Crystallogr. A* **53**, 603 (1997).
15. I. G. Voigt-Martin, G. Li, A. A. Yakimanski, J. J. Wolff, and H. Gross, *J. Phys. Chem. A* **101**, 7265 (1997).
16. C. Lin and K. Wu, *Chem. Phys. Lett.* **321**, 83 (2000).

17. X. L. Zhu, X. Z. You, Y. Zhong, Z. Yu, and S. L. Guo, *Chem. Phys.* **253**, 241 (2000).

18. See Sections 3.1 and 3.3.1 of B. Champagne and B. Kirtman, in *Handbook of Advanced Electronic and Photonic Materials and Devices*, Vol. 9, *Nonlinear Optical Materials*, H. S. Nalwa, ed., Academic Press, San Diego, 2001, Chapter 2, p. 63.

19. G. J. B. Hurst, M. Dupuis, and E. Clementi, *J. Chem. Phys.* **89**, 385 (1988).

20. T. Hamada, *J. Chem. Soc. Faraday Trans.* **92**, 3165 (1996).

21. T. Hamada, *J. Chem. Soc. Faraday Trans.* **94**, 509 (1998).

22. E. A. Perpète, B. Champagne, and B. Kirtman, *J. Chem. Phys.* **107**, 2463 (1997).

23. E. A. Perpète, J. M. André, and B. Champagne, *J. Chem. Phys.* **109**, 4624 (1998).

24. F. L. Gu, D. M. Bishop, and B. Kirtman, *J. Chem. Phys.* **115**, 10548 (2001).

25. S. Yamada, M. Nakano, I. Shigemoto, S. Kiribayashi, and K. Yamaguchi, *Chem. Phys. Lett.* **267**, 438 (1997).

26. Y. Luo, P. Norman, P. Macak, and H. Ågren, *Phys. Rev. B* **61**, 3060 (2000).

27. W. D. Cheng, J. S. Huang, and J. X. Lu, *Phys. Rev. B* **57**, 1527 (1998).

28. R. K. Li, N. Ye, and Y. C. Wu, *Phys. Rev. B* **62**, 7654 (2000).

29. B. Kirtman, J. L. Toto, K. A. Robins, and M. Hasan, *J. Chem. Phys.* **102**, 5350 (1995).

30. B. Champagne, D. Jacquemin, J. M. André, and B. Kirtman, *J. Phys. Chem. A* **101**, 3158 (1997).

31. E. K. Dalskov, J. Oddershede, and D. M. Bishop, *J. Chem. Phys.* **108**, 2152 (1998).

32. E. J. Weniger and B. Kirtman, in *Numerical Methods in Physics*, T. E. Simos, G. Avdelas, and J. Vigo-Aguiar, eds., *Chemistry and Engineering. Computers & Mathematics with Applications*, Elsevier, Holland, 2002.

33. See Fig. 7 of B. Champagne and B. Kirtman, in *Handbook of Advanced Electronic and Photonic Materials and Devices*, Vol. 9, *Nonlinear Optical Materials*, H. S. Nalwa, ed., Academic Press, San Diego, 2001, Chapter 2, p. 63.

34. F. Castet and B. Champagne, *J. Phys. Chem. A* **105**, 1366 (2001).

35. E. Botek and B. Champagne, *Appl. Phys. B* **74**, 627 (2002).

36. Q. Zhu, J. E. Fischer, R. Zusok, and S. Roth, *Solid State Commun.* **83**, 179 (1992); K. Pressl, K. D. Aichholzer, G. Leising, and H. Kahlert, *Synth. Metals* **55**, 4432 (1993).

37. D. M. Bishop and F. L. Gu, *Chem. Phys. Lett.* **317**, 322 (2000).

38. B. Jansik, B. Schimmelpfennig, P. Norman, Y. Mochizuki, Y. Luo, and H. Ågren, *J. Phys. Chem. A* **106**, 395 (2002).

39. G. H. Wagnière and J. B. Hutter, *J. Opt. Soc. Am. B* **6**, 693 (1989).

40. A. D. Buckingham, *Adv. Chem. Phys.* **12**, 107 (1967).

41. V. Magnasco and R. McWeeny, in *Theoretical Models of the Chemical Bonding*, Z. B. Maksic, ed., Springer-Verlag, Berlin, 1991, p. 133; G. Chalasinski and M. M. Szczesniak, *Chem. Rev.* **100**, 4227 (2000).

42. See, for instance, A. D. Buckingham, E. P. Concannon, and I. D. Hands, *J. Phys. Chem.* **98**, 10455 (1994); X. Li, K. L. C. Hunt, J. Pipin, and D. M. Bishop, *J. Chem. Phys.* **105**, 10594 (1996).

43. D. M. Bishop and M. Dupuis, *Mol. Phys.* **88**, 887 (1996).

44. J. D. Auspurger and C. E. Dykstra, *Int. J. Quantum Chem.* **43**, 135 (1992).

45. B. Kirtman, in *Nonlinear Optical Materials: Theory and Modeling*, S. P. Karna and A. T. Yeates, eds., ACS Symposium Series, New York, 1995, Vol. 628, Chapter 3, p. 58.

46. B. Kirtman, C. E. Dykstra, and B. Champagne, *Chem. Phys. Lett.* **305**, 132 (1999).

47. C. W. Dirk, R. J. Twieg, and G. Wagnière, *J. Am. Chem. Soc.* **108**, 5387 (1986).

48. J. Waite and M. G. Papadopoulos, *Z. Naturforsch.* **45a**, 189 (1990).

49. J. Waite and M. G. Papadopoulos, *Z. Naturforsch.* **43a**, 253 (1988).

50. T. Yasukawa, T. Kimura, and M. Uda, *Chem. Phys. Lett.* **169**, 259 (1990).

51. S. Di Bella, M. A. Ratner, and T. J. Marks, *J. Am. Chem. Soc.* **114**, 5842 (1992).

52. S. Di Bella, G. Lanza, I. Fragalá, S. Yitzchaik, M. A. Ratner, and T. J. Marks, *J. Am. Chem. Soc.* **119**, 3003 (1997).

53. Y. Okuno, S. Yokoyama, and S. Mashiko, *J. Phys. Chem. B* **105**, 2163 (2001).

54. J. Perez and M. Dupuis, *J. Phys. Chem.* **95**, 6525 (1991).

55. V. Moliner, P. Escribano, and E. Peris, *New J. Chem.* **22**, 387 (1998).

56. M. Guillaume, E. Botek, B. Champagne, F. Castet, and L. Ducasse, *Int. J. Quantum Chem.*, in press (2002).

57. B. Kirtman and B. Champagne, *Int. Rev. Phys. Chem.* **16**, 389 (1997); D. M. Bishop, *Adv. Chem. Phys.* **104**, 1 (1998).

58. P. C. M. McWilliams and Z. G. Soos, *J. Chem. Phys.* **95**, 2127 (1991).

59. D. Guo and S. Mazundar, *J. Chem. Phys.* **97**, 2170 (1992); P. C. M. McWilliams, Z. G. Soos, and G. W. Hayden, *J. Chem. Phys.* **97**, 2172 (1992).

60. See Section 3.4.2 of B. Champagne and B. Kirtman, in *Handbook of Advanced Electronic and Photonic Materials and Devices*, Vol. 9, *Nonlinear Optical Materials*, H. S. Nalwa, ed., Academic Press, San Diego, 2001, Chapter 2, p. 63.

61. S. Y. Chen and H. A. Kurtz, *J. Mol. Struct.* **388**, 79 (1996).

62. B. Champagne and B. Kirtman, *J. Chem. Phys.* **109**, 6450 (1998).

63. B. Kirtman, B. Champagne, F. L. Gu, and D. M. Bishop, *Int. J. Quantum Chem.* **90**, 709 (2002).

64. O. Xie and C. W. Dirk, *J. Phys. Chem. B* **102**, 9378 (1998).

65. B. Champagne, M. Spassova, J. B. Jadin, and B. Kirtman, *J. Chem. Phys.* **116**, 3935 (2002); J. Zyss, I. Ledoux, S. Volkov, V. Chernyak, S. Mukhamel, G. P. Bartholomew, and G. C. Bazan, *J. Am. Chem. Soc.* **122**, 11956 (2000).

66. M. Tomonari, N. Ookubo, T. Takada, and H. Hirayama, *Chem. Phys. Lett.* **272**, 199 (1997).

67. T. T. Rantala, M. I. Stockman, D. A. Jelski, and T. F. George, *J. Chem. Phys.* **93**, 7427 (1990).

68. P. P. Korambath and S. P. Karna, *J. Phys. Chem. A* **104**, 4801 (2000).

69. A. Banerjee and M. K. Harbola, *J. Chem. Phys.* **113**, 5614 (2000).

70. H. A. Lorentz, *The Theory of Electrons*, Leipzig, 1952, p. 138.

71. C.J. F. Böttcher and P. Bordewijk, *Theory of Electric Polarization*, Vol. 2, *Dielectrics in Time-Dependent Fields*, Elsevier, Amsterdam, 1978, Chapter XV.

72. R. W. Munn, *Mol. Phys.* **64**, 1 (1988).

73. P. J. Bounds and R. W. Munn, *Chem. Phys.* **24**, 343 (1977).

74. A. J. Stone, *Mol. Phys.* **56**, 1065 (1985).

75. R. F. W. Bader, T. A. Keith, K. M. Gough, and K. E. Laidig, *Mol. Phys.* **75**, 1167 (1992).

76. C. G. Giribet, M. D. Demarco, M.Ruiz de Azúa, and R. H. Contreras, *Mol. Phys.* **91**, 105 (1997).

77. J. M. Stout and C. E. Dykstra, *J. Phys. Chem. A* **102**, 1576 (1998).

78. M. B. Ferraro, M. C. Caputo, and P. Lazzeretti, *J. Chem. Phys.* **109**, 2987 (1998).

79. N. Celebi, J. G. Ángyán, F. Dehez, C. Millot, and C. Chipot, *J. Chem. Phys.* **112**, 2709 (2000).

80. J. R. Maple and C. S. Ewig, *J. Chem. Phys.* **115**, 4981 (2001).

81. R. F. Bader and C. F. Matta, *Int. J. Quantum Chem.* **85**, 592 (2001).

82. L. Silberstein, *Philos. Mag.* **33**, 92 (1917); J. Applequist, J. R. Carl, and K. K. Fung, *J. Am. Chem. Soc.* **94**, 2952 (1972); M. L. Olson and K. R. Sundberg, *J. Chem. Phys.* **69**, 5400 (1978); B. T. Thole, *Chem. Phys.* **59**, 341 (1981); P. T. van Duijnen and M. Swart, *J. Phys. Chem.* **102**, 2399 (1998); L. Jensen, P. O. Åstrand, K. O. Sylvester-Hvid, and K. V. Mikkelsen, *J. Phys. Chem. A* **104**, 1563 (2000).

83. L. Jensen, P. O. Åstrand, A. Osted, J. Kongsted, and K. V. Mikkelsen, *J. Chem. Phys.* **116**, 4001 (2002).

84. M. Hurst and R. W. Munn, *J. Mol. Electron.* **2**, 35 (1986).

85. M. Hurst and R. W. Munn, *J. Mol. Electron.* **3**, 75 (1987).

86. T. Zhou and C. E. Dykstra, *J. Phys. Chem. A* **104**, 2204 (2000).

87. V. Geskin and J. L. Brédas, *J. Chem. Phys.* **109**, 6163 (1998); R. Zanasi, *Chem. Phys. Lett.* **315**, 217 (1999).

88. M. Hurst and R. W. Munn, *J. Mol. Electron.* **2**, 43 (1987).

89. M. Hurst and R. W. Munn, *J. Mol. Electron.* **2**, 139 (1986).

90. M. Hurst, R. W. Munn, and J. O. Morley, *J. Mol. Electron.* **6**, 15 (1990).

91. R. W. Munn, *Int. J. Quantum Chem.* **43**, 159 (1992).

92. P. G. Cummins, D. A. Dunmur, and R. W. Munn, *Chem. Phys. Lett.* **36**, 199 (1975).

93. M. Hurst and R. W. Munn, *J. Mol. Electron.* **2**, 101 (1986).

94. C. E. Dykstra, *J. Comput. Chem.* **9**, 476 (1988).

95. C. E. Dykstra, S. Y. Liu, and D. J. Malik, *J. Mol. Struct (Theochem.)* **135**, 357 (1986).

96. R. W. Munn, *Mol. Phys.* **89**, 555 (1996).

97. R. W. Munn, Z. Shuai, and J. L. Brédas, *J. Chem. Phys.* **108**, 5975 (1998).

98. R. W. Munn, *J. Chem. Phys.* **106**, 3870 (1997).

99. M. Malagoli and R. W. Munn, *J. Chem. Phys.* **107**, 7926 (1997).

100. H. Reis, S. Raptis, M. G. Papadopoulos, R. H. C. Janssen, D. N. Theodorou, and R. W. Munn, *Theor. Chem. Acc.* **99**, 384 (1998).

101. H. Reis, M. G. Papadopoulos, and R. W. Munn, *J. Chem. Phys.* **109**, 6828 (1998).

102. H. Reis, M. G. Papadopoulos, C. Hättig, J. G. Ángyán, and R. W. Munn, *J. Chem. Phys.* **112**, 6161 (2000).

103. M. in het Panhuis and R. W. Munn, *J. Chem. Phys.* **112**, 6763 (2000).

104. M. Malagoli and R. W. Munn, *J. Chem. Phys.* **112**, 6757 (2000).

105. H. Reis, M. G. Papadopoulos, P. Calaminici, K. Jug, and A. M. Köster, *Chem. Phys.* **261**, 359 (2000).

106. M. in het Panhuis and R. W. Munn, *J. Chem. Phys.* **113**, 10685 (2000).

107. M. in het Panhuis and R. W. Munn, *J. Chem. Phys.* **113**, 10691 (2000).

108. H. Reis, S. Raptis, and M. G. Papadopoulos, *Chem. Phys.* **263**, 301 (2001).

109. R. W. Munn, *J. Chem. Phys.* **114**, 5607 (2001).

110. R. W. Munn, *J. Chem. Phys.* **114**, 5404 (2001).

111. H. Nobutoki and H. Koezuka, *J. Phys. Chem. A* **101**, 3762 (1997).

112. J. Knoester and S. Mukamel, *Phys. Rev. A* **39**, 1899 (1989).

113. J. Knoester and S. Mukamel, *J. Chem. Phys.* **91**, 989 (1989).

114. S. Mukamel in *Molecular Nonlinear Optics, Materials, Physics, and Devices*, J. Zyss, ed., Academic Press, New York 1994, Chapter 1, p. 1.

115. K. Kunc and R. Resta, *Phys. Rev. Lett.* **51**, 686 (1983); R. Resta and K. Kunc, *Phys. Rev. B* **34**, 7146 (1986).

116. D. E. Aspnes, *Phys. Rev. B* **6**, 4648 (1972).

117. D. J. Moss, J. E. Sipe, and H. M. van Driel, *Phys. Rev. B* **36**, 9708 (1987).

118. D. J. Moss, E. Ghahramani, J. E. Sipe, and H. M. van Driel, *Phys. Rev. B* **41**, 1542 (1990).

119. E. Ghahramani, D. J. Moss, and J. E. Sipe, *Phys. Rev. B* **43**, 8990 (1991).

120. V. N. Genkin and P. M. Mednis, *Sov. Phys. JETP* **27**, 609 (1968) [*Zh. Eksp. Teor. Fiz.* **54**, 1137 (1968)].

121. J. E. Sipe and E. Ghahramani, *Phys. Rev. B* **48**, 11705 (1993).

122. C. Aversa, J. E. Sipe, M. Sheik-Bahae, and E. W. Van Stryland, *Phys. Rev. B* **50**, 18073 (1994).

123. C. Aversa and J. E. Sipe, *Phys. Rev. B* **52**, 14636 (1995).

124. J. E. Sipe and A. I. Shkrebtii, *Phys. Rev. B* **61**, 5337 (2000).

125. M. I. Bell, in *Electronic Density of States*, L. H. Bennett, ed., National Bureau of Standards (U.S.) Special Publication No. 323, U. S. GPO, Washington, D.C., 1971, p. 757.

126. C. Y. Fong and Y. R. Shen, *Phys. Rev. B* **12**, 2325 (1975).

127. E. Ghahramani, D. J. Moss, and J. E. Sipe, *Phys. Rev. Lett.* **64**, 2815 (1990); E. Ghahramani and J. E. Sipe, *Phys. Rev. B* **46**, 1831 (1992).

128. E. Ghahramani, D. J. Moss, and J. E. Sipe, *Phys. Rev. B* **43**, 9700 (1991).

129. M. Z. Huang and W. Y. Ching, *Phys. Rev. B* **45**, 8738 (1992).

130. M. Z. Huang and W. Y. Ching, *Phys. Rev. B* **47**, 9464 (1993).

131. W. Y. Ching and M. Z. Huang, *Phys. Rev. B* **47**, 9479 (1993).

132. J. L. P. Hugues and J. E. Sipe, *Phys. Rev. B* **53**, 10751 (1996).

133. J. L. P. Hugues, Y. Wang, and J. E. Sipe, *Phys. Rev. B* **55**, 13630 (1997).

134. J. L. P. Hugues and J. E. Sipe, *Phys. Rev. B* **58**, 7761 (1998).

135. R. W. Godby, M. Schlüter, and L. J. Sham, *Phys. Rev. B* **37**, 10159 (1988); V. Fiorentini and A. Baldereschi, *Phys. Rev. B* **51**, 17196 (1995).

136. B. M. Deb and S. K. Ghosh, *J. Chem. Phys.* **77**, 342 (1982); E. Runge and E. K. U. Gross, *Phys. Rev. Lett.* **52**, 997 (1984); for a recent review, see R. van Leeuwen, *Int. J. Mod. Phys. B* **15**, 1969 (2001).

137. A. Zangwill and P. Soven, *Phys. Rev. A* **21**, 1561 (1980).

138. See the review by X. Gonze, *Phys. Rev. A* **52**, 1096 (1995) and references therein.

139. Z. H. Levine, *Phys. Rev. B* **42**, 3567 (1990); *Phys. Rev. B* **44**, 5981 (1991).

140. Z. H. Levine and D. C. Allan, *Phys. Rev. Lett.* **66**, 41 (1991); *Phys. Rev. B* **44**, 12781 (1991).

141. H. Zhong, Z. H. Levine, D. C. Allan, and J. W. Wilkins, *Phys. Rev. Lett.* **69**, 379 (1992); *Phys. Rev. B* **70**, 1032 (1993).

142. Z. H. Levine and D. C. Allan, *Phys. Rev. B* **48**, 7783 (1993).

143. J. Chen, Z. H. Levine, and J. W. Wilkins, *Phys. Rev. B* **50**, 11514 (1994).

144. W. G. Aulbur, Z. H. Levine, J. W. Wilkins, and D. C. Allan, *Phys. Rev. B* **51**, 10691 (1995).

145. J. Chen, Z. H. Levine, and J. W. Wilkins, *Appl. Phys. Lett.* **66**, 1129 (1995).

146. Z. H. Levine, *Phys. Rev. B* **49**, 4532 (1994).

147. E. K. Chang, E. L. Shirley, and Z. H. Levine, *Phys. Rev. B* **65**, 35205 (2001).

148. X. Gonze and J. P. Vigneron, *Phys. Rev. B* **39**, 13120 (1989).

149. A. Dal Corso and F. Mauri, *Phys. Rev. B* **50**, 5756 (1994).

150. R. W. Nunes and D. Vanderbilt, *Phys. Rev. Lett.* **73**, 712 (1994).

151. A. Dal Corso, F. Mauri, and A. Rubio, *Phys. Rev. B* **53**, 15638 (1996).

152. J. Chen, L. Jönsson, J. W. Wilkins, and Z. H. Levine, *Phys. Rev. B* **56**, 1787 (1997).

153. R. W. Nunes and X. Gonze, *Phys. Rev. B* **63**, 155107 (2001).

154. R. D. King-Smith and D. Vanderbilt, *Phys. Rev. B*. **47**, 1651 (1993). See also R. Resta, *Rev. Mod. Phys.* **66**, 899 (1994).

155. B. Kirtman, F. L. Gu, and D. M. Bishop, *J. Chem. Phys.* **113**, 1294 (2000).

156. D. M. Bishop, F. L. Gu, and B. Kirtman, *J. Chem. Phys.* **114**, 7633 (2001).

157. B. Adolph and F. Bechstedt, *Phys. Rev. B* **57**, 6519 (1998).

158. B. Adolph and F. Bechstedt, *Phys. Rev. B* **62**, 1706 (2000).

159. S. N. Rashkeev, W. R. L. Lambrecht, and B. Segall, *Phys. Rev. B* **57**, 3905 (1998).

160. S. N. Rashkeev, W. R. L. Lambrecht, and B. Segall, *Phys. Rev. B* **57**, 9705 (1998).

161. S. N. Rashkeev, S. Limpijumnong, and W. R. L. Lambrecht, *Phys. Rev. B* **59**, 2737 (1999).

162. S. N. Rashkeev and W. R. L. Lambrecht, *Phys. Rev. B* **63**, 165212 (2001).

163. M. Rérat, W. D. Cheng, and R. Pandey, *J. Phys. Condens. Matter* **13**, 343 (2001).

164. X. L. Zhu, X. Z. You, and Y. Zhang, *Chem. Phys.* **254**, 287 (2000).

165. X. Gonze, Ph. Ghosez, and R. W. Godby, *Phys. Rev. Lett.* **74**, 4035 (1995).

166. R. M. Martin and G. Ortiz, *Phys. Rev. B* **56**, 1124 (1997).

167. W. G. Aulbur, L. Jönsson, and J. W. Wilkins, *Phys. Rev. B* **54**, 8540 (1996).

168. B. Champagne, E. A. Perpète, S. J. A. van Gisbergen, J. G. Snijders, E. J. Baerends, C. Soubra-Ghaoui, K. A. Robins, and B. Kirtman, *J. Chem. Phys.* **109**, 10489 (1998); *erratum* **110**, 11664 (1999); B. Champagne, E. A. Perpète, D. Jacquemin, S. J. A. van Gisbergen, E. J. Baerends, C. Soubra-Ghaoui, K. A. Robins, and B. Kirtman, *J. Phys. Chem. A* **104**, 4755 (2000).

169. X. Gonze, Ph. Ghosez, and R. W. Godby, *Phys. Rev. Lett.* **78**, 294 (1997).

170. S. J. A. van Gisbergen, P. R. T. Schipper, O. V. Gritsenko, E. J. Baerends, J. G. Snijders, B. Champagne, and B. Kirtman, *Phys. Rev. Lett.* **83**, 694 (1999).

171. P. L. de Boeij, F. Kootstra, J. A. Berger, R. van Leeuwen, and J. G. Snijders, *J. Chem. Phys.* **115**, 1995 (2001).

172. M. van Faassen, P. L. de Boeij, R. van Leeuwen, J. A. Berger, and J. G. Snijders, *Phys. Rev. Lett.* **88**, 186401 (2002).

173. C. P. Agrawal, C. Cojan, and C. Flytzanis, *Phys. Rev. B* **17**, 776 (1978).

174. B. Champagne, D. Jacquemin, and J. M. André, in *Nonlinear Optical Properties of Organic Materials VIII*, G. R. Möhlmann, ed., SPIE Proceedings, Vol. 2527, 1995, p. 71, Washington D.C.

175. P. Otto, *Phys. Rev. B* **45**, 10876 (1992).

176. J. Ladik, in *Nonlinear Optical Materials: Theory and Modeling* S. P. Karna and A. T. Yeates, eds., ACS, Washington, D.C., 1996, Vol. 628, Chapter 10, p. 174.

177. P. Otto, F. L. Gu, and J. Ladik, *J. Chem. Phys.* **110**, 2717 (1999).

178. K. Schmidt and M. Springborg, *Phys. Chem. Chem. Phys.* **1**, 1743 (1999).

179. F. L. Gu, Y. Aoki, and D. M. Bishop, *J. Chem. Phys.*, **117**, 385 (2002).

180. C. Flytzanis and J. Ducuing, *Phys. Rev.* **178**, 1218 (1969).

181. S. S. Jha and N. Bloembergen, *Phys. Rev.* **171**, 891 (1968).

182. B. F. Levine, *Phys. Rev. Lett.* **22**, 787 (1969).

183. J. C. Phillips, *Phys. Rev. Lett.* **20**, 550 (1968).

184. J. A. Van Vechten, *Phys. Rev.* **182**, 891 (1969).

185. J. C. Phillips and J. A. Van Vechten, *Phys. Rev.* **183**, 709 (1969).

186. D. A. Kleinman, *Phys. Rev. B* **2**, 3139 (1970).

187. B. F. Levine, *Phys. Rev. Lett.* **25**, 440 (1970).

188. C. Flytzanis, *Phys. Rev. Lett.* **23**, 1336 (1969).

189. W. L. Faust and C. H. Henry, *Phys. Rev. Lett.* **17**, 1265 (1966).

190. C. Flytzanis, *Phys. Lett. A* **34**, 99 (1971).

191. C. C. Shih and A. Yariv, *Phys. Rev. Lett.* **44**, 281 (1980).

192. C. L. Tang and C. Flytzanis, *Phys. Rev. B* **4**, 2520 (1971).

193. B. F. Levine, *Phys. Rev. B* **7**, 2600 (1973); *Phys. Rev. B* **8**, 4046 (1973).

194. V. G. Tsirelson, O. V. Korolkova, I. S. Rez, and R. P. Ozerov, *Phys. Status Solidi* (b) **122**, 599 (1984).

195. M. E. Lines, *Phys. Rev. B* **41**, 3383 (1990).

196. D. Xue and S. Zhang, *J. Phys. Condens. Matter* **8**, 1949 (1996).

197. D. Xue and S. Zhang, *J. Solid State Chem.* **128**, 17 (1997).

198. D. Xue and S. Zhang, *J. Phys. Chem. A* **101**, 5547 (1997).

199. D. Xue and S. Zhang, *Chem. Phys. Lett.* **301**, 449 (1999).

200. D. Xue and S. Zhang, *Chem. Phys. Lett.* **287**, 503 (1998).

201. D. Xue, K. Betzler, and H. Hesse, *J. Phys. Condens. Matter* **12**, 6245 (2000).

202. D. Xue and D. M. Bishop, unpublished.

203. P. W. Fowler and P. A. Madden, *Phys. Rev. B* **30**, 6131 (1984).

204. P. W. Fowler and P. A. Madden, *Mol. Phys.* **49**, 913 (1983).

205. M. D. Johnson, K. R. Subbaswamy, and G. Senatore, *Phys. Rev. B* **36**, 9202 (1987).

206. M. D. Johnson and K. R. Subbaswamy, *Phys. Rev. B* **39**, 10275 (1989).

207. R. Adair, L. L. Chase, and S. A. Payne, *Phys. Rev. B* **39**, 3337 (1989).

208. M. M. Mestechkin, *J. Phys. Condens. Matter* **9**, 157 (1997).

209. R. C. Miller, *Appl. Phys. Lett.* **5**, 17 (1964).

210. C. W. Wang, *Phys. Rev. B* **2**, 2045 (1970).

211. N. L. Boling, A. J. Glass, and A. Owyoung, *IEEE J. Quantum Electron.* **QE-14**, 601 (1978).

BRIDGING THE GAP BETWEEN LONG TIME TRAJECTORIES AND REACTION PATHWAYS

RON ELBER, ALFREDO CÁRDENAS, AVIJIT GHOSH, and HARRY A. STERN

Department of Computer Science, Cornell University, Ithaca, New York, U.S.A.

CONTENTS

I. INTRODUCTION

Molecular Dynamics (MD) simulations provide an atomically detailed description of complex systems on a wide range of temporal and spatial scales. The MD approach has had numerous successes and has provided many insightful observations, but a clear limitation is the restriction to short time scales. While some heroic efforts have reached the microsecond time scale [1], routine

Advances in Chemical Physics, Volume 126, Edited by I. Prigogine and Stuart A. Rice.
ISBN 0-471-23582-2.   © 2003 John Wiley & Sons, Inc.

simulations of complex and large molecular systems at the atomic level of detail are restricted to nanoseconds. This time scale is too short to address many interesting processes in biophysics, such as conformational transitions of proteins, transport phenomena, and reactions.

An alternative approach to MD for modeling kinetics of molecular systems is based on (a) statistical theories of the rate and (b) the concept of a reaction coordinate [2]. The latter is usually identified as the steepest descent path (SDP). The SDP is a curve with a low-energy barrier connecting two minima, a reactant and a product. It provides a qualitative view of the progress of the reaction, along with useful input to theories of rate such as the transition state theory [3].

The focus of this chapter is a recently developed methodology [4–6] that enables us to calculate approximate MD trajectories at extended time scales. The method serves as a bridge between exact trajectories, reaction coordinates, and statistical theories of rate. It has been applied to the investigations of numerous systems [5–7], and we have computed trajectories for durations of nanoseconds [5], microseconds [6], and milliseconds [7]. Of course, it should be noted that the millisecond trajectories are highly approximate, since a very significant fraction of the motions was filtered out (see Section III.C). Nevertheless, they provide a view of the reaction pathway that is useful in interpretations of experimental data [7].

After presenting the algorithm, we show that the proposed numerical protocol interpolates (as a function of the approximation quality) from a steepest descent path (a poor approximation to the dynamics) to an exact Newtonian trajectory. Finally, we present an efficient algorithm to compute relative rates that can be used with our formulation to compute experimental observables.

II. MOLECULAR DYNAMICS

We start with a brief review of existing well-established approaches while emphasizing the current limitations. We use the term MD for a simulation technique that solves the classical equations of motion *at the atomic level of detail*. We assume that an atomically detailed potential is available (we use empirical potentials, but other approaches can be used as well). The dynamics on the energy surface is described by classical mechanics (Newton's law). The discussion is limited to dynamical models and differential equations that directly follow from microscopic parameters and can be made to approach the exact solution.

A. Initial Value Formulation

The initial value formulation of classical dynamics (Newton's equations of motion) is the most widely used in numerical solutions of classical mechanic problems:

$$M\frac{d^2X}{dt^2} = -\nabla U \tag{1}$$

Throughout the text the lowercase x, y, or z variables denote scalar coordinates, while the uppercase X, Y, and Z denote vectors of coordinates. In the above formula, M is the mass matrix, X is the coordinate vector, t is the time, and U is the potential energy. A successful (and simple) algorithm to solve Eq. (1) is the Verlet algorithm [8]. In the "velocity" form, it reads

$$\begin{aligned} X_{i+1} &= X_i + V_i\Delta t - (\Delta t^2/2)M^{-1}dU/dX_i \\ V_{i+1} &= V_i - (\Delta t/2)M^{-1}[dU/dX_i + dU/dX_{i+1}] \end{aligned} \tag{2}$$

A trajectory is obtained after specifying two initial conditions, the coordinate vector, $X(t = 0) \equiv X_0$, and the velocity vector, $V(t = 0) \equiv V_0$. The size of the time step is restricted because steps larger than a few femtoseconds result in numerical instabilities. To obtain long time dynamics, many small time steps of size Δt are required (e.g., to reach a few nanoseconds, millions of steps are required). The necessity of using small time steps is the major obstacle in the computation of long time dynamics with the initial value formulation.

B. A Boundary Value Formulation in Time

Another well-established formulation of classical mechanics (which is, of course, equivalent) is based on a boundary value problem. We seek a stationary solution of a functional of the path, S [9]:

$$\begin{aligned} S[X(t')] &= \int_0^t L \cdot dt' \\ L &= \frac{1}{2}\left(\frac{dX}{dt'}\right)^T M\left(\frac{dX}{dt'}\right) - U(X) \end{aligned} \tag{3}$$

In Eqs. (3) the two end points, $X(0)$ and $X(t)$, and the total time t are held fixed. It is possible (in principle) to solve Eqs. (3) numerically by discretizing the integral and computing the stationary discrete path (S now is a function of the set $\{X_i\}_{i=1}^N$):

$$S[\{X_i\}_{i=1}^N] = \Delta t \cdot \sum_{i=0,\dots,N} \frac{1}{2\Delta t^2}(X_{i+1} - X_i)^T M(X_{i+1} - X_i) - U(X_i) \tag{4}$$

Solving Eq. (4) (i.e., finding a trajectory that makes S stationary) is an alternative to solving the initial value problem [Eqs. (2)]. Equation (4) has a certain philosophical appeal since it provides a global solution of the whole path instead of Eqs. (2), which provide a sequence of local solutions in time. There is a hope that having a global view of the classical trajectory will yield more robust and stable solutions as a function of step size.

In support of the "philosophical appeal" is the fact that Eq. (4) is an integral while in Eqs. (2) we use derivatives. Numerical estimates of integrals are, in general, more accurate and more stable compared to estimates of derivatives. On the other hand, computations of the whole path are more expensive than the calculation of one temporal slice of the trajectory at a time. The computational effort is larger in the boundary value formulation by at least a factor of N, where N is the number of time slices, compared to the calculation of a step in the initial value approach. To make the global approach computationally attractive (assuming that it does work), the gain in step size must be substantial.

The hope is then that Eq. (4) may be a useful alternative to initial value solvers if *approximate* long time trajectories are desirable. However, the use of Eq. (4) "as is" is problematic. One problem is that even in Eq. (4) we need an estimate of (first order) derivatives of the coordinates with respect to time. Here, the estimate is based on a finite difference. The finite difference estimate is of poor quality as the time step increases, and it leads to numerical instabilities even in the integral formulation.

If we are after large time steps, the functional of Eq. (4) changes from being a minimum to being a maximum. This is not a desirable property for a function to be minimized!

To exemplify the above problems with a simple example, it is instructive to use one-dimensional harmonic oscillator:

$$S[\{x_j\}_{j=1}^N] = \Delta t \cdot \sum_{j=0,\ldots,N} \left(\frac{1}{2} M \frac{(x_{j+1} - x_j)^2}{\Delta t^2} - \frac{K}{2} x_j^2 \right) \tag{5}$$

The stationary condition on the function S is

$$\frac{1}{\Delta t} \partial S/\partial x_k = \frac{M}{\Delta t^2} [(x_k - x_{k+1}) + (x_k - x_{k-1})] - K x_k = 0$$

$$\left(\frac{2M}{\Delta t^2} - K \right) x_k - \frac{M}{\Delta t^2} (x_{k+1} + x_{k-1}) = 0 \tag{6}$$

Guessing a solution of the type $x_k = x_0 \exp[-ik\omega]$ ($i = \sqrt{-1}$), we have

$$\left(\frac{2M}{\Delta t^2} - K \right) - \frac{2M}{\Delta t^2} \cos(\omega) = 0$$

$$\left(1 - \frac{K \cdot \Delta t^2}{2M} \right) = \cos(\omega) \tag{7}$$

It is obvious that we obtain a stability condition that is not much different from the stability condition of the initial value equation. If Δt is larger than $2 \cdot \sqrt{M/K}$, (the cosine is smaller than -1), the solution grows exponentially and is numerically unstable. Hence, in the straightforward boundary value formulation of classical mechanics, we gain very little in terms of stability and step size compared to the solution of the initial value differential equation. The difficulty is not in the "philosophical" view (global or local) but in the estimate of the time derivative, which is approximated by a local finite difference expression.

It is also possible to demonstrate a shift from a maximum to a minimum for S. The diagonal elements of the second derivative matrix of the action, $\partial^2 S/\partial x_k^2 = \left(\frac{2M}{\Delta t^2} - K\right)$, are changing their sign. They are positive at small Δt and become negative at sufficiently large Δt.

Of course, it is not reasonable to expect that oscillations with a frequency ν will be reproduced accurately with a time step Δt larger than π/ν. However, there is a difference between an accurate representation and being blown out of the roof. An attractive alternative (if possible) is the removal of modes that change significantly on a time scale shorter than Δt.

Consider, for example, a two-dimensional harmonic oscillator. Instead of Eq. (5), we now have

$$S[\{x_j, y_j\}_{j=1}^N] = \Delta t \cdot \sum_{j=1,\ldots,N} \left(\frac{1}{2}M\frac{(x_{j+1}-x_j)^2}{\Delta t^2} + \frac{1}{2}M\frac{(y_{j+1}-y_j)^2}{\Delta t^2} - \frac{K_x}{2}x_j^2 - \frac{K_y}{2}y_j^2\right)$$

$$K_x \gg K_y \qquad (8)$$

The frequency along the x direction, $\sqrt{K_x/M}$, is set to be much larger than the frequency of the oscillation along the y coordinate, $\sqrt{K_y/M}$. However, we cannot increase the time step, Δt. The use of a time step appropriate for the y direction will cause numerical instability in the x direction. An oscillatory solution of the type $\exp[i\omega t]$ will start growing exponentially like $\exp[\omega t]$.

In the above trivial case simple, filtering of x (set $x_j = 0$ for all j) enables us to obtain an exact solution of y. In Eq. (8) there is no coupling between the two coordinates, whereas in real-life applications, coupling is usually present. For coupled degrees of freedom, freezing high-frequency modes provides only an approximate solution. Nevertheless, the freezing is essential in maintaining the numerical stability of other slow motions.

It is possible to use multiple sizes of time steps to integrate separately along x and y axes [10–12]. However, this procedure requires the prior identification of the fast and the slow modes; that is, we need to know which modes to integrate with small time steps and for which modes it is possible to employ the more economic larger time steps.

The identification of fast coordinates can be difficult in simulations of condensed phases. Some of the fast modes are bond or angle vibrations that can be identified and integrated separately. However, other fast modes are transient. They are fast for a short duration of time and slow otherwise. The transient fast modes are collisions—for example, two atoms that are close and feel strong repulsive forces due to excluded volume interactions. The relevant degree of freedom (the distance between the atoms) is a fast mode during the collision event and a slow mode before or after the short collision period. The fundamental complication in the treatment of these modes is the "identity crisis" of these fast/slow coordinates.

Earlier studies suggest a special treatment to a collision coordinate as a fast coordinate that is turned on and off [13]. A transformation of a collision coordinate, to a form more appropriate for strongly or weakly interacting particles, was used in the framework of a mean field approximation. However, tracking down collision events and treating them in a special way is computationally expensive. It cannot be done in practice for more than a few collision events at the same time. Since the number of collisions at a given time slice is proportional to the number of atoms in the system, it is difficult to come with a general scalable tracking scheme that will be independent of the system size. Moreover, as the size and the density of the system increase, three- and four-body collisions (that are not considered in the above scheme [13]) may be relevant as well.

Our goals are therefore twofold:

1. We seek a "stable" treatment of the fast modes (on the scale of Δt) and approximate description of the slow modes. We hope that the simplification done for the rapid displacements will still produce sound description of the slow motions.
2. We seek a formulation in which we will not need to identify (to begin with) what are the slow and the fast degrees of freedom. Hence we seek an automated "stabilizing" algorithm.

To achieve 1 and 2 a new model is required.

We note that we cannot expect to do better than the goals outlined above when a large step is used. Without detailed small-step integration, it is indeed impossible to follow fast motions. So "stabilizing" the modes with frequencies higher than $\pi/\Delta t$ is the best we could hope for.

Before describing the formulation of an alternative model, we consider yet another boundary value formulation that will be of considerable interest to us later on.

C. A Boundary Value Formulation in Length

A classical trajectory is a curve in Q-dimensional space (where Q is the number of degrees of freedom). In the discussion so far we consider the parameterization of the curve as a function of time, $X(t)$. However, this parameterization is clearly not unique, and we may choose to distribute the points along the curve in other ways that have computational benefits. Another well-established protocol in classical mechanics, which we consider here, parameterizes the curve as a function of the path length, l [9].

It is convenient at this point to change to mass weighted coordinates $Z = \sqrt{M}X$. The Lagrangian in Eqs. (3) is now modified to

$$L = \frac{1}{2}\left(\frac{dZ}{dt}\right)^2 - U(Z) \tag{9}$$

The usual formulation of the action as a function of length has fixed end points, fixed total energy, E, and variable length and time:

$$S_L[Z(l')] = \int_{Z(0)}^{Z(l)} \sqrt{2(E - U)} \cdot dl' \tag{10}$$

One of the advantages of Eq. (10) with respect to Eqs. (3) is that the total time of the trajectory is an output (versus an input). The energy can be estimated from equilibrium considerations. For example, the total kinetic energy $K \equiv E - U$ may be set to $Qk_BT/2$ (or sampled from the Boltzmann distribution), where Q is the number of degrees of freedom. The total time is recovered from the stationary path with length parameterization as $t = \int_{Z(0)}^{Z(l)} \frac{dl'}{\sqrt{2(E-U)}}$.

A discrete version of Eq. (10) can be optimized in a way similar to that of Eq. (4).

$$S_L[\{Z_j\}_{j=1}^N] = \sum_{j=1,\dots,N-1}\left(\frac{1}{\sqrt{2}}\left[\sqrt{E - U(Z_j)} + \sqrt{E - U(Z_{j+1})}\right] \cdot |Z_j - Z_{j+1}|\right) \tag{11}$$

The first and the last coordinate sets are fixed. Another advantage of this parameterization (besides switching from constant time to constant energy) is the elimination of time derivatives. A finite difference estimate remains (the length element, $\Delta l' \cong |Z_j - Z_{j+1}|$). However, since the configurations are equally distributed along the path, the length element behaves better than the finite difference estimate of the velocity. For example, the distance between two points provides a lower bound to the true length of the path. No such bound is available when estimating velocity.

The action, S_L, in Eq. (11) is not necessarily a minimum making the computation of the stationary path a nontrivial challenge. We have tried to use Eq. (11) in a straightforward way but finally gave up [A. Ghosh and R. Elber, unpublished]. A variant on Eq. (11) within the framework of the stochastic difference equation (Section III.F) is at present our most promising approach.

To conclude the discussion on the length-dependent action, we note that an initial value differential equation as a function of length also exists [6] and is

$$\frac{d^2Z}{dl^2} = -\frac{1}{2(E-U)}(\nabla U - (\nabla U \cdot \hat{\eta})\hat{\eta}) \tag{12}$$

where $\hat{\eta}$ is a unit vector in the current direction of the trajectory. This second-order differential equation resembles the Newton's equation of motion. It includes a force component only in the direction perpendicular to the path and an effective "mass" of $2(E-U)$. The length parameterization provides a constant density of points along the path, similarly to the constant density of points in time for the time parameterization. The scalar product of the first derivative of the path with a unit vector along the path direction, $dZ/dl \cdot \hat{\eta}$, is therefore a constant. Alternatively [as is evident from Eq. (12)], we note that the scalar product $(d^2Z/dl^2) \cdot \hat{\eta}$ is zero.

III. THE STOCHASTIC DIFFERENCE EQUATION

A. Stochastic Difference in Time: Definition

The expression we derive below leads to an action and to a stationary (minimum) condition on the classical path. The optimal path is a discrete approximation to a classical trajectory. Interestingly, in the integral limit (an infinitesimal time step), the action below was used already by Gauss (!) to compute classical trajectories [14]. At variance with Gauss we keep a finite Δt.

Despite the similarity to the Gauss approach to classical mechanics, there is a key difference between the classical actions described above and the corresponding action of the stochastic difference equation. The classical actions are deterministic mechanical models; the SDE is a nondeterministic approach that is based on stochastic modeling of the numerical errors introduced by the finite difference formula.

Computer simulations (obviously) use a finite time step. Consider a finite difference approximation to the Newton's equations of motion [Eq. (1)]:

$$M\frac{X_{i+1} + X_{i-1} - 2X_i}{\Delta t^2} + \frac{dU}{dX_i} = \varepsilon_{i+1} \tag{13}$$

The new feature in Eq. (13) is the use of an error vector, ε_{i+1}. Even if the trajectory we have at hand $\{X_i\}_{i=1}^N$ is exact, the left-hand side of Eq. (13) will not be zero, as finite difference is used to approximate the second derivative of the coordinate vector as a function of time. In straightforward Molecular Dynamics simulations the time step Δt is taken to be small with the hope that the errors (which we here denoted by ε_{i+1}) can be neglected. Usually we do not know what the errors are, since we do not have the exact trajectory. Therefore, ignoring the errors for small Δt seems like a reasonable idea. In the SDE approach we attempt to model these errors. This allows us (in principle) the use of larger time steps.

It is important to emphasize that the errors so calculated are with respect to the exact trajectory. The computational procedure for the errors is as follows: We first compute an exact solution of the Newton's equation of motion, $\{X_i\}_{i=1}^{N}$ (an analytical or a numerical solution). The exact solution is plugged into Eq. (13) to produce estimates of the errors due to the use of an approximate formula (a finite difference estimate of second derivatives). The errors so obtained are therefore not the deviation of an approximate trajectory from the exact trajectory. Here the errors estimate the accuracy of a finite difference formula, tested on the exact trajectory. Of course if we solve Eq. (13) we will have a trajectory that is different from the exact solution, and we may use Eq. (13) to measure this difference.

The distribution of errors is a property of the exact trajectories. If we generate an approximate trajectory based on the finite difference formula, we should generate an error distribution that is consistent with what we know about the true solution.

Note also that the error is evaluated at the upper edge of the interval. The use of the edge for the error calculation is for computational convenience (as discussed below) and should not affect the results. There is an ambiguity about the placement of the error vector that adds interesting complications to the derivation below, and it had led to some unnecessary anxiety by other researchers in the field. Since we are developing a new statistical model, we have the liberty of defining our model at our convenience. Why this choice is indeed convenient will become clearer in Section III.B.

We have performed numerous numerical experiments on the properties of the errors. Some of these studies are described in Ref. 5. A typical experiment is presented in Fig. 1.

Based on the numerical experiments, we suggest the following two basic assumptions to be used in the error modeling:

(i) The errors are considered stochastic variables, not correlated in time. Or, more explicitly,

$$\langle \varepsilon_i \rangle = 0, \qquad \langle \varepsilon_i \varepsilon_j \rangle = C \delta_{ij} \qquad (14)$$

(ii) The probability density of the norms of the error vectors is assumed Gaussian:

$$P(\varepsilon_i) = \sqrt{1/2\pi\sigma^2} \, \exp[-\varepsilon_i^2/2\sigma^2] \qquad (15)$$

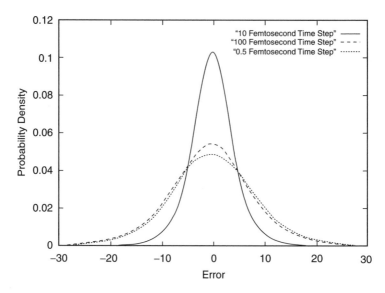

Figure 1.   Distribution of norms of the error vectors computed by the finite difference formula [Eq. (13)] from exact trajectories of valine dipeptide. The dipeptide was initially equilibrated at 300 K. The largest errors are significant and are of the same order of magnitude as the forces.

Assumption (ii) is for convenience only. In fact the formalism can be used for any functional form of the probability density of errors, provided that the first assumption is satisfied.

We have no "proof" that the above two assumptions are true. All we have are numerical experiments on systems that vary from a dipeptide to a small solvated protein. Our results suggest that the above assumptions are sound for sufficiently large time steps. We note that at the limit of small Δt, in which we obtain a nearly exact trajectory, the (much smaller) errors become correlated. For the statistical assumption to be valid the time step needs to be sufficiently large so that correlations will decay rapidly. A few numerical experiments for different step sizes are presented in Fig. 2.

Are the results of the numerical experiments surprising? Let us examine first the second assumption and assume for the moment that the correlation is lost rapidly. Is the normal distribution a surprise? It is not. It is a simple demonstration of the Central Limit Theorem (CLT). For sufficiently large systems and after ensemble averaging, the addition of the (nearly uncorrelated) elements of the error vector leads to a normal distribution. Note also that the first and second moments of the errors are bound if the coordinates of the exact trajectory are

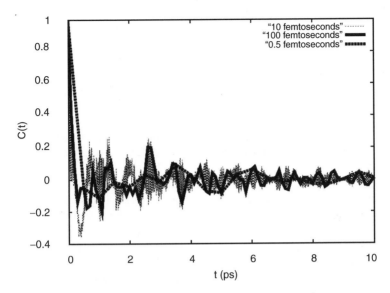

Figure 2. The correlation of errors $\langle (\varepsilon(t) \cdot \varepsilon(0))/\varepsilon(0) \cdot \varepsilon(0)) \rangle$ estimated from exact trajectories of valine dipeptide and Eq. (13) for three different time steps.

bound (as they are in practical condensed phase simulations). Therefore the conditions for the application of the CLT are satisfied.

Is it a surprise that the correlations diminished rapidly? If Δt is very small, then the two terms—the estimate for the second derivative of the coordinate with respect to time and the force—are comparable (along the exact trajectory). As Δt is made larger, the force contribution and the errors are made larger as well. At very large Δt (maximum errors) the error correlations become the correlation of the forces computed at time separation Δt (without short time integration). The forces are changing significantly on the time scale of Δt and their discrete average $C_{dis}(\tau) = (1/N)\sum_i F(\tau + i\Delta t) \cdot F(i\Delta t)$ decays quickly, much faster than the corresponding integral $C_{int}(\tau) = (1/t')\int_0^{t'} F(t) \cdot F(t + \tau)\, dt$. Even the integral is known from statistical mechanics to decay quite rapidly. It is sometimes used to estimate the friction kernel and the memory function of the generalized Langevin equation. Our noise models extra degrees of freedom, which are the high-frequency modes that were filtered out (see Section III.C). The multiple trajectories of the stochastic difference equation are approximations to Newtonian trajectories that share similar motions along the slow coordinates and may deviate significantly along the fast degrees of freedom. In the Langevin formulation the extreme view is taken that the correlation function is $C_{int}(\tau) \approx C_0 \delta(\tau)$. This is somewhat similar to our (weaker) assumption for $C_{dis}(\tau)$.

It is important to emphasize, however, that our model is different from the Langevin equation, which is a stochastic differential equation. Our model has no noise in the limit of small time steps in which the numerical errors approach zero. The "noise" we introduce is numerical. Once we filter the rapid oscillations, it is impossible for us to recover the true trajectory using only the low-frequency modes. The noise in the SDE approach is introduced when we approximate a differential equation by a finite difference formula and filter out high-frequency motions.

An obvious limitation of the above argument is that it applies only for "sufficiently" large systems and a "sufficiently" large time step. How large is sufficient requires numerical experiments. In our experience, the dynamics of valine dipeptide with 42 degrees of freedom and a time step of 10 femtoseconds already shows the desired properties (Figs. 1 and 2). Considerably "nicer" Gaussian curves were obtained for yet larger systems (folding of C peptide [5]). It is therefore pointless to test the above assumption on model systems that are not ergodic even if we love them very much (like the one-dimensional harmonic oscillator). The conditions for applying the CLT are clearly not satisfied in these cases.

A final comment on the model: In the applications that were pursued so far [4–7], we assume that the variance, σ^2, is a time-independent scalar. There is no theoretical or computational restriction to make it so, and a potential extension to the model may make the variance a time-dependent tensor. The current choice is based on insufficient data to fit, rather than on true conviction of simplicity.

B. A Stochastic Model for a Trajectory

If the above model of errors is accepted, we can proceed to use it to generate an ensemble of appropriate trajectories. A sequence of errors, $\{\varepsilon_i\}_{i=1}^{N}$, is generated with probability

$$\bar{P}(\{\varepsilon_i\}_{i=1}^{N}) \prod_j d\varepsilon_j = \prod_i P(\varepsilon_i) \, d\varepsilon_i \qquad (16)$$

If the errors are zero, we obtain the most probable trajectory within the framework of the stochastic difference equation. This trajectory is *not* exact and is within σ from the exact trajectory [Eq. (13)]. What are the approximations made? In Section III.C we argue that the approximate trajectory is a solution of the slow modes in the system where the high-frequency modes are filtered out.

Focusing on Eq. (16), it is more useful to write the probability in terms of coordinates (instead of errors):

$$\bar{P}(\{\varepsilon_i\}_{i=1}^N) \prod_j d\varepsilon_j = \prod_i P\left(\varepsilon_i \equiv M \frac{X_i + X_{i-2} - 2X_{i-1}}{\Delta t^2} + \frac{dU}{dX_{i-1}}\right)$$

$$\times \left[\prod_k dX_k\right] \cdot \det\left[J_{ij} \equiv \frac{d\varepsilon_i}{dX_j}\right] \tag{17}$$

The determinant on the right-hand side is the Jacobian of transformation from the error vector to the coordinate vector. More explicitly, the Jacobian of the transformation is

$$\det\left[\frac{d\varepsilon_i}{dX_j}\right] = \begin{bmatrix} \partial\varepsilon_1/\partial X_1 & \partial\varepsilon_1/\partial X_2 & \partial\varepsilon_1/\partial X_3 & \cdots \\ \partial\varepsilon_2/\partial X_1 & \partial\varepsilon_2/\partial X_2 & \partial\varepsilon_2/\partial X_3 & \cdots \\ \partial\varepsilon_3/\partial X_1 & \partial\varepsilon_3/\partial X_2 & \partial\varepsilon_3/\partial X_3 & \cdots \\ \cdots & \cdots & \cdots & \cdots \end{bmatrix}$$

$$= \begin{bmatrix} M/\Delta t^2 & 0 & 0 & \cdots \\ \left[\frac{-2M}{\Delta t^2} + \frac{d^2 U}{dX_1^2}\right] & M/\Delta t^2 & 0 & \cdots \\ M/\Delta t^2 & \left[\frac{-2M}{\Delta t^2} + \frac{d^2 U}{dX_2^2}\right] & M/\Delta t^2 & \cdots \\ \cdots & \cdots & \cdots & \cdots \end{bmatrix} = \left(\frac{M}{\Delta t^2}\right)^N \tag{18}$$

Expressing the determinant in Eq. (18), we find that only the diagonal terms remain, since the upper off-diagonal part of the determinant is zero. The final result is coordinate-independent. This is the most convenient choice and is the motivation behind the error placement at the edge of the interval. Placing the error at the center would introduce second derivatives of the potential to the Jacobian, which would make it coordinate-dependent and more expensive to compute.

The ambiguity in the choice of the Jacobian is well known from path integral studies [15]. There the choice is made based on the physics that we wish to present. There is no "correct" or "wrong" choice before a concrete physical model is introduced. Our freedom in defining the errors allows us to make the most convenient choice.

Note that the boundary conditions implicitly written into the determinant requires the knowledge of (fixed) X_0 and X_{N+1}. Our model as outlined below leads, at the limit of small Δt, to a fourth-order differential equation in time that requires four initial or boundary values. In our experience, fixing only a pair of coordinates (optimizing also the velocities at the boundaries) affects the overall results only very little when a large time step is used. The noise overwhelms subtle differences in the initial conditions.

The probability density of a trajectory can then be written as

$$
\begin{aligned}
P(\{X\}_{i=1}^{N}) &= A \prod_{i} \exp\left[\frac{\varepsilon_i^2}{2\sigma^2}\right] = A \prod_{i} \exp\left[-\frac{1}{2\sigma^2}\left(M\frac{X_{i+1}+X_{i-1}-2X_i}{\Delta t^2}+\frac{dU}{dX_i}\right)^2\right] \\
&= A \exp\left[-\frac{1}{2\sigma^2}\sum_{i}\left(M\frac{X_{i+1}+X_{i-1}-2X_i}{\Delta t^2}+\frac{dU}{dX_i}\right)^2\right] \\
&= A \exp\left[-\frac{1}{2[\sigma^2\cdot\Delta t]}\sum_{i}\Delta t\left(M\frac{X_{i+1}+X_{i-1}-2X_i}{\Delta t^2}+\frac{dU}{dX_i}\right)^2\right]
\end{aligned}
\tag{19}
$$

where A is the normalization factor that is coordinate-independent. The last trick of dividing and multiplying by Δt is not necessary from a computational viewpoint but makes the sum approach a limit at small Δt. We define now an "action" S_{SDET} for the stochastic difference equation in time. A trajectory that minimizes S_{SDET} is an approximation to the true classical trajectory:

$$
\begin{aligned}
S_{\text{SDET}}(\{X_i\}_{i=1}^{N}) &\equiv \sum_{i}\Delta t\left(M\frac{X_{i+1}+X_{i-1}-2X_i}{\Delta t^2}+\frac{dU}{dX_i}\right)^2 \\
S_{\text{SDET}} &\xrightarrow[\substack{\Delta t\to 0 \\ N\to\infty}]{} \int_{0}^{(N-1)\Delta t} dt'\left(M\frac{d^2X}{dt'^2}+\frac{dU}{dX}\right)^2 \equiv S_G
\end{aligned}
\tag{20}
$$

The last limiting expression is the Gauss action, S_G, for classical mechanics. It clearly has a minimum that satisfies the equations of motion (when the action is zero). The action is non-negative, which makes it easier to identify the true minimum. The non-negativity is an important difference from the classical action formulation that we introduced at the beginning and makes the calculations with the S_{SDET} and S_G significantly more stable. The approximate trajectories that are produced by optimization of S_{SDET} are stable. An exponential solution of the type $\exp[\omega t]$ (ω positive) cannot be obtained in a reasonable formulation of a boundary value problem if the boundaries are not "explosive". However, an "explosive" solution can be obtained with initial value formulation.

A complication we should keep in mind when comparing S_G to the usual classical action is that the Newtonian trajectory is not the only stationary solution of the Gauss action. A standard variation of Eq. (20) leads to a fourth order differential equation and hence to two more solutions in addition to the true classical trajectory (the two additional solutions are related by a time reversal operation). An example was discussed in details in Ref. 4 [see discussions and Eqs. (15) and (16) of Ref. 4]. The good news is that the true trajectory is still the global minimum (when the action is zero), which is a clear computational guideline. However, the possibility of being trapped at a wrong minimum solution exists.

C. Weights of Trajectories and Sampling Procedures

At this point we may continue in one of two directions. We may use a single approximate trajectory at the neighborhood of the exact trajectory—that is, the trajectory that was obtained by the minimization of the discrete action. Alternatively, we recognize that the exact trajectory deviates from the optimal trajectory by errors distributed normally (keep in mind that the error distribution is a property of the exact trajectory). We may sample errors (and plausible trajectories) from the appropriate distribution of coordinates in the neighborhood of the trajectory with filtered high-frequency modes. The sampling in the neighborhood of the optimized trajectory should be normalized (we approximate one trajectory):

$$\int P(\{X\}_{i=1}^N) \prod_{k=2,\ldots,N} dX_k = \int A \exp\left[-\frac{S_{\mathrm{SDET}}}{2[\sigma^2 \cdot \Delta t]}\right] \prod_{k=2,\ldots,N} dX_k = 1 \qquad (21)$$

where A is the normalization constant. The above expression suggests that the weight of a single trajectory with fixed boundaries is $\exp\left[-\frac{S_{\mathrm{SDET}}}{2\sigma^2\Delta t}\right]$ for trajectories with the same energy. Additional weighting may be required according to the desired statistical mechanics ensemble. This weight opens the way for the use of Molecular Dynamics or the Monte Carlo procedure to sample *probable solutions* to the boundary value problem. We note that the exact trajectory should be within a distance $\sigma\sqrt{\Delta t}$ from the most probable trajectory with a large time step, Δt, underlining the need for trajectory sampling.

Define the vector $Y \equiv [X_1, X_2, \ldots, X_N]$ that includes the complete discrete approximation for the trajectory. The action, S_{SDET}, is a function of Y. To generate the trajectory distribution with the above weight, we create a canonical distribution with S_{SDET} for an energy function and $2\sigma^2\Delta t$ replacing the usual thermal energy ($k_B T$). In our code MOIL [16] we implemented a Molecular Dynamics protocol that solves the following Newton-like equation:

$$\frac{d^2 Y}{d\xi^2} = -\nabla S_{\mathrm{SDET}} \qquad (22)$$

The parameter ξ is a fictitious time, and the fictitious masses are set uniformly to 1. The solution of (22) conserves the total energy of the system where S_{SDET} is the potential. To obtain a temperature of $2\sigma^2 \Delta t$, we scale the velocities, $dY/d\xi$, periodically by a single factor λ [17]:

$$\lambda \cdot \sum_i \left(\frac{dy_i}{d\xi}\right)^2 = N \cdot Q \cdot (2\sigma^2 \cdot \Delta t) \qquad (23)$$

The scaling factor λ is chosen to satisfy Eq. (23). The number of time slices is N, and Q is the number of degrees of freedom in a single time slice. The code in MOIL is producing complete trajectories in any of the ξ steps following the dynamics of Eqs. (22) and (23). The trajectories so obtained have the Boltzmann-like weight for a trajectory, expected from formula (21).

We note that the algorithm can be parallelized efficiently. Implementations for loosely coupled clusters of PCs are available for LINUX and Windows operating systems using the MPI communication library. In practice, the scaling with the number of processors (linear), as well as the load balancing, is excellent [6]. This is, of course, not surprising since the parallelization protocol is exceedingly simple and consists of assigning different time slices to different CPUs [6]. Only nearest-neighbor communications between the CPUs are required.

The above argument for the weight of trajectories holds for sampling trajectories with fixed end points. The argument becomes subtler if we wish to vary the starting and ending points and sample alternative trajectories. In straightforward classical mechanics, all trajectories with the same energy have the same weight. In the present approximate formulation the total time (and not the energy) is made fixed. The weight of different classical trajectories (different boundary conditions) in S_{SDET} can be estimated only approximately. There are two basic assumptions we have to make: The first is on the total energy, and the second is on the kinetic energy.

In the first assumption we set the total energy, E, to be a constant during the whole trajectory. This is correct in a true classical trajectory, but is not exact in an S_{SDET} path. In the same sense that we have errors in our coordinates, we are likely to have errors in the energy as well. Nevertheless, this is a useful constraint to have and is likely to make the trajectories more realistic. Using the first assumption, the trajectory weight is (for a thermal system with an inverse temperature β), $\exp\left[-\beta E - \frac{S_{SDET}}{2\sigma^2 \Delta t}\right]$.

Note that we change our philosophy here. If we allow different initial conditions while keeping the total time fixed, we also must allow different energies. We maintain, however, the same energy for one trajectory. The total energy is then written as a sum of kinetic and potential energies $(K = E - U)$ and computed for every time slice during the trajectory. The calculation of the energy for each time slice instead of the starting point makes the functional more symmetric, and it should not matter if the energy is indeed conserved. The functional becomes

$$S_{SDET}^{\beta} \equiv -\beta E - \frac{S_{SDET}}{2\sigma^2 \Delta t} = -\frac{1}{2\sigma^2 \Delta t} \sum_i [2\sigma^2 \cdot (\beta/N)(K_i + U_i) + \varepsilon_i^2] \Delta t \quad (24)$$

Estimating the total energy, E, of the current S_{SDET}^{β} trajectory is, however, difficult. The large time step that we employ makes it difficult to estimate time derivatives of the type dX/dt (and the corresponding kinetic energy). The paths we compute do not have enough information to estimate the kinetic and therefore the total energy. We are therefore making a (second) assumption. This time the assumption is on the kinetic energy.

The second assumption is that the *average* kinetic energy is roughly thermal. We assume that

$$\sum_i (\beta/N) K_i \cong Q/2 \quad (25)$$

This assumption is not bad at all for sufficiently large microcanonical systems and long trajectories. It is consistent with the numerical observation of the similarity between canonical and microcanonical ensembles for systems with a few hundred coupled degrees of freedom.

Using Eq. (25), we note that the kinetic energy makes no contribution to the weight because the constant factor will disappear in the normalization. The log of the weight of a trajectory with arbitrary starting and ending points is therefore reduced to

$$S_{SDET}^{\beta} \equiv -\beta E - \frac{S_{SDET}}{2\sigma^2 \Delta t} \cong -\frac{1}{2\sigma^2 \Delta t} \sum_i [2\sigma^2 \cdot (\beta/N)U_i + \varepsilon_i^2] \Delta t \quad (26)$$

Note that Eq. (26) leads to a very similar trajectory sampling protocol as outlined in Eqs. (22) and (23). We only need to change S_{SDET} by the modified action, S_{SDET}^{β} of Eq. (26).

Of the above two assumptions, (i) constant total energy E and (ii) thermally averaged kinetic energy, the first one is more difficult to justify because the distribution of errors in the coordinates should affect the energy as well. It would be nice if we could enforce the energy conservation and still maintain the simple properties of the S_{SDET} formulation. The other formulation that we discuss (the stochastic difference equation in length, S_{SDEL}, Section III.F) fixes the energy and is therefore a more natural procedure to sample alternate initial conditions. There is no need to enforce energy conservation because this property is already built in. In the next section we consider more sampling concerns.

D. Mean Field Approach, Fast Equilibration, and Molecular Labeling

So far we have discussed algorithms that with the addition of more computational resources (more time slices) approach the exact answer. It is useful at this point to introduce one physically based approximation that significantly reduces the computational resources required. At the least we can have it as an option when the computational resources are limited. Another useful feature of the approximation below is the ability to solve the problem of molecular labeling and proper solvent sampling.

The molecular labeling problem is as follows. Consider a solvated system (e.g., a protein immersed in a box of water). To compute a S_{SDET} path, we need to specify the initial and the final coordinate sets, X_1 and X_N. Some of the coordinates are the spatial locations of water molecules. The coordinates are required by classical mechanics, true; but the exact labeling of the different water molecules creates a huge labeling degeneracy. All the permutations of water molecules will create identical trajectories. Moreover, a slight perturbation in the solvent coordinate will create alternative trajectories that we are not interested in. Our prime interest is in the dynamics of the protein rather than the dynamics of the water molecules.

Here we proposed a physically based approximation to get around this problem. We separate the coordinate set X into two domains: X^{slow} and X^{fast}, where "slow" and "fast" are with respect to the rate of approaching equilibrium. For example, we argue that the translation and rotation degrees of freedom of a bulk water molecule relax to equilibrium more rapidly than the protein dihedral angles $(\phi, \psi)_i$ of an amino acid i.

We assume separation of (equilibration) time scale that can be done only if the properties of the system are reasonably well understood. Proceeding with the example of a solvated protein, we set the water coordinates to be X^{fast}, and the protein coordinates to be X^{slow}. Consider a time step, Δt, that is significantly longer than the relaxation time to equilibrium of the fast part but is still slow on the time scale of relaxation to equilibrium of the slow part.

In the above-mentioned example, the time scale of 100 picoseconds is probably in that range. It is significantly longer than the local orientation and translation relaxation of the water molecules but too short to allow complete relaxation of the protein dihedral angles. If such a time step, Δt, is used in S_{SDET} calculation, it eliminates the need to follow the explicit dynamics of the water molecules. On this time scale, in a single step, the water molecules will already relax to equilibrium (with a "frozen" configuration of the slow protein). Their explicit dynamics will become irrelevant.

The consequences of the above picture for the S_{SDET} calculation are as follows: Instead of following the explicit dynamics of all the degrees of freedom in X, we follow the explicit dynamics only of X^{slow} and we thermally average the action (for each time slice) over the X^{fast} coordinates.

$$\langle S_{SDET}\rangle_{X^{fast}} = \sum_i \Delta t \left\langle \left(M^{slow} \frac{X_{i+1}^{slow} + X_{i-1}^{slow} - 2X_i^{slow}}{\Delta t^2} + \frac{dU}{dX_i^{slow}} \right)^2 \right\rangle_{X^{fast}} \quad (27)$$

In the above average, only the force depends on the fast coordinates. So we are required to perform averages of the type $\langle \frac{dU}{dX^{slow}} \rangle_{X^{fast}}$ and $\langle (\frac{dU}{dX^{slow}})^2 \rangle_{X^{fast}}$. These averages are performed in practice by short molecular dynamics trajectories for the fast components while keeping the slow components fixed at their current time slice configuration. The average $\langle G(X^{fast}, X^{slow}) \rangle_{X^{fast}}$ of the function $G(X^{fast}, X^{slow})$ over X^{fast} is computed as follows:

$$\langle G(X^{fast}; X^{slow}) \rangle_{X^{fast}} = \frac{1}{\delta t} \int_0^{\delta t} G(X^{fast}(t); X^{slow}) \cdot dt$$
$$M^{fast} \frac{d^2 X^{fast}(t)}{dt^2} = -\nabla U(X^{fast}; X^{slow}) \quad (28)$$

The molecular dynamics trajectory for X^{fast} is computed at fixed slow coordinates and for a time duration, δt, that is significantly smaller than Δt. In the calculations of folding of C peptide [5], of a conformational transition in hemoglobin [6], and of ion migration through the gramicidin channel [Koneshan Siva and Ron Elber, *Protein, Structure Function, and Genetics*, in press], the above averaging was used for a selected set of slow coordinates. In these calculations we average the fast coordinates for only tens of steps between sequential optimizations of the slow coordinates. While more extensive averaging of the fast coordinates could help, this is what we can afford at present.

We also note that there is more than one choice of a function to average. Besides the specific choice made in Eq. (27), S_{SDET}, it is also possible to average (over the fast coordinates) the weight, $\exp[-S_{SDET}/2\sigma^2 \cdot \Delta t]$, or the force, $-dU/dX$. Of the three possibilities the last choice is equivalent to generating a potential of mean force prior to the calculations of dynamics. The direct use of a potential of mean force for peptides and proteins is another direction that we are currently pursuing [18].

E.   Stochastic Difference Equation in Length

The stochastic difference equation in length is conceptually similar to the stochastic difference in time. We therefore do not repeat all of the arguments and discussions above. Rather we briefly list the main formulas and focus on the differences between the two approaches.

Instead of starting from the Newton's equation, we use the action formalisms. In the time formulation, we obtain the equation of motion by requiring that the action is stationary—that is, $\delta S/\delta X(\tau) = 0$—or in the discrete approximation to the path $\{\partial S/\partial X_i = 0\}_{i=1}^{N}$ [see Eq. (4)]. The S_{SDET} action can be written in that case as

$$S_{\text{SDET}} = \sum_i \left(\frac{1}{\Delta t} \partial S/\partial X_i \right)^2 \Delta t \qquad (29)$$

We are using a similar approach to define the analogous action for the stochastic difference equation in length, S_{SDEL} [S_L is defined in Eq. (10)]:

$$S_{\text{SDEL}} = \sum_i \left(\frac{1}{\Delta l_{i,i+1}} \partial S_L/\partial Z_i \right)^2 \Delta l_{i,i+1} \qquad (30)$$

The optimization of S_{SDEL} is performed subject to the constraint that all the lengths of the path segments, $\Delta l_{i,i+1}$, are the same. This is (of course) equivalent to the requirement that the points are equally distributed along the path. The constraint is conveniently formulated as a penalty function [19]:

$$\text{Constraint} = \frac{\lambda}{(N-1)Q} \sum_i (\Delta l_{i,i+1} - \langle \Delta l \rangle)^2, \qquad \langle \Delta l \rangle = \frac{1}{N-1} \sum_i \Delta l_{i,i+1}$$
$$(31)$$

In the above expression, λ is a constant that is determined by experimentation. Formally the constraints are not necessary. They are added for computational convenience. Moreover, since the length element is computed in mass-weighted Cartesian coordinates, $\Delta l_{i,i+1} = |Z_i - Z_{i+1}|$, it is necessary to avoid overall translations and rotations of individual length slices, Z_i. Imposing linear constraints and solving for the corresponding Lagrange multipliers is a convenient way of removing the rigid body motions [20]. The following constraints are imposed on each length slice:

$$\sum_k m_k (R_{ik} - R_{Rk}) = 0$$
$$\sum_k m_k (R_{ik} - R_{Rk}) \times R_{Rk} = 0 \qquad (32)$$

The k index is running over the atoms in a single length slice i. The vector R_{ik} is of rank 3 [$R_k \equiv (x, y, z)$], and it provides the Cartesian coordinates of a single atom; m_k is the mass of the k atom. The vector R_{Rk} is a reference coordinate system (the coordinates of the middle intermediate structure) that is used to determine the absolute orientation.

The errors connected with the length formulation are defined as before. An exact trajectory assesses the accuracy of the finite difference formula. The choice of the finite difference formula to use is biased (as before) by the convenience of a constant Jacobian of transformation from the errors to coordinates [see also Eq. (13)]:

$$[\varepsilon_{i+1}^{(l)}]^T = \left(\frac{1}{\Delta l_{i,i+1}} \partial S_L / \partial Z_i\right)^T = 2(E - U(Z_i)) \cdot \frac{(Z_{i+1} + Z_{i-1} - 2Z_i)^T}{\Delta l_{i,i-1}^2}$$
$$- (dU/dZ_i)^T \cdot (1 - (Z_i - Z_{i-1})(Z_i^T - Z_{i-1}^T)/\Delta l_{i,i-1}^2) \qquad (33)$$

The statistical properties of the "length" errors are similar to the statistical properties of the "time" errors. This is demonstrated in Fig. 3.

The overall similar behavior of the time-dependent and the length-dependent errors suggest that a related modeling of trajectories can be used in the last case. This is what we have done. It is also possible to show that the high-frequency motions (as a function of path length) are filtered out similarly to the removal of rapid motions as a function of time in S_{SDET}. A legitimate question is then, Why do we need yet another stochastic formulation? What did we gain (or lose) by the alternative representation?

Below we list the favorable and less favorable features:

One of the difficulties in the time formulation of the errors, discussed earlier, is the filtering of fast transitional motions (e.g., a transition over an energy barrier). These are rare and rapid motions that may be of considerable interest. In the time formulation they are removed because they are fast. In the length formulation the situation is different. These motions are *not* of small amplitude, and the constraint of uniform distribution of points along the path [Eq. (31)] ensures proper sampling of spatially significant transitions.

A second point in favor is the use of energy conservation explicitly instead of specifying the total time. The energy, as argued before in Section III.D, is easier to estimate because it requires only equilibrium observations (like the temperature of the system). It is also nice to have the total time of the process as an output (Section II.C) instead as an input like in S_{SDET}.

Moreover, the sampling arguments of Section III.D become simpler conceptually, and there is no need to make additional assumptions (beyond the SDE formulation). The energy conservation of the trajectory is already built in. The usual classical mechanics weight applies: Trajectories with the same energy will have the same weight.

There is also some bad news. The integral that determines the time is a weighted sum over the spatial path. Each Δl is the distance between two sequential points and is the shortest path between them. If the step in length, Δl, is large (i.e., only a small number of grid points is used), then the overall length

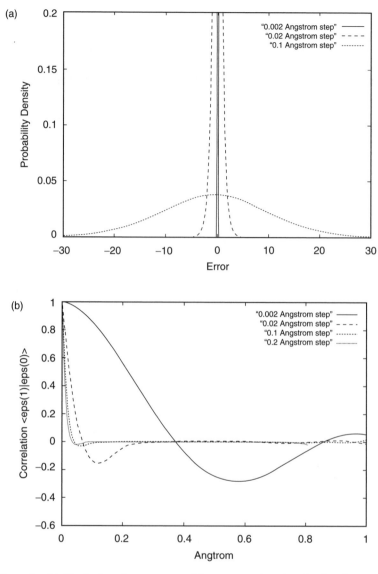

Figure 3. (a) Distribution of errors extracted from exact trajectories of valine dipeptide and Eq. (33) at three different length steps. Note that the narrowest distribution corresponds to essentially exact trajectory. (b) Correlation function of errors extracted from exact trajectories of valine dipeptide and Eq. (33) at four different length steps. Note that at the smallest length step (in which the correlation persists for more than one angstrom) the errors are very small and the trajectory is very close to exact. Hence, as we approach the exact result, no filtering of high-frequency modes occur, the errors become correlated, and the first assumption of the SDE approach is invalid. In practice, it means that there is a minimal size of the step that can be used in the calculation if the statistical assumptions are to be satisfied. Nevertheless, for small length step there is no need for a statistical model, and we obtain the exact trajectory.

of the path is bound to be shorter than the length of the true trajectory. The total time of the trajectory, which is an integral over the path and is an important variable, is likely to be too short as well. A way to get around this problem and obtain a sound estimate of the total time of the trajectory is by statistical refinement, a procedure that is described in the next section.

Another interesting feature of path parameterized by length is the nonuniqueness of the solution. In the length parameterization we fix the end points and the total energy. If we consider periodic motions, the addition of complete periods will return us to the same end points (and of course the same energy). Hence, it is possible to obtain very short (one period) trajectories or infinitely long trajectories with infinitely many repeats of the periodic motions. Depending on our initial guess and the extent of our annealing, we may find trajectories with different total length and time; all of them are legitimate solutions by the conditions we set.

F. "Fractal" Refinement of Trajectories Parameterized by Length

The argument below is similar in spirit to the estimation of the length of the coast of England [21]. The calculated length depends on the yardstick in which it is measured. The smaller the yardstick, the longer the observed length. The coast of England is therefore considered to be a fractal, for which the total distance measured as described above will increase without limit. For the classical trajectory at hand there is a limit and we define it to be the total length of the path that is obtained when a length element, Δl_{limit}, leads to the same solution of the initial value differential equation and the boundary value problem. It is typically $0.01 \, \text{Å}$.

The algorithm for trajectory refinement with the prime purpose of determining the time is as follows:

(i) *Initiate.* Optimize S_{SDEL} to obtain a path parameterized by length of N intermediate points. We denote the path segment by $\Delta l_{i,i+1}^{(1)}$.

(ii) *Sample Intervals.* From the calculated path, sample at random a few pairs of points (intervals), $\{Z_i, Z_{i+1}\}_i$. The number of sampled intervals, ξ, is much smaller than N to ensure computational efficiency.

(iii) *Refine Selected Intervals.* For each of the ξ intervals, compute an interpolating trajectory with N intermediate points [i.e., compute length-dependent trajectories for all the $\{Z_i, Z_{i+1}\}_i$ pairs that are used for boundary conditions and sampled in (ii)]. The length of the path segment in this refinement is denoted by $\Delta l_{i,i+1}^{(n)}$, where (n) is the index of the refinement cycle.

(iv) *Examine the Convergence of the Newly Generated Path Segments.* Convergence is assumed when the path computed with the initial value

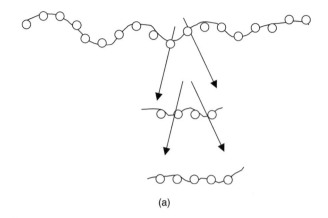

(a)

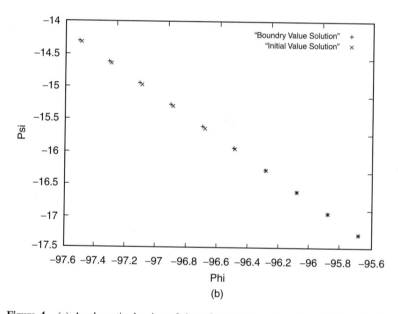

(b)

Figure 4. (a) A schematic drawing of the refinement procedure of an SDEL path. Length segments are sampled and length-dependent paths are computed (with more intermediate points) for the selected segments. Convergence is assumed when the criterion discussed in part b is satisfied. (b) Comparing the boundary value solution and the initial value formulation for a segment of a refined SDEL path for valine dipeptide. At this level of agreement we considered the refinement process complete.

formulation [Eq. (12)] agrees within a threshold to the path created by the optimization of S_{SDEL}. If converged, go to (vi).

(v) *If Convergence Was Not Reached, Return to (ii).* Sampling is now done from the *refined* ξ segments. The sequential points that we sample, $\{Z_i, Z_{i+1}\}$, are from the segments of the segments.

(vi) *Convergence Was Reached.* We have an estimate for the value of the time integral (Section II.C) for a path of a short length $\{\Delta l_{i,i+1}^{(n)} \equiv |Z_i, Z_{i+1}|\}$. Use this estimate (and estimates from other length intervals) to estimate the time length for an earlier refinement cycle with a length step $\Delta l_{i,i+1}^{(n-1)}$. Repeat until the original length step, $\Delta l_{i,i+1}^{(1)}$, is reached.

After a few refinement cycles (a typical number is 5) the protocol above converges in some examples [18]. The path segments calculated with S_{SDEL} or with the initial value formulation (equation (12) with a step size, $\Delta l_{i,i+1}^{(n)}$) are essentially the same (Figs. 4 and 5). Sometimes, however, the process is difficult to converge and the step size in sequential optimizations does not decrease in a substantial and systematic way [18].

An intriguing question is, How much does the spatial distribution of the trajectories change upon refinement? Do we need to refine the trajectories as described above for all types of studies? Or is it possible to extract useful information from paths with significantly lower resolution? We discuss this

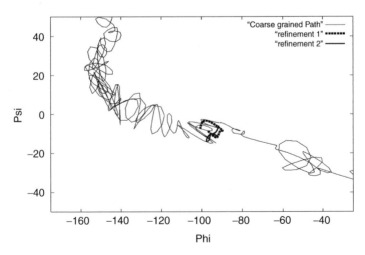

Figure 5. Refinement example of a trajectory along a coarse-grained path for valine dipeptide.

question first formally (by deriving the asymptotic behavior of the approximate trajectory) and then provide a few numerical examples.

IV. SDEL TRAJECTORIES APPROACH THE STEEPEST DESCENT PATH

The question we address in the present section is the asymptotic behavior of a trajectory when the step becomes quite large. We are unable to answer that question in a useful way for the SDET algorithm; however, an intriguing result is obtained for SDEL. In the present section we will show a connection between SDEL trajectories and the usual definition of the reaction coordinate, the steepest descent path (SDP).

In the SDEL formulation the most probable trajectory is the solution of the following finite difference formula [Eq. (33)]

$$[\varepsilon_{i+1}^{(l)}]^T = \left(\frac{1}{\Delta l_{i,i+1}} \partial S_L / \partial Z_i\right)^T = 2(E - U(Z_i)) \cdot \frac{(Z_{i+1} + Z_{i-1} - 2Z_i)^T}{\Delta l_{i,i-1}^2}$$
$$- (dU/dZ_i)^T \cdot (1 - (Z_i - Z_{i-1})(Z_i^T - Z_{i-1}^T)/\Delta l_{i,i-1}^2) = 0$$

The solution, $\{Z_i\}_{i=1}^N$, is subject to the additional constraints of equal distances between sequential configurations and the avoidance of rigid body motions. Consider a specific limit for the length element; Δl is set sufficiently large so the absolute value of the first term on the left-hand side of the equation, $2(E - U(Z_i)) \cdot (Z_{i+1} + Z_{i-1} - 2Z_i)^T/\Delta l_{i,i-1}^2$, is significantly smaller than the second term $(dU/dZ_i)^T \cdot (1 - (Z_i - Z_{i-1})(Z_i^T - Z_{i-1}^T)/\Delta l_{i,i-1}^2)$. The first term is inversely proportional to square of the length element, while the second term is independent of it, so this limit can be achieved [note that $(Z_i^T - Z_{i-1}^T)/\Delta l_{i,i-1} \approx \hat{\eta}$ is a unit vector along the direction of the path and is independent of the size of the length element]. Hence, in this limit the finite difference equation of SDEL becomes

$$[\varepsilon_{i+1}^{(l)}]^T = \left(\frac{1}{\Delta l_{i,i+1}} \partial S_L / \partial Z_i\right)^T = -(dU/dZ_i)^T \cdot$$
$$\times (1 - (Z_i - Z_{i-1})(Z_i^T - Z_{i-1}^T)/\Delta l_{i,i-1}^2) = 0 \qquad (34)$$

We also assume that Δl is not too large so the estimate of $\hat{\eta}$ by a finite difference can be made accurately. This limit is easy to realize in practice because a trajectory with its high frequency motions filtered out is a slowly varying function of the path length and an accurate estimate of derivatives can be made with a reasonably large step. However, this observation may depend on the specific system at hand. Nevertheless, as is demonstrated below, this trajectory is the steepest descent path (SDP). Past experience [20] in calculations of the SDP clearly demonstrate that $\hat{\eta}$ can be approximated quite well by a small number of slices.

Note that the equidistance constraint keeps the different slices close to each other and correlated. This is a useful feature (which we miss at SDET) at the limit of very large steps. A solution of Eq. (34) provides a set of configurations that are equally distributed along the path and satisfy

$$\{(dU/dZ_i)^T \cdot (1 - \hat{\eta}_i\hat{\eta}_i^T) = 0\}_{i=1}^{N} \tag{35}$$

Hence, for each configuration, the force is minimized in all directions excluding the direction of the path. This is one of the definitions of the steepest descent path. The above definition was used numerically in the past [22] to compute the SDP while avoiding the use of second derivatives. The other common approach to compute SDP is based on the Hessian matrix and the propagation along (at most) one negative mode [2]. The conclusion is that the SDEL approach provides the SDP at large steps and therefore interpolates (depending on the quality of the approximation and the length of the step) between the steepest descent path and a classical trajectory.

V. NUMERICAL EXPERIMENTS

The present chapter is mostly methodological, presenting the conceptual framework behind the new technique of the stochastic difference equation. It is therefore appropriate to discuss numerical examples of small systems for which different aspects can be tested in greater details. On the other hand the numerical examples should be sufficiently complex so that nontrivial effects could be observed. So, despite the fact that the techniques were already applied to investigate much larger systems, we focus here on conformational transitions of smaller systems: dipeptides.

Dipeptides are useful models for conformational transitions of proteins. They form "minimal models" on which protein backbone conformational changes can be investigated. A schematic picture of valine dipeptide is shown in Fig. 6. An extended atom model (CH_n groups are treated as a point mass) is used.

Glycine dipeptide has 11 extended atoms; and valine dipeptide, with its larger side chain, has 14 atoms. Nevertheless the backbone of the two peptides is essentially identical. It includes only two soft degrees of freedom, the φ, ψ

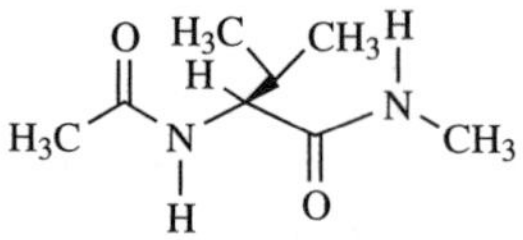

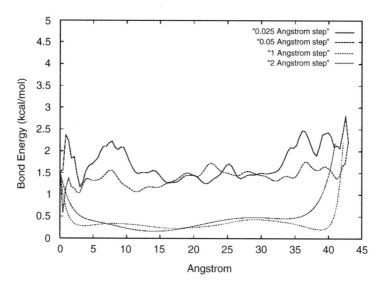

Figure 6. A schematic drawing of valine dipeptide. Note that the only soft degrees of freedom of this system are the rotations (φ, ψ) around the bonds as indicated by the arrows.

dihedral angles. Other backbone modes are too stiff to be significantly excited at room temperature. It is therefore a common practice to describe the dynamics (and thermodynamics) of dipeptides on a two-dimensional φ, ψ energy map.

The AMBER/OPLS force field is implemented in MOIL [16] and is used throughout the calculations. No cutoffs were used for this small system, and the 1–4 scaling factor was 2 and 8 for electrostatic and van der Waals interactions. No constraints on fast vibration were used. However, the stochastic difference equation filters the bond vibrations anyway. In Fig. 7 we compare the energy content of the bond vibrations in S_{SDEL} optimization with different step sizes.

Another question of interest that we can examine in this small system is the dependence of the paths on the number of length slices. In Fig. 8 we present a

Figure 7. Bond energy distribution along the path for four different sizes of length steps. The data are extracted from SDEL calculations of valine dipeptide. Note the significant reduction of bond energies (filtering) as the step size increases.

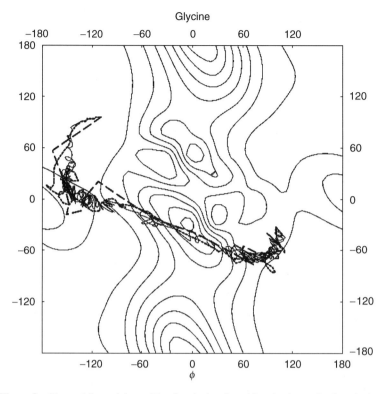

Figure 8. Equatorial-to-axial transition for glycine dipeptide using increasing length size with SDEL. Note that the spatial distribution of configurations along the trajectories remains similar at all resolutions. The main difference for a refined trajectory is the significantly larger density of configurations near minima associated with incubation periods.

comparison for S_{SDEL} trajectories computed with 80, 320, and 640 number of grid points. The trajectories are shown on (φ, ψ) map, the two relevant degrees of freedom, though the complete 33 degrees of freedom were used in the calculations. The trajectories with different resolution in length cover the same domains in conformation space.

Our trajectories are sampled with the help of a simulated annealing protocol. But how can we test that the sampling is appropriate? One measure that can help us to assess the quality of the simulation is the distribution of the orientation of the initial momentum vector. If we sample effectively the space of initial conditions (by sampling *complete* trajectories), then the momentum vectors should cover all of the orientation space. Alternatively, the vectors of the initial direction of the momentum behave as random vectors (with norm of one). In

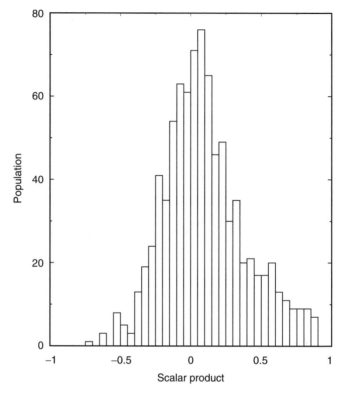

Figure 9. Histogram showing the distribution function of the scalar product of normalized initial momentum vectors for 21 reactive and 21 nonreactive trajectories for valine dipeptide conformational change.

Fig. 9, we show the distribution function of the scalar products of different (normalized) momentum vectors. The distribution for valine dipeptide is a combination of reactive and nonreactive trajectories. While the distribution is not exactly a Gaussian, it is not too far from it.

VI. THE KINETIC CYCLE

We wish to conclude with a future direction. The majority of the applications pursued so far were focused on individual trajectories and their analysis. Here we propose a practical approach for the calculation of relative rates. A theoretical argument is presented in full, together with a simple example.

We define two states: A, the state of the reactants, and B, the state of the product. Both states are assumed to be in local equilibrium; the transition time between the two states is considerably longer than relaxation times to local equilibrium within any of the states. The reaction progress is described by the conditional probability that the system will be at state B after time t, given that it was at the reactant state, A, at time $t = 0$

$$C(t) = P[B(t)|A(0)] \tag{36}$$

From a phenomenological point of view, a rate law can summarize the transitions between the two states:

$$\begin{aligned} dA/dt &= -k_+ A + k_- B \\ dB/dt &= k_- A - k_+ B \end{aligned} \tag{37}$$

with the initial condition $B(0) = 0$. We have

$$C(t) = (B_{eq}/A_{eq})(1 - \exp[-kt]) \tag{38}$$

The constants B_{eq} and A_{eq} are the equilibrium populations of states B and A respectively, and $k = k_+ + k_-$.

It is also possible to write the conditional probability from a molecular perspective. The coordinate vector is X and the following notation is used:

$$A(X) = \begin{cases} 1 & \text{if } X \in A \\ 0 & \text{otherwise} \end{cases}$$
$$B(X) = \begin{cases} 1 & \text{if } X \in B \\ 0 & \text{otherwise} \end{cases} \tag{39}$$

The conditional probability is now written in the canonical ensemble. We also use the notations $X_0 \equiv X(t = 0)$ and $X_t = X(t)$:

$$C(t) = \frac{P(A(0), B(t))}{P(A(0))} = \frac{\int \exp[-\beta E(X_0)]A(X_0) \cdot B(X_t)dX_0}{\int \exp[-\beta E(X_0)]A(X_0)dX_0} \tag{40}$$

The denominator is the thermal probability of being at state A at time zero. The numerator is the joint probability of being at state A at time zero and at state B at time t. The present correlation function can be computed with ordinary trajectories or with the SDE approach. For the SDE formulation we write

$$P(A(0), B(t)) = \int \exp[-\beta E(X_0) - S(X_0, X_t, X(\tau))/2\sigma^2]A(X_0) \cdot$$
$$B(X_t) \cdot dX_0 dX_t DX(\tau) \tag{41}$$

The notation $DX(\tau)$ means a path integral (summation over all trajectories, $X(\tau)$), that starts at X_0 and ends at X_t. The action depends on the boundary coordinates as well as on the trajectory. At the limit of a small time step the distribution of the trajectories, $X(\tau)$, is sharply peaked at the exact trajectory. In fact, in the numerical examples discussed below we only consider exact trajectories. The effects of large steps and the corresponding errors on the conditional probability will be the topics of future work.

The calculation above for the conditional probability is hard to perform because it is equivalent to the computation of a partition function. However, similarly to tricks in equilibrium statistical mechanics (the free-energy perturbation method [23]), we can compute the ratios of the conditional probabilities for slightly different Hamiltonians. For example, we may compare the diffusion of different ions: a sodium ion and a potassium ion permeating through the gramicidin channel [Koneshan Siva and Ron Elber, *Protein, Structure Function, and Genetics*, in press].

For the ratio of two correlation functions, $C_1(t)$ and $C_2(t)$, with corresponding energy functions E_1 and E_2 we have

$$\frac{C_1(t)}{C_2(t)} = \frac{Q_{A_2}}{Q_{A_1}} \cdot \frac{P(A_1(0), B_1(t))}{P(A_2(0), B_2(t))} \tag{42}$$

The Q_{A_x} are the local partition functions of the reactants, for system 1 or system 2, respectively.

We comment that not only the Hamiltonians are changing from state 1 to state 2 but also the spatial boundaries of the reactants and the products may differ, For example, A_1 and A_2 are not necessarily the same, which complicates the averages that we are after. The following identity proves useful

$$(A_1 \cup A_2)A_x = A_x, \qquad x = 1, 2 \tag{43}$$

The union of reactant spaces is equal to one if the initial coordinate is either at A_1 or at A_2. It is zero otherwise. We now have

$$\frac{Q_{A_2}}{Q_{A_1}} = \frac{\int \exp[-\beta H_1](A_1 \cup A_2)\exp[-\beta \Delta H]A_2 \, dX}{\int \exp[-\beta H_1](A_1 \cup A_2)A_1 \, dX} \cdot \frac{\int \exp[-\beta H_1](A_1 \cup A_2) \, dX}{\int \exp[-\beta H_1](A_1 \cup A_2) \, dX}$$
$$= \frac{\langle \exp[-\beta \Delta H] \cdot A_2 \rangle_{e^{-\beta H_1}, A_1 \cup A_2}}{\langle A_1 \rangle_{e^{-\beta H_1}, A_1 \cup A_2}} \tag{44}$$

A similar procedure follows for the calculation of the ratio of the time-dependent joint probability densities.

An effective "Hamiltonian" is defined as

$$H_x \equiv E_x + S_x/2\beta\sigma^2, \qquad \Delta H = H_2 - H_1 \qquad (45)$$

We consider a related union of starting and end coordinates, $A_1 B_1 \cup A_2 B_2$. This union is one if the trajectory starts at the reactant of Hamiltonian 1 and ends at the product of the same Hamiltonian, or if it starts at the reactant of Hamiltonian 2 and ends at the product state of Hamiltonian 2, and is zero otherwise. The ratio of the joint probabilities is now

$$\frac{P(A_{10}, B_{1t})}{P(A_{20}, B_{2t})} = \frac{\int \exp[-\beta H_1](A_1 B_1 \cup A_2 B) A_{10} B_{1t} \, dX_0}{\int \exp[-\beta H_1](A_1 B_1 \cup A_2 B_2) \exp[-\beta \Delta H] A_{10} B_{1t} \, dX_0} \cdot$$

$$\frac{\int \exp[-\beta H_1](A_1 B_1 \cup A_2 B_2) \, dX_0}{\exp[-\beta H_1](A_1 B_1 \cup A_2 B_2) \, dX_0}$$

$$= \frac{\langle A_{10} B_{1t} \rangle_{\exp[-\beta H_1], A_1 B_1 \cup A_2 B_2}}{\langle \exp[-\beta \Delta H] A_{20} B_{2t} \rangle_{\exp[-\beta H_1], A_1 B_1 \cup A_2 B_2}} \qquad (46)$$

The expressions for the ratios of the (local) partition functions and the joint probabilities can therefore be computed as averages over ensembles of reactive trajectories only. This is clearly an advantage in comparison to the usual trajectory calculations that include both reactive and nonreactive trajectories. The average above also converges quickly if the two Hamiltonians are not very different from each other. It is a similar gain to the calculations of relative free energies compared to absolute free energies.

To demonstrate the efficiency of the numerical protocol on a small model system, we consider the two-dimensional energy surface:

$$U(x, y) = \exp[-(x/\gamma_x)^2] \cdot [1 - \exp[-(y/\gamma_y)^2]] + x^6 + y^6 \qquad (47)$$

The surface is a symmetric double well and the parameters, γ_x, and γ_y, determine the width of the barriers (Fig. 10).

For the two Hamiltonians the different parameters were ($\gamma_x = 10^{-3}, \gamma_y = 10^{-3}$) and ($\gamma_x = 10^{-3}, \gamma_y = 1.2 \times 10^{-3}$), respectively. The mass of the particle was set to 1, and kT was set to 0.1. The barrier separating reactants and products is narrow, making the probability of a reactive trajectory smaller and our procedure more favorable. If the probability of a reactive trajectory is higher, then the benefits of the kinetic cycle are smaller.

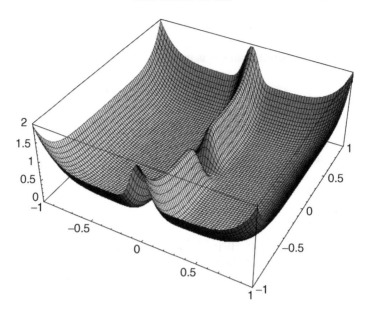

Figure 10. Surface plot for the potential energy function with $\gamma_x = \gamma_y = 0.1$. See text for more details.

The ratio of the correlation functions was computed in two ways:

(i) Sampling initial values for the trajectories at state A and histogramming reactive and nonreactive trajectories. The process was repeated for different times and for the different Hamiltonians. This is the most straightforward calculation, and the results serve as an exact comparison to the second calculation.

(ii) Computing the averages in Eq. (43). Note that only reactive trajectories need to be considered. A reactive trajectory is needed to initiate the sampling. This may be a hard task for initial value formulation (finding the first reactive trajectory) but is considerably easier with the boundary value calculation.

In Fig. 11 we compared the ratio of the correlation functions computed by method (i) and method (ii).

It is obvious that the results of the two numerical methods are indistinguishable. Note that the procedure we outlined requires separate calculation at different times. An efficient method for calculating the probability at any other time t' from the probability at a single time t is the transition path sampling method, which was developed by Dellago et al. [24]. We have employed that procedure to complete graph in Fig. 11.

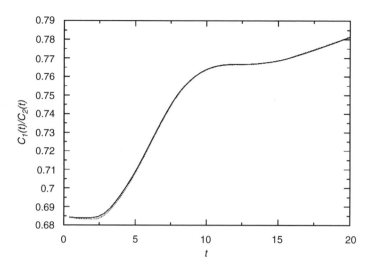

Figure 11. Comparison of the ratio of the reactive correlation functions for two Hamiltonians, $C_1(t)/C_2(t)$, as a function of time t. Hamiltonian H_1 is the model potential with $\gamma_x = 0.1, \gamma_y = 0.2$. Hamiltonian H_2 is the model potential with $\gamma_x = 0.1$, $\gamma_y = 0.3$. The solid line is the computation with the kinetic cycle method; the dotted line is the computation done in the conventional manner (running many trajectories with different initial conditions and counting how many are in region B at time t.

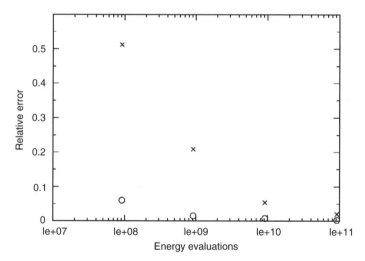

Figure 12. The ratio C_1/C_2 computed for a very rare event in contrast to the system in Fig. 10. For the rare event it is difficult to do the exact calculations. Therefore, accuracy check was done for not-so-rare events. Error bars are calculated by adding the standard errors for each factor in the numerator and denominator, each of which is given from the standard error of the mean from 20 uncorrelated runs. Here the potential is $\gamma_x = 10^{-3}, \gamma_y = 10^{-3}$ (for H_1), and $\gamma_y = 1.2 \times 10^{-3}$ (for H_2). kT is 0.1. Since γ_x is 10^{-3} and kT is 0.1, it takes only $t \sim 0.01$ to get across the channel. But after time $t = 1$, the probability of a reaction is only ~ 0.0001.

127

Comparison of the efficiency of the calculation is provided in Fig. 12, in which we compare the rate of convergence for the two procedures. We estimate it as the number of energy evaluation that is required to reach desired error bars. The kinetic cycle is more efficient for the problem at hand by about two orders of magnitude.

VII. CONCLUDING REMARKS

We have outlined a new numerical approach to compute approximate long-time molecular dynamics trajectories. We have explained the underlying assumptions and the limitations of the present approach as well as its promise. Numerical examples were shown for relatively small system for which detailed and extensive calculations can be performed. A future direction, the calculation of relative rates, was outlined. The research described in this chapter was supported by grants from the NIH GM59796 and the NSF Grant No. 9982524 to Ron Elber.

References

1. Y. Duan and P. A. Kollman, Pathways to a protein folding intermediate observed in a microsecond simulation in aqueous solution. *Science* **282**, 740–744 (1998).

2. R. Elber, Reaction path studies of biological molecules, in *Recent Developments in Theoretical Studies of Proteins*, Ron Elber, ed., World Scientific, Singapore, 1996, pp. 65–136.

3. See, for instance, D. G. Truhlar and B. C. Garrett, Multidimensional transition state theory and the validity of Grote Hynes theory. *J. Phys. Chem. B* **104**, 1069–1072 (2000).

4. R. Olender and R. Elber, Calculation of Classical Trajectories with a Very Large Time Step: Formalism and Numerical Examples. *J. Chem. Phys.* **105**, 9299–9315 (1996).

5. R. Elber, J. Meller, and R. Olender, A stochastic path approach to compute atomically detailed trajectories: Application to the folding of C peptide. *J. Phys. Chem. B*, **103**, 899–911 (1999).

6. V. Zaloj and R. Elber, Parallel computations of molecular dynamics trajectories using stochastic path approach. *Comput. Phys. Commun.* **128**, 118–127 (2000).

7. J. C. M. Uitdehaag, B. A. van der Veen, L. Dijkhuizen, R. Elber, and B. W. Dijkstra, The enzymatic circularization of a malto-octaose linear chain studied by stochastic reaction path calculations on cyclodextrin glycosyltransferase. *Proteins Struct. Funct. Genet.* **43**, 327–335 (2001).

8. L. Verlet, Computer experiments on classical fluids I. Thermodynamics properties of Lennard Jones molecules. *Phys. Rev.* **98**, 159 (1967).

9. L. D. Landau and E. M. Lifshitz, *Mechanics*, Butterworth-Heinenann, Oxford, 1993.

10. P. F. Batcho, D. A. Case, and T. Schlick, Optimized particle-mesh Ewald/multiple-time step integration for molecular dynamics simulations. *J. Chem. Phys.* **115**, 4003–4018 (2001).

11. R. H. Zhou, E. Harder, H. F. Xu, and B. J. Berne, Efficient multiple time step method for use with Ewald and particle mesh Ewald for large biomolecular systems. *J. Chem. Phys.* **115**, 2348–2358 (2001).

12. T. C. Bishop, R. D. Skeel, K. Schulten, Difficulties with multiple time stepping and fast multipole algorithm in molecular dynamics. *J. Comp. Chem.* **18**, 1785–1791 (1997).

13. A. Ulitsky and R. Elber, The equilibrium of the time dependent Hartree and the locally enhanced sampling approximations: Formal properties, corrections, and computational studies of rare gas clusters. *J. Chem. Phys.* **98**, 3380 (1993).

14. C. Lanczos, *The Variational Principles of Mechanics*, University of Toronto Press, 1970.

15. H. Kleinert, *Path Integrals in Quantum Mechanics, Statistics and Polymer Physics*, World Scientific, Singapore, 1995, Chapters 18.5 and 18.6.

16. R. Elber, A. Roitberg, C. Simmerling, R. Goldstein, H. Li, G. Verkhivker, C. Keasar, J. Zhang, and A. Ulitsky, Moil: A program for simulations of macromolecules. *Comput. Phys. Commun.* **91**, 159–189 (1995).

17. J. M. Haile and S. Gupta, Extensions of the molecular dynamics simulation method 2. Isothermal systems. *J. Chem. Phys.* **79**, 3067 (1983).

18. R. Elber, A. Ghosh, and A. Cárdenas, *Account of Chemical Research* **33**, 396–403 (2002).

19. R. Czerminski and R. Elber, Self avoiding walk between two fixed points as a tool to calculate reaction paths in large molecular systems. *Int. J. Quant. Chem.* **24**, 167–186 (1990).

20. R. Elber, Calculation of the potential of mean force using molecular dynamics with linear constraints: An application to a conformational transition in a solvated dipeptide. *J. Chem. Phys.* **93**, 4312 (1990).

21. B. M. Mandelbort, *The Fractal Geometry of Nature*, Freeman, San Francisco, 1977.

22. A. Ulitsky and R. Elber, A new technique to calculate steepest descent paths in flexible polyatomic systems. *J. Chem. Phys.* **96**, 1510 (1990).

23. See, for instance, R. J. Radmer and P. A. Kollman, Free energy calculation methods: A theoretical and empirical comparison of numerical errors and a new method for qualitative estimates of free energy changes. *J. Comp. Chem.*, **18**, 902–919 (1997).

24. C. Dellago, P. G. Bolhuis, and D. Chandler, On the calculation of reaction rate constants in the transition path ensemble. *J. Chem Phys.* **110**, 6617–6625 (1999).

ITINERANT OSCILLATOR MODELS OF FLUIDS

WILLIAM T. COFFEY

Department of Electronic and Electrical Engineering, School of Engineering, Trinity College, Dublin, Ireland

YURI P. KALMYKOV

Centre d'Etudes Fondamentales, Université de Perpignan, Perpignan, France

SERGEY V. TITOV

Institute of Radio Engineering and Electronics of the Russian Academy of Sciences, Fryazino, Moscow Region, Russian Federation

CONTENTS

Advances in Chemical Physics, Volume 126, Edited by I. Prigogine and Stuart A. Rice.
ISBN 0-471-23582-2. © 2003 John Wiley & Sons, Inc.

I. INTRODUCTION

The itinerant oscillator introduced by Hill [1] and Sears [2] is a model for the dynamical (rotational or translational) behavior of a molecule in a fluid embodying the suggestion that a typical molecule of the fluid is capable of vibration in a temporary equilibrium position (cage), which itself undergoes Brownian motion. Sears [2] used a translational version of the model (see Fig. 1) to evaluate the velocity correlation function for liquid argon, while a two-dimensional rotational version was applied in Refs. 1 and 3 to explain the relaxational (Debye) and far-infrared (Poley) absorption spectrum of dipolar fluids: The Brownian motion of the cage gives rise to the Debye absorption; the librational motion of the molecule gives rise to the resonance absorption in the far-infrared (FIR) region. The treatment of Refs. 1 and 3 based on the small oscillation (harmonic potential) approximation was further expanded upon by Coffey et al. [4–7], where a preliminary attempt was made to generalize the model to an anharmonic (cosine) potential. Yet another recent application of the model is relaxation of ferrofluids (colloidal suspensions of single-domain ferromagnetic particles) [8]. A three-dimensional rotational version excluding inertial effects, termed an "egg model," has been used by Shliomis and Stepanov [9] to simultaneously explain the Brownian and Néel relaxation in ferrofluids, which are due to the rotational diffusion of the particles and random reorientations of the magnetization inside the particles, respectively [8]. The analysis in the context of ferrofluids yields similar equations of motion to those given by Damle et al. [10] and van Kampen [11] for the translational itinerant oscillator. This formulation has the advantage that, in the noninertial limit, the equations of motion automatically decouple into those of the molecule and its surroundings (cage). Such a decoupling of the exact equations of motion is also possible in the inertial case if we assume a massive cage.

The first objective of this review is to describe a method of solution of the Langevin equations of motion of the itinerant oscillator model for rotation about a fixed axis in the massive cage limit, discarding the small oscillation approximation; in the context of dielectric relaxation of polar molecules, this solution may be obtained using a matrix continued fraction method. The second

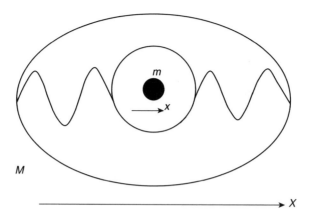

Figure 1. Itinerant oscillator model; that is, a body of mass M moves in a fluid and contains in its interior a damped oscillator of mass m. The displacement of the body is X relative to the fluid and the displacement of the oscillator is x.

objective is to show, using the Langevin equation method, how the model may be easily adapted to explain the resonance and relaxation behavior of a ferrofluid [12]. In the final section of the review, we shall demonstrate how it is possible to extend the model to the fractal time random walk process where the characteristic waiting time between encounters of molecules is divergent [13].

II. APPLICATION OF THE ITINERANT OSCILLATOR (CAGE) MODEL TO THE CALCULATION OF THE DIELECTRIC ABSORPTION SPECTRA OF POLAR MOLECULES

The far-infrared (FIR) absorption spectrum of low-viscosity liquids contains a broad peak of resonant character with a resonant frequency and intensity which decreases with increasing temperature [14,15]. This phenomenon is known as the Poley absorption. It takes its name from the work of Poley [16], who observed that the difference $\varepsilon_\infty - n_{ir}^2$ between the high-frequency dielectric permittivity ε_∞ and the square of the infrared refractive index n_{ir}^2 of several dipolar liquids was proportional to the square of the dipole moment of a molecule, leading him to predict a significant power absorption in these liquids in the -70-cm^{-1} region. (A detailed review of the problem and of various theoretical approaches to its solution is given in Refs. 14 and 15, where an interested reader can find many related references.)

As far as other physical systems are concerned, the FIR absorption has also been observed in supercooled viscous liquids and glasses (for a detailed review see Ref. 17). A similar resonance phenomenon seems to occur [17] in the crystalline structures of ice clathrates where dipolar (and nondipolar) molecules are trapped in the small-size symmetrical and rigid cages formed by tetrahedral hydrogen-bounded water molecules. The encaged molecules undergo both torsional oscillations and orientational diffusion [17]. Thus, like liquids and glasses, the spectra of ice clathrates exhibit a resonant-type peak in the infrared region, a broad relaxation spectra due to the motion of the encaged molecules at intermediate frequencies (in this instance occurring in the MHz region, unlike the microwave region characteristic of simple polar liquids), and a very-low-frequency broad-band relaxation process due to orientational motion of the water molecules confined at the lattice sites. The latter relaxation process has small activation energy [17], and its relaxation rate follows the Arrhenius law. Yet another phenomenon that appears to be related to those described above is the peak in the low-frequency (in the region of acoustic phonons of the corresponding crystals) Raman spectra for a variety of glasses which appears to be an intrinsic feature of the glassy state [17]. Since the scattering intensity associated with this feature could be explained by the Bose–Einstein statistics, the peak in the low-energy Raman spectra has been called the *Boson peak*. Referring to the opening part of our discussion, dielectric loss spectra of several viscous liquids and glasses also exhibit a peak in the THz frequency range (particularly, in the 40- to 80-cm^{-1} range of the infrared spectra) [17]. That peak has again been identified as the Boson peak, because the scattering function for the light is approximately proportional to the magnitude of the dielectric loss, ε''.

The apparent similarities between the low-frequency Raman scattering peak and Poley FIR absorption peak has led Johari [17] on an analysis of the experimental data, to the conclusion that these are simply two different names for the same underlying molecular process which may be regarded as analogous to the torsional oscillations of molecules confined to the cage-like structures of an ice clathrate which is mathematically tractable [14,15]. Moreover, the overriding advantage of such a representation of the relaxation process is that it leads to the modeling of the process as the inertial effects [14,15,18,19] of the rotational Brownian motion of a rigid dipolar molecule undergoing torsional oscillations about a temporary equilibrium position in the potential well created by its cage of neighbors. It follows from the foregoing discussion that the schematic description of the process corresponds to the itinerant oscillator or cage model of polar fluids originally suggested by Hill [1,20]. Hill considered a specific mechanism whereby at any instant an individual polar molecule may be regarded as confined to a temporary equilibrium position in a cage formed by its neighbors where the potential energy surface may in general have several minima. The molecule is considered to librate (i.e., execute oscillations about temporary equilibrium positions) in this cage. Using an approximate analysis of this model based on the Smoluchowski equation [21] for the rotational diffusion of an assembly of noninteracting dipolar molecules proposed by Debye [22],

Hill demonstrated that the frequency of the Poley absorption peak is inversely proportional to the square root of the moment of inertia of a molecule. Hill's model was later rigorously generalized to take full account of inertial effects and evaluated in the small oscillation approximation, using the Langevin equation of the theory of the Brownian motion, in a series of papers [3–7,23] beginning with the work of Wyllie [3,23]. These results are summarized in Refs. [14,15,21]. The disadvantage of all these analyses, however, is that they invariably rely on a small oscillation approximation because no reliable method of treating the finite oscillations of a pendulum when the Brownian torques are included had existed. Thus some of the most important nonlinear aspects of the relaxation processes were omitted, an example being the dependence of the frequency of oscillation of the dipole on the amplitude of the oscillation. A preliminary attempt to include nonlinear or anharmonic effects has been made in Refs. 7 and 24; only very recently [25], however, has it become possible to treat the Brownian motion in a potential other than a parabolic one in a general fashion using matrix continued fractions. The solution has been illustrated [25] by considering the Brownian motion of a rotator in a cosine potential with a *fixed* equilibrium position rather than the *temporary* one of Hill's model.

It is one of the purposes of this section to illustrate how Hill's model may be solved exactly for the first time in the limit of a very large cage and small dipole using the existing results [24–26] for the free Brownian rotator [19] and the matrix continued fraction method for the fixed center of oscillation cosine potential model [25]. Thus we shall summarize how in the limit of large cage moment of inertia the equations of motion of cage and dipole will, in general, decouple. Moreover, we shall give the exact solution for rotation about a fixed axis utilizing the matrix continued fraction method of Ref. 25. In addition, we shall indicate how the problem may also be solved for rotation in space. Furthermore, we shall remark on the similarity of the cage or itinerant oscillator model to the problem of generalizing [27,28] the Onsager model of the static permittivity of polar fluids to calculate the frequency-dependent complex susceptibility. Thus for the purpose of physical justification of the cosine potential we shall pose our discussion of the Hill model in the context of a generalization of the Onsager model, which suggests that the origin of the Poley absorption may lie in the long-range dipole–dipole interaction of the encaged molecule with its neighbors.

We remark that in its original static form the Onsager model constitutes the first attempt to take into account the contribution of the long-range dipole–dipole interactions to the static permittivity of a polar fluid. The key difference between the dynamic Onsager model and its static counterpart is that when a time-varying external field is applied, the reaction field produced by the action of the orienting dipole on its surroundings will lag [27,28] behind the dipole. The net effect of this is to produce a torque on the dipole. If the inertia of the dipole is taken into account, the torque will naturally give rise to a resonance absorption peak with peak frequency in the FIR region, thus explaining the Poley absorption (as well as the microwave Debye absorption). The whole process is analogous to the behavior of a driven damped pendulum in a uniform gravitational field with a time-varying center of oscillation [15,29].

In order to proceed, we first describe the problem of generalizing the Onsager model to include the frequency dependence of the relative permittivity and how this problem may be linked to the cage or itinerant oscillator model.

A. Generalization of the Onsager Model— Relation to the Cage Model

The Onsager model [14,28,30] consists essentially of a very large spherical dielectric sample of static relative permittivity $\varepsilon(\omega)$ placed in a spatially uniform electric field that is applied along the polar axis. The effect of the long-range dipole–dipole coupling is taken account of by imagining that a particular (reference or tagged) dipole of the sample is situated within an empty spherical cavity at the center of the sample (now a shell). The polarizing effect of the dipole then creates a reaction field that exerts a torque on the dipole if the dipole direction is not the same as the reaction field direction as is so in the time-varying case [28,29]. In calculating the polarization of the sample when placed in the uniform external field, it is assumed that the shell may be treated macroscopically (i.e., by continuum electrostatics) while the orientation of the (tagged) dipole in the cavity is treated using classical statistical mechanics. The treatment in the static case then yields the famous Onsager equation [14] (whence the static permittivity may be calculated), which was subsequently generalized to a macroscopic cavity by Kirkwood [30] and Fröhlich [30]. We remark that in the static case the reaction field is a uniform field which is parallel to the dipole direction so that it cannot orient the dipole.

In order to attempt a generalization of the Onsager model to a time-varying applied field for the purpose of the explanation of the FIR absorption peak, we shall suppose, following a suggestion of Fröhlich [31], that the surroundings of a tagged dipole—that is, the shell or cage—may be treated as an inertia corrected Debye dielectric [14,15]. Thus the electrical interactions between cage dipoles are ignored; the only interaction taken account of is that between the cage and the tagged dipole. We shall also suppose that a weak external uniform applied field parallel to the polar axis is switched off at an instant $t = 0$ so that we may utilize the methods of linear response theory. The Langevin equation of motion of the surroundings is then

$$I_s \dot{\omega}_s(t) + \zeta_\mu [\omega_s(t) - \omega_\mu(t)] + \zeta_s \omega_s(t) + \mathbf{R}(t)$$
$$\times \frac{\partial}{\partial \mathbf{R}} V[\boldsymbol{\mu}(t) \cdot \mathbf{R}(t)] = -\lambda_\mu(t) + \lambda_s(t) \qquad (1)$$

while the Langevin equation of motion of the dipole $\boldsymbol{\mu}$ in the cavity is

$$I_\mu \dot{\boldsymbol{\omega}}_\mu(t) + \zeta_\mu [\boldsymbol{\omega}_\mu(t) - \boldsymbol{\omega}_s(t)] + \boldsymbol{\mu}(t) \times \frac{\partial}{\partial \boldsymbol{\mu}} V[\boldsymbol{\mu}(t) \cdot \mathbf{R}(t)] = \boldsymbol{\lambda}_\mu(t) \qquad (2)$$

where we note that by Newton's third law

$$\boldsymbol{\mu} \times \frac{\partial V}{\partial \boldsymbol{\mu}} + \mathbf{R} \times \frac{\partial V}{\partial \mathbf{R}} = 0 \qquad (3)$$

Note that we do not have to introduce the dipole moment of the surroundings explicitly because their influence is represented by $\mathbf{R}(t)$. In the Langevin equations (1) and (2), which reflect the balance of the torques acting on the cage and dipole, respectively, I_s is the moment of inertia of the surroundings (i.e., the cage) of the cavity which are supposed to rotate with angular velocity $\boldsymbol{\omega}_s$, $\zeta_s \boldsymbol{\omega}_s$ and $\boldsymbol{\lambda}_s$ are the stochastic torques on the surroundings which are generated by the heat bath, I_μ is the moment of inertia of the cavity dipole, and $\frac{\partial V}{\partial \boldsymbol{\mu}} \times \boldsymbol{\mu}$ is the torque on $\boldsymbol{\mu}$. The torque arises because in a time-varying applied field, $\boldsymbol{\mu}$ will not be parallel to the reaction field $\mathbf{R}(t)$, unlike in a static field. The terms $\zeta_\mu(\boldsymbol{\omega}_\mu - \boldsymbol{\omega}_s)$ and $\boldsymbol{\lambda}_\mu$ represent the dissipative Brownian torques acting on $\boldsymbol{\mu}$ due to the heat bath, and $\boldsymbol{\omega}_\mu$ is the angular velocity of the dipole. Equations (1) and (2) are recognizably a form of the equations of motion of the itinerant oscillator model [21,32] with a time-dependent potential. In the above equations the white noise torques $\boldsymbol{\lambda}_s$ and $\boldsymbol{\lambda}_\mu$ are centered Gaussian random variables so that they obey Isserlis's theorem [21] and have correlation functions

$$\overline{\lambda_s^{(i)}(t)\lambda_s^{(j)}(t')} = 2kT\zeta_s \delta_{ij}\delta(t - t') \qquad (4)$$

$$\overline{\lambda_\mu^{(i)}(t)\lambda_\mu^{(j)}(t')} = 2kT\zeta_\mu \delta_{ij}\delta(t - t') \qquad (5)$$

where δ_{ij} is Kronecker's delta, $\delta(t - t')$ is the Dirac delta function, and $i, j = 1, 2, 3$ refer to distinct Cartesian axes fixed in the system of mutually coupled rotators represented by Eqs. (1) and (2). It is also assumed that $\boldsymbol{\lambda}_s$ and $\boldsymbol{\lambda}_\mu$ are uncorrelated, with the overbars denoting the statistical average over the realizations in a small time $|t - t'|$ of the Gaussian white noise processes $\boldsymbol{\lambda}_s$ and $\boldsymbol{\lambda}_\mu$. Thus the Langevin equations (1) and (2) are stochastic differential equations. We also note that the system is, in general, governed by the time-dependent Hamiltonian [since $\mathbf{R}(t)$ involves dissipation]

$$H = \frac{1}{2}I_s\omega_s^2 + \frac{1}{2}I_\mu\omega_\mu^2 + V(\boldsymbol{\mu} \cdot \mathbf{R}) \qquad (6)$$

The instantaneous orientation of the dipole $\mathbf{\mu}$ is governed by the kinematic relation [14]

$$\frac{d\mathbf{\mu}}{dt} = \mathbf{\omega}_\mu \times \mathbf{\mu} \tag{7}$$

Our objective is to calculate from the system of Eqs. (1)–(7) the dipole autocorrelation function of an assembly of encaged dipoles:

$$C_\mu(t) = \frac{\langle \mathbf{\mu}(0) \cdot \mathbf{\mu}(t) \rangle_0}{\langle \mathbf{\mu}(0) \cdot \mathbf{\mu}(0) \rangle_0} \tag{8}$$

Hence the complex susceptibility of such an assembly from the linear response theory formula [27,28]

$$\frac{\chi(\omega)}{\chi'(0)} = 1 - i\omega \int_0^\infty C_\mu(t) e^{-i\omega t}\, dt \tag{9}$$

In Eq. (8) the zero inside the angular braces which denote equilibrium averages refers to the instant at which the small external uniform electric field $\mathbf{F}_0$ is switched off, and the subscript zero outside the braces indicates that the average is to be evaluated in the absence of that field.

The stochastic differential Eqs. (1) and (2) cannot be integrated to yield Eq. (8) in explicit form as they stand. The reason is that they are mutually coupled via the term $I_\mu \dot{\mathbf{\omega}}_\mu$ in Eq. (2) and $\mathbf{\omega}_s$ in Eq. (1); moreover, we have no knowledge of the functional form of the time-dependent reaction field $\mathbf{R}(t)$, except that it should still be spatially uniform and that by quasi-electrostatics its Fourier transform $\tilde{\mathbf{R}}(\omega)$ should be of the order of magnitude [14]

$$\frac{\mu(\varepsilon(\omega) - 1)}{2\pi\varepsilon_0 a^3(2\varepsilon(\omega) + 1)} \equiv \tilde{g}(\omega) \tag{10}$$

where a is the radius of the cavity. In the Onsager model the radius of the cavity a is determined from

$$v = \frac{4}{3}\pi a^3 N \tag{11}$$

so that $\frac{4}{3}\pi a^3$ is the mean volume per molecule and N is the number of molecules in the spherical sample. Since we do not know the functional form of $\mathbf{R}(t)$, it is apparent that no further progress can be made unless we make an assumption concerning the time-varying amplitude of $\mathbf{R}(t)$. Thus we shall assume that the amplitude $\tilde{g}(\omega)$ of the reaction field factor, Eq. (10), is only a very slowly varying function of the frequency (corresponding to a *quasi-stationary* function of the time) so that we may replace it by a constant. Thus the time dependence of $\mathbf{R}(t)$ arises solely from the time-varying angle between μ and $\mathbf{R}$. We shall now demonstrate how the variables may be separated in Eqs. (1) and (2) by making the plausible assumption that the moment of inertia of the tagged dipole is much less than that of the surroundings or cage of dipoles. If this is so, the dipole autocorrelation function $C_\mu(t)$ will automatically factor into the product of the autocorrelation function of the surrounding inertia corrected Debye dielectric and the autocorrelation function of the orientation of the dipole relative to its surroundings. We remark that the treatment closely resembles that of the egg model of orientation of a single domain ferromagnetic particle in a ferro-fluid proposed by Shliomis and Stepanov [9], which will be referred to in Section III.

B.   Calculation of the Dipole Correlation Function

By addition and subtraction of Eqs. (1) and (2), we have

$$I_\mu \dot{\omega}_\mu(t) + I_s \dot{\omega}_s(t) + \zeta_s \omega_s(t) = \lambda_s(t) \tag{12}$$

$$\dot{\omega}_\mu(t) - \dot{\omega}_s(t) + (\omega_\mu(t) - \omega_s(t))\left(\frac{\zeta_\mu}{I_\mu} + \frac{\zeta_s}{I_s}\right)$$

$$+ \mu(t) \times \frac{\partial V}{\partial \mu}\left(\frac{1}{I_\mu} + \frac{1}{I_s}\right) - \frac{\zeta_s}{I_s}\omega_s(t) = \lambda_\mu(t)\left(\frac{1}{I_\mu} + \frac{1}{I_s}\right) - \frac{\lambda_s(t)}{I_s} \tag{13}$$

In the limit $I_\mu \ll I_s$ Eqs. (12) and (13) become

$$I_s \dot{\omega}_s(t) + \zeta_s \omega_s(t) = \lambda_s(t) \tag{14}$$

and

$$I_\mu \dot{\Omega}_R(t) + \zeta_\mu \Omega_R(t) + \mu(t) \times \frac{\partial}{\partial \mu} V[\mu(t) \cdot \mathbf{R}(t)] = \lambda_\mu(t) \tag{15}$$

where the relative angular velocity of the dipole and cage is

$$\Omega_R = \omega_\mu - \omega_s \tag{16}$$

so that the equations of motion decouple in Ω_R and ω_s. Thus the cage and dipole orientation processes may be considered as *statistically independent* in the limit of a large cage and small dipole. Hence, the autocorrelation function of the tagged dipole from Eq. (8) factors into the product of the autocorrelation function of the inertia-corrected Debye process corresponding to the behavior of the cage or surroundings of the dipole and the longitudinal $\langle \cos \vartheta(0) \cos \vartheta(t) \rangle_0$ and transverse $\langle \sin \vartheta(0) \cos \varphi(0) \sin \vartheta(t) \cos \varphi(t) \rangle_0$, $\langle \sin \vartheta(0) \sin \varphi(0) \sin \vartheta(t) \sin \varphi(t) \rangle_0$ autocorrelation functions of the motion of the dipole in the cosine potential

$$V = -\mu R \cos \vartheta(t) \tag{17}$$

(ϑ and φ are the polar and azimuthal angles, respectively).

The calculation of orientational autocorrelation functions from the free rotator Eq. (14) which describes the rotational Brownian motion of a sphere is relatively easy because Sack [19] has shown how the one-sided Fourier transform of the orientational autocorrelation functions (here the longitudinal and transverse autocorrelation functions) may be expressed as continued fractions. The corresponding calculation from Eq. (15) for the three-dimensional rotation in a potential is very difficult because of the nonlinear relation between ω_μ and μ [33] arising from the kinematic equation, Eq. (7).

A considerable simplification of the problem can be achieved in two particular cases, however: (a) in the noninertial limit and (b) for rotation about a fixed axis. In case (a), a general method of solution proposed by Kalmykov [21,34] for the calculation of the noninertial response of three-dimensional rotators and described in the preliminary approach to the present problem given in Ref. 26 may be used to write differential recurrence relations for the longitudinal and transverse autocorrelation functions. This method is valid even when the form of $R(t)$ is not explicitly given. Moreover, if the quasi-stationary assumption for the amplitude of $R(t)$ holds, the autocorrelation function from Eq. (8) may be calculated numerically. Preliminary analysis of this case, based on the work of Waldron et al. [21,35] on noninertial dielectric relaxation in a strong uniform field, indicates that for very strong fields such as the reaction field, both longitudinal and transverse relaxation times decrease as the inverse of the field strength. Moreover, the corresponding spectra exhibit marked Debye-like relaxation behavior, effectively with a single relaxation time that may be an explanation for the nearly pure Debye-like behavior of strongly associated liquids at low frequencies.

In case (b), the kinematic relation Eq. (7) reduces to the linear equation

$$\frac{d\mu}{dt} = \omega_\mu \mathbf{k} \times \mu \tag{18}$$

Thus Eqs. (14) and (15) reduce in the limit $I_\mu \ll I_s$ to the itinerant oscillator equations [32]

$$I_s \ddot{\psi}(t) + \zeta_s \dot{\psi}(t) = \lambda_s(t) \tag{19}$$

$$I_\mu \ddot{\theta}(t) + \zeta_\mu \dot{\theta}(t) + \mu R(t) \sin \theta(t) = \lambda_\mu(t) \tag{20}$$

where

$$\Omega_R = \dot{\theta} = \dot{\phi}_\mu - \dot{\phi}_s, \qquad \omega_s = \dot{\psi} = \dot{\phi}_s \tag{21}$$

Here the tagged dipole autocorrelation function is [32]

$$C_\mu(t) = 2\wp_s(t)\wp_\theta(t) = 2\langle\cos\psi(0)\cos\psi(t)\rangle_0\langle\cos\Delta\theta(t)\rangle_0 \tag{22}$$

and the cage autocorrelation function is

$$\wp_s(t) = \langle\cos\psi(0)\cos\psi(t)\rangle_0 = \frac{1}{2}\exp\left\{-\left[\frac{t}{\tau_D} - \gamma + \gamma e^{\frac{-t}{\gamma\tau_D}}\right]\right\} \tag{23}$$

Moreover,

$$\Delta\theta = \theta(t) - \theta(0), \quad \gamma = kTI_s/\zeta_s^2 \quad \text{and} \quad \tau_D = \zeta_s/(kT) \tag{24}$$

The angle $\psi(t)$ is taken as the instantaneous angle of rotation of the cage. The cage autocorrelation function, Eq. (23), is the inertia-corrected Debye result for an assembly of noninteracting fixed axis rotators giving rise to the characteristic low-frequency (microwave) Debye peak and a return to transparency at high frequencies without any resonant behavior (see, for example, Fig. 33 in Ref. 27). Here γ is the inertial parameter introduced by Gross and Sack [18,19] and τ_D is the Debye relaxation time of the surroundings. The autocorrelation function $\wp_\theta(t) = \langle\cos\Delta\theta(t)\rangle_0$ comprising the sum of the sine and cosine (longitudinal and transverse) autocorrelation functions of the angle $\theta(t)$ between the dipole and the reaction field has been examined in detail [6,21,36] in the small oscillation (itinerant oscillator) approximation. Invariably, $\wp_\theta(t)$ gives rise to a pronounced far-infrared absorption peak in the frequency domain. The characteristic frequency ω_0 of this peak on making the quasistationary assumption $R = \text{const}$ for $R(t)$ is given by

$$\omega_0^2 = \frac{\xi}{2\eta^2} \tag{25}$$

and the complete expression for $C_\mu(t')$ is

$$C_\mu(t') = \exp\left\{-\left[\frac{\eta t'}{\tau_D} - \gamma + \gamma e^{\frac{-\eta t'}{\gamma\tau_D}}\right]\right\}\exp\left\{-\frac{1}{\xi}\left[1 - \exp\left(-\frac{\beta't'}{2}\right)\right.\right.$$
$$\left.\left. \times \left(\cos\omega_1't' + \frac{\beta'}{2\omega_1'}\sin\omega_1't'\right)\right]\right\} \tag{26}$$

The damped natural angular frequency ω_1 is given by

$$\omega_1'^2 = \omega_0'^2 - \frac{\beta'^2}{4} = \frac{\xi}{2} - \frac{\beta'^2}{4} \tag{27}$$

and

$$t' = \frac{t}{\eta}, \qquad \eta = \sqrt{\frac{I_\mu}{2kT}}, \qquad \beta' = \frac{\zeta_\mu \eta}{I_\mu}, \qquad \xi = \frac{\mu R}{kT} \tag{28}$$

It is apparent that Eq. (26) represents a discrete set of damped resonances and Debye relaxation mechanisms [5,21]. We remark in passing that an equation, very similar in mathematical form but not identical to the small oscillation solution, Eq. (26), may be obtained [14] by applying Mori theory [37] to the angular velocity correlation function of the tagged dipole. Here the notion of a cage is dispensed with, and instead the angular velocity is supposed to obey a generalized Langevin equation incorporating memory effects which is truncated after the third convergent (three-variable Mori theory [14]). The solution of that equation in the frequency domain is then represented by a continued fraction. A comparison of the two approaches has been made in Ref. 36. Moreover, the Mori three-variable theory has been extensively compared with experiment by Evans et al. [14]. We now return to the behavior of the complex susceptibility as yielded by Eq. (22) when the small oscillation approximation is discarded.

C. Exact Solution for the Complex Susceptibility Using Matrix Continued Fractions

The complex susceptibility $\chi(\omega)$ yielded by Eq. (9), combined with Eq. (22) when the small oscillation approximation is abandoned, may be calculated using the shift theorem for Fourier transforms combined with the matrix continued fraction solution for the fixed center of oscillation cosine potential model treated in detail in Ref. 25. Thus we shall merely outline that solution as far as it is needed here and refer the reader to Ref. 25 for the various matrix manipulations, and so on. On considering the orientational autocorrelation function of the surroundings $\wp_s(t)$ and expanding the double exponential, we have

$$\wp_s(t) = \frac{1}{2} e^\gamma \sum_{n=0}^{\infty} \frac{(-\gamma)^n}{n!} e^{-(1+n/\gamma)t/\tau_D} \tag{29}$$

Now in view of the shift theorem for Fourier transforms as applied to one-sided Fourier transforms [38], namely,

$$\mathscr{I}\{e^{-at}f(t)\} = \int_0^\infty e^{-i\omega t - at} f(t)\, dt = \tilde{f}(i\omega + a) \tag{30}$$

we have the following expression for the Fourier transform of the composite expression given by Eq. (22):

$$\mathscr{I}\{2\wp_s(t)\wp_\theta(t)\} = e^\gamma \sum_{n=0}^{\infty} \frac{(-\gamma)^n}{n!} \tilde{\wp}_\theta(i\omega + \tau_D^{-1} + n/(\gamma\tau_D)) \qquad (31)$$

where the Fourier transform $\tilde{\wp}_\theta(i\omega)$ is to be determined from Eq. (20) by the matrix continued fraction method [25]. Equation (31) is useful for the calculation of the complex susceptibility χ, which is given by

$$\frac{\chi(\omega)}{\chi'(0)} = 1 - i\omega e^\gamma \sum_{n=0}^{\infty} \frac{(-\gamma)^n}{n!} \tilde{\wp}_\theta(i\omega + \tau_D^{-1} + n/(\gamma\tau_D)) \qquad (32)$$

and now only involves the quantity $\tilde{\wp}_\theta(i\omega)$, which is determined from the Langevin equation, Eq. (20). This equation, in turn, is clearly the same as Eq. (4) of Coffey et al. [25], which has already been solved by using the matrix continued fraction technique (the method of solution of the Langevin equation without recourse to the underlying Fokker–Planck equation has been described in detail in Ref. 21). On applying the method of Coffey et al. [25], we may derive the differential-recurrence relation, which describes the dynamics of the model under consideration. The corresponding recurrence relation is [25]

$$\eta \frac{d}{dt}(H_n e^{-iq\theta}) = -\beta' n H_n e^{-iq\theta} - \frac{iq}{2}(H_{n+1}e^{-iq\theta} + 2n H_{n-1}e^{-iq\theta})$$
$$- \frac{in\xi}{2}(H_{n-1}e^{-i(q+1)\theta} - H_{n-1}e^{-i(q-1)\theta}) \qquad (33)$$

where $H_n = H_n(\eta\dot{\theta})$ are the Hermite polynomials, $n = 0, 1, 2, \ldots, q = 0, \pm 1, \pm 2, \ldots$.

Our interest as dictated by Eq. (22) is in the equilibrium average of $\langle \cos \Delta\theta(t) \rangle_0$ so that it is advantageous to obtain a hierarchy that will allow us to calculate $\langle \cos \Delta\theta(t) \rangle_0$ directly. Hence let us introduce the functions

$$c_{n,q}(t) = \langle H_n(\eta\dot{\theta}(t))e^{-i[q\theta(t)-\theta(0)]} \rangle(t) \qquad (34)$$

Now on multiplying Eq. (33) by $e^{-i\theta_0}$ [$\theta_0 = \theta(0)$ denotes a sharp initial value] and averaging over a Maxwell–Boltzmann distribution of θ_0 and $\dot{\theta}_0$ (that procedure is denoted by the angular braces) we have a differential recurrence relation for $c_{n,q}(t)$—that is, the equilibrium averages given by Eq. (34), namely,

$$\eta \dot{c}_{n,q}(t) = -\beta' n c_{n,q}(t) - \frac{iq}{2} [c_{n+1,q}(t) + 2n c_{n-1,q}(t)]$$
$$- \frac{in\xi}{2} [c_{n-1,q+1}(t) - c_{n-1,q-1}(t)] \qquad (35)$$

In the present problem, according to Eq. (32), we are interested in

$$\wp_\theta(t) = \langle \cos \Delta\theta(t) \rangle_0 = \frac{1}{2}(c_{0,1}(t) + c_{0,-1}(t)) \qquad (36)$$

so that

$$\frac{\chi(\omega)}{\chi'(0)} = 1 - \frac{i\omega e^\gamma}{c_{0,1}(0) + c_{0,-1}(0)} \sum_{n=0}^\infty \frac{(-\gamma)^n}{n!} [\tilde{c}_{0,1}(i\omega + \tau_D^{-1} + n(\gamma\tau_D)^{-1})$$
$$+ \tilde{c}_{0,-1}(i\omega + \tau_D^{-1} + n(\gamma\tau_D)^{-1})] \qquad (37)$$

Now, the scalar recurrence Eq. (35) can be transformed into the matrix *three-term* recurrence equation

$$\eta \frac{d}{dt} \mathbf{C}_n(t) = \mathbf{Q}_n^- \mathbf{C}_{n-1}(t) + \mathbf{Q}_n \mathbf{C}_n(t) + \mathbf{Q}_n^+ \mathbf{C}_{n+1}(t) + \mathbf{R}\delta_{n,2} \qquad (n \geq 1) \quad (38)$$

where the column vectors $\mathbf{R}$ and $\mathbf{C}_n(t)$ are given by

$$\mathbf{R} = \frac{i\xi I_1(\xi)}{2 I_0(\xi)} \begin{pmatrix} \vdots \\ 0 \\ -1 \\ 0 \\ 1 \\ 0 \\ \vdots \end{pmatrix} \qquad (39)$$

$$\mathbf{C}_0(t) = \mathbf{0}, \quad \mathbf{C}_1(t) = \begin{pmatrix} \vdots \\ c_{0,-2} \\ c_{0,-1} \\ c_{0,1} \\ c_{0,2} \\ \vdots \end{pmatrix}, \quad \mathbf{C}_n(t) = \begin{pmatrix} \vdots \\ c_{n-1,-2} \\ c_{n-1,-1} \\ c_{n-1,0} \\ c_{n-1,1} \\ c_{n-1,2} \\ \vdots \end{pmatrix} \quad (n \geq 2) \quad (40)$$

and the matrices $\mathbf{Q}_n^-$, $\mathbf{Q}_n$, and $\mathbf{Q}_n^+$ are defined by

$$
\mathbf{Q}_n^+ = -\frac{i}{2}
\begin{pmatrix}
\ddots & \vdots & \vdots & \vdots & \vdots & \vdots & \iddots \\
\cdots & -2 & 0 & 0 & 0 & 0 & \cdots \\
\cdots & 0 & -1 & 0 & 0 & 0 & \cdots \\
\cdots & 0 & 0 & 0 & 0 & 0 & \cdots \\
\cdots & 0 & 0 & 0 & 1 & 0 & \cdots \\
\cdots & 0 & 0 & 0 & 0 & 2 & \cdots \\
\iddots & \vdots & \vdots & \vdots & \vdots & \vdots & \ddots
\end{pmatrix}
\tag{41}
$$

$$
\mathbf{Q}_n^- = -\frac{i(n-1)}{2}
\begin{pmatrix}
\ddots & \vdots & \vdots & \vdots & \vdots & \vdots & \iddots \\
\cdots & -4 & \xi & 0 & 0 & 0 & \cdots \\
\cdots & -\xi & -2 & \xi & 0 & 0 & \cdots \\
\cdots & 0 & -\xi & 0 & \xi & 0 & \cdots \\
\cdots & 0 & 0 & -\xi & 2 & \xi & \cdots \\
\cdots & 0 & 0 & 0 & -\xi & 4 & \cdots \\
\iddots & \vdots & \vdots & \vdots & \vdots & \vdots & \ddots
\end{pmatrix}
\tag{42}
$$

$$
\mathbf{Q}_n = -(n-1)\beta'\mathbf{I}
\tag{43}
$$

and $\mathbf{I}$ is the unit matrix of infinite dimension. The exceptions are the matrices $\mathbf{Q}_1^+$ and $\mathbf{Q}_2^-$, which are given by

$$
\mathbf{Q}_1^+ = -\frac{i}{2}
\begin{pmatrix}
\ddots & \vdots & \vdots & \vdots & \vdots & \vdots & \iddots \\
\cdots & -2 & 0 & 0 & 0 & 0 & \cdots \\
\cdots & 0 & -1 & 0 & 0 & 0 & \cdots \\
\cdots & 0 & 0 & 0 & 0 & 0 & \cdots \\
\cdots & 0 & 0 & 0 & 1 & 0 & \cdots \\
\cdots & 0 & 0 & 0 & 0 & 2 & \cdots \\
\iddots & \vdots & \vdots & \vdots & \vdots & \vdots & \ddots
\end{pmatrix}
\tag{44}
$$

$$
\mathbf{Q}_2^- = -\frac{i}{2}
\begin{pmatrix}
\ddots & \vdots & \vdots & \vdots & \vdots & \iddots \\
\cdots & -4 & \xi & 0 & 0 & \cdots \\
\cdots & -\xi & -2 & 0 & 0 & \cdots \\
\cdots & 0 & -\xi & \xi & 0 & \cdots \\
\cdots & 0 & 0 & 2 & \xi & \cdots \\
\cdots & 0 & 0 & -\xi & 4 & \cdots \\
\iddots & \vdots & \vdots & \vdots & \vdots & \ddots
\end{pmatrix}
\tag{45}
$$

By applying the one-sided Fourier transform to Eq. (38), we obtain the matrix recurrence relations

$$i\eta\omega\,\tilde{\mathbf{C}}_1(i\omega) - \eta\mathbf{C}_1(0) = \mathbf{Q}_1^+\tilde{\mathbf{C}}_2(i\omega) \tag{47}$$

$$i\eta\omega\tilde{\mathbf{C}}_n(i\omega) = \mathbf{Q}_n\tilde{\mathbf{C}}_n(i\omega) + \mathbf{Q}_n^+\tilde{\mathbf{C}}_{n+1}(i\omega)$$
$$+ \mathbf{Q}_n^-\tilde{\mathbf{C}}_{n-1}(i\omega) + \frac{\mathbf{R}}{i\omega}\delta_{n,2} \qquad (n \geq 2) \tag{48}$$

where the initial value vector is

$$\mathbf{C}_1(0) = \frac{1}{I_0(\xi)}\begin{pmatrix} \vdots \\ I_3(\xi) \\ I_2(\xi) \\ I_0(\xi) \\ I_1(\xi) \\ \vdots \end{pmatrix} \tag{49}$$

and we have used the fact that $\mathbf{C}_n(0) = 0$ for $n \geq 2$ because

$$\langle H_n \rangle = 0 \quad (n \geq 1) \tag{50}$$

for the (equilibrium) Maxwell–Boltzmann distribution function. By invoking the general method for solving the matrix recursion Eq. (38) [21], we have the exact solution for the spectrum $\tilde{\mathbf{C}}_1(i\omega)$ in terms of matrix continued fractions, namely,

$$\tilde{\mathbf{C}}_1(i\omega) = \Delta_1(\eta\mathbf{C}_1(0) + \mathbf{Q}_1^+\Delta_2\mathbf{R}/i\omega) \tag{51}$$

where the matrix continued fractions Δ_n are given by

$$\Delta_n = \cfrac{\mathbf{I}}{i\eta\omega\mathbf{I} - \mathbf{Q}_n - \mathbf{Q}_n^+ \cfrac{\mathbf{I}}{i\eta\omega\mathbf{I} - \mathbf{Q}_{n+1} - \mathbf{Q}_{n+1}^+ \cfrac{\mathbf{I}}{i\eta\omega\mathbf{I} - \mathbf{Q}_{n+2} \ddots}\mathbf{Q}_{n+2}^-}\mathbf{Q}_{n+1}^-} \tag{52}$$

and the fraction lines designate the matrix inversions. The foregoing equations allow us to evaluate the complex susceptibility. Before analyzing the numerical results, we shall first describe various approximate treatments of the complex susceptibility.

D. Approximate Expressions for the Complex Susceptibility

We first remark that the small oscillation (itinerant oscillator) approximation has the advantage of being very easy to use for the purpose of comparison with experimental observations because the solution in the time domain is available in closed form [Eq. (26)]. Moreover, if the inertial parameter γ is sufficiently small (≤ 0.1) and $I_\mu \omega_0^2 \ll kT$, the complex susceptibility for small oscillations [i.e., Eq. (26)] may be closely approximated in the frequency domain [21,36] by the simple expression [32]

$$\frac{\chi(\omega)}{\chi'(0)} = \frac{1}{(i\omega\tau_D + 1)(i\omega\gamma\tau_D + 1)} + \frac{kT}{I_\mu \omega_0^2} \frac{i\omega\tau_D}{i\omega\tau_D + 1} \left[\left(1 - \frac{\omega^2}{\omega_0^2}\right) + \frac{\zeta_\mu}{I_\mu} \frac{i\omega}{\omega_0^2} \right]^{-1} \quad (53)$$

a form of which was originally given in Ref. 4 for the single friction itinerant oscillator model (see also Ref. 14). The first term in Eq. (53) is essentially the Rocard equation [9] of the inertia-corrected Debye theory of dielectric relaxation and thus is due to the cage motion, and the second damped harmonic oscillator term represents in our picture the high-frequency effects due to the cage–tagged-dipole interaction.

Another approximate formula may also be derived for the cosine potential as follows. On expansion of $\wp_s$ and $\wp_\theta$ in Taylor's series, we have from Eqs. (9) and (22)

$$\frac{\chi(\omega)}{\chi'(0)} \approx 1 - i\omega \int_0^\infty \left(1 - \frac{1}{2}\langle(\Delta\psi)^2\rangle_0\right)\left(1 - \frac{1}{2}\langle(\Delta\theta)^2\rangle_0\right)e^{-i\omega t}\, dt$$

$$\approx \frac{1}{2} i\omega \mathscr{I}\{\langle(\Delta\psi)^2\rangle_0 + \langle(\Delta\theta)^2\rangle_0\} \quad (54)$$

if we assume that both series may be truncated after the second term and the product $\langle(\Delta\psi)^2\rangle_0\langle(\Delta\theta)^2\rangle_0$ may be ignored. Now, for a random variable $y(t)$ obeying the Langevin equation

$$\mathscr{I}\{\langle(\Delta y)^2\rangle_0\} = -\frac{2}{\omega^2}\mathscr{I}\{\langle\dot{y}(0)\dot{y}(t)\rangle_0\} \quad (55)$$

thus, we have

$$\frac{\chi(\omega)}{\chi'(0)} \approx \frac{1}{i\omega}\mathscr{I}\{\langle\dot{\psi}(0)\dot{\psi}(t)\rangle_0 + \langle\dot{\theta}(0)\dot{\theta}(t)\rangle_0\} = \frac{1}{i\omega}[\tilde{C}_{\dot\psi}(i\omega) + \tilde{C}_{\dot\theta}(i\omega)] \quad (56)$$

where we note that the Fourier transform of the angular velocity autocorrelation function of the cage is from the Ornstein–Uhlenbeck theory of the Brownian motion [21]:

$$\tilde{C}_{\dot{\psi}}(i\omega) = \frac{kT}{I_s} \frac{1}{i\omega + \zeta_s/I_s} \tag{57}$$

We now suppose just as in the derivation of Eq. (53) that we may shift the term $(i\omega)^{-1}$ multiplying the whole expression in Eq. (56) by $1/\tau_D$, which is the assumption used to derive the harmonic oscillator approximate formula Eq. (53); thus we have

$$\frac{\chi(\omega)}{\chi'(0)} \approx \frac{1}{(i\omega + 1/\tau_D)} \left[\frac{kT}{I_s} \frac{1}{i\omega + \zeta_s/I_s} + \tilde{C}_{\dot{\theta}}(i\omega) \right] \tag{58}$$

We remark that the quantity

$$\tilde{C}_{\dot{\theta}}(i\omega) = \mathscr{M}'(i\omega) - i\mathscr{M}''(i\omega) \tag{59}$$

is the dynamic mobility in the cosine potential model.

In order to calculate $\tilde{C}_{\dot{\theta}}(i\omega)$, we again introduce functions

$$f_{n,q}(t) = \langle H_1(\eta\dot{\theta}(0))H_n(\eta\dot{\theta}(t))e^{-iq\theta(t)}\rangle(t) \tag{60}$$

Multiplying Eq. (33) by $H_1(\eta\dot{\theta}_0)$ (θ_0 denotes a sharp initial value) and averaging over a Maxwell–Boltzmann distribution of θ_0 and $\dot{\theta}_0$, we automatically have a differential recurrence relation for $f_{n,q}(t)$, namely,

$$\eta\dot{f}_{n,q}(t) = -\beta'nf_{n,q}(t) - \frac{iq}{2}[f_{n+1,q}(t) + 2nf_{n-1,q}(t)]$$
$$- \frac{in\xi}{2}[f_{n-1,q+1}(t) - f_{n-1,q-1}(t)] \tag{61}$$

with

$$f_{1,q}(0) = \langle [H_1(\eta\dot{\theta})]^2 e^{-iq\theta} \rangle = \frac{\int_{-\infty}^{\infty} [H_1(\eta\dot{\theta})]^2 e^{-(\eta\dot{\theta})^2} d\dot{\theta} \int_0^{2\pi} e^{-iq\theta} e^{\xi\cos\theta} d\theta}{\int_{-\infty}^{\infty} e^{-(\eta\dot{\theta})^2} d\dot{\theta} \int_0^{2\pi} e^{\xi\cos\theta} d\theta}$$
$$= \frac{2I_q(\xi)}{I_0(\xi)} \tag{62}$$

Thus we have for $\tilde{C}_{\dot{\theta}}(\omega)$

$$\tilde{C}_{\dot{\theta}}(i\omega) = \frac{1}{4\eta^2}\tilde{f}_{1,0}(i\omega) \tag{63}$$

The function $\tilde{f}_{1,0}(\omega)$ also can be obtained by using the matrix continued fraction method. The scalar recurrence equation, Eq. (61), can once again be transformed into the matrix *three-term* recurrence equation

$$\eta\frac{d}{dt}\mathbf{C}_n(t) = \mathbf{Q}_n^-\mathbf{C}_{n-1}(t) + \mathbf{Q}_n\mathbf{C}_n(t) + \mathbf{Q}_n^+\mathbf{C}_{n+1}(t) \quad (n \geq 1) \tag{64}$$

where the column vectors are

$$\mathbf{C}_0(t) = \mathbf{0}, \quad \mathbf{C}_1(t) = \begin{pmatrix} \vdots \\ f_{0,-2} \\ f_{0,-1} \\ f_{0,1} \\ f_{0,2} \\ \vdots \end{pmatrix}, \quad \mathbf{C}_n(t) = \begin{pmatrix} \vdots \\ f_{n-1,-2} \\ f_{n-1,-1} \\ f_{n-1,0} \\ f_{n-1,1} \\ f_{n-1,2} \\ \vdots \end{pmatrix} \quad (n \geq 2) \tag{65}$$

and the matrices $\mathbf{Q}_n^-$, $\mathbf{Q}_n$, and $\mathbf{Q}_n^+$ are defined by Eqs. (41)–(45). By invoking the general method for solving the matrix recursion equation, Eq. (64) [21], we have the exact solution for the spectrum $\tilde{C}_1(i\omega)$ in terms of matrix continued fractions, namely,

$$\tilde{C}_2(i\omega) = \eta(\Delta_2\mathbf{Q}_2^-\Delta_1\mathbf{Q}_1^+ + \mathbf{I})\Delta_2\mathbf{C}_2(0) \tag{66}$$

where the matrices Δ_1 and Δ_2 are defined from Eq. (52) and

$$\mathbf{C}_2(0) = \frac{2}{I_0(\xi)}\begin{pmatrix} \vdots \\ I_1(\xi) \\ I_0(\xi) \\ I_1(\xi) \\ \vdots \end{pmatrix} \tag{67}$$

E. Numerical Results and Comparison with Experimental Data

In Fig. 2, we show plots of the imaginary part $\chi''(\omega)$ of the complex susceptibility from the matrix continued fraction solution, Eqs. (37) and (51), for $\beta' = 2$ and $\gamma = 0.01$ and illustrate the effect of varying the reaction field parameter ξ. The real part of the susceptibility $\chi'(\omega)$ is shown in Fig. 3. Superimposed on these are $\chi'(\omega)$ and $\chi''(\omega)$ as yielded by the harmonic oscillator

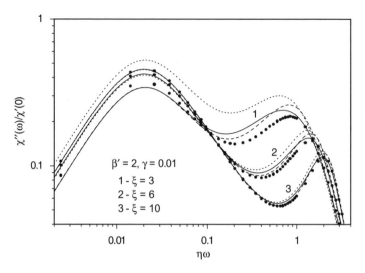

Figure 2. Imaginary part of the complex susceptibility $\chi''(\omega)$ versus normalized frequency $\eta\omega$ for various values of the reaction field parameter. Solid lines correspond to the matrix continued fraction solution, Eqs. (37) and (51); circles correspond to the small oscillation solution, Eq. (26); dashed lines correspond to the approximate small oscillation solution Eq. (53); and dotted lines correspond to the solution based on the dynamic mobility, Eq. (58).

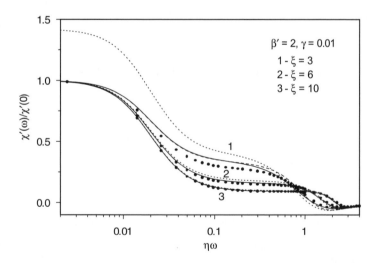

Figure 3. The same as in Fig. 2 for the real part of the complex susceptibility $\chi'(\omega)$. Note the failure of Eq. (58) for small ξ because it then becomes impossible to normalize that equation. Equation (58) is accurate for large ξ because the behavior of $\tilde{C}_\theta$ is like that of the harmonic oscillator which vanishes as $\omega = 0$ cf. Eqs. (3.3.5, 3.3.15) of Ref. [21] or Eq. (50c) of Ref. [86].

approximation [Eq. (26)], which is the conventional itinerant oscillator model. The approximate formula (58) involving the dynamic mobility is compared both with its harmonic oscillator counterpart Eq. (53) and the matrix continued fraction solution, Eqs. (37) and (51). These results are also shown in Figs. 2 and 3. The advantage of doing this is that it demonstrates the extent to which one is justified in using the small oscillation approximation, Eq. (26) or (53), in order to describe the relaxation process. As far as Figs. 2 and 3 are concerned, it appears that the simple analytic equation [Eq. (53)], which is the harmonic oscillator dynamic mobility approximation, can adequately describe the response for curves 2 and 3 where the reaction field parameter is of the order of 6 or more. The agreement for curve 1—that is, $\xi = 3$—is not so good. In particular, the cosine potential dynamic mobility approximation, Eq. (58), fails in this region. The above behavior is entirely in accord with intuitive considerations because one would expect the harmonic or simple pendulum approximation to become less and less accurate as the reaction field parameter is decreased. Nevertheless, it appears that Eq. (53), which is simple to implement numerically, provides a reasonable approximation for a large reaction field parameter.

In Fig. 4, we demonstrate the effect of varying the cage inertial parameter γ for fixed dipole friction parameter β' and large reaction field parameter ξ. It

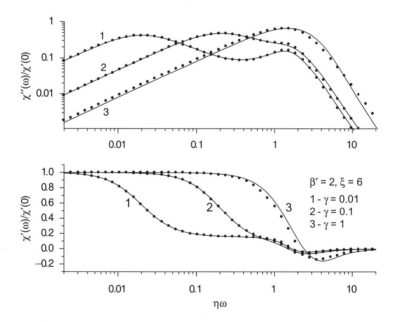

Figure 4. Effect of varying the cage inertial parameter γ: real $\chi'(\omega)$ and imaginary $\chi''(\omega)$ parts of the complex susceptibility versus $\eta\omega$. Solid lines correspond to the matrix continued fraction solution, Eqs. (37) and (51); and circles correspond to the small oscillation solution, Eq. (53).

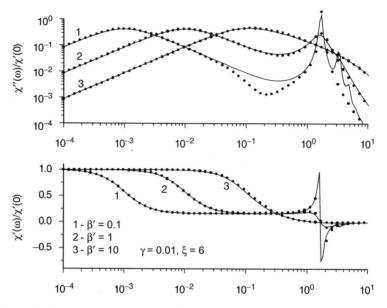

Figure 5. Effect of varying the dipole friction parameter β': real $\chi'(\omega)$ and imaginary $\chi''(\omega)$ parts of the complex susceptibility vs $\eta\omega$. Solid lines correspond to the matrix continued fraction solution, Eqs. (37) and (51); and circles correspond to the small oscillation solution, Eq. (53).

appears that if $\gamma < 1$, then the small oscillation approximation given by of Eq. (53) again provides a very close approximation to the exact solution based on numerical evaluations of Eqs. (37) and (51). The approximate Eq. (53) begins to fail at $\gamma = 1$ and higher. This is to be expected because one is now dealing with the region in which the Rocard approximation for the complex susceptibility is no longer valid.

In Fig. 5, we fix the cage inertial parameter and the reaction field parameter and we demonstrate the effect of varying the dipole friction parameter β' for a reasonably high value of the reaction field parameter so that one can expect the harmonic approximation of Eq. (53) to be valid. Again the agreement with the exact cosine potential solution, Eqs. (37) and (51), is good for relatively high values of β' (see curves 2 and 3). For relatively small values of β', however, there is some discrepancy in the intermediate frequency region. At high frequencies the harmonic approximation, Eq. (53), cannot predict the comb-like peak structure of the exact solution. (The peaks occur at the harmonics of the fundamental frequency.) This to be expected because Eq. (53) results from an expansion of the correlation function, Eq. (26) (a double transcendental function with sines and cosines in its argument), which is truncated at the fundamental frequency term so that the comb-like structure may not be reproduced. Nevertheless, Eq. (53) provides a reasonable description of the fundamental frequency peak in the FIR region as is obvious by inspection of Fig. 5. We remark that a similar comb-like structure occurs in the theory of ferromagnetic resonance [39]

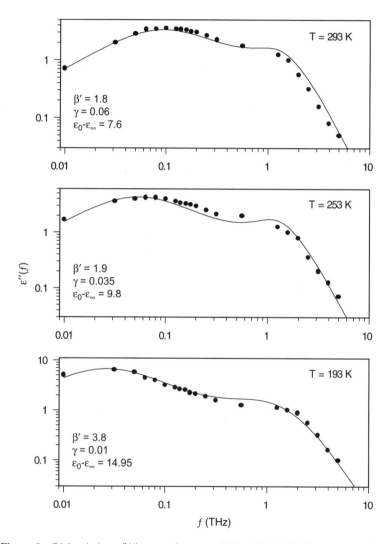

Figure 6. Dielectric loss $\varepsilon''(f)$ versus frequency f ($f = \omega/2\pi$). Solid lines correspond to the matrix continued fraction solution, Eqs. (37) and (51); and circles are experimental data for the symmetrical top molecule CH_3Cl [14,21] ($I_r = 28.5 \times 10^{-40}$ g·cm^2 and $\xi = 193\xi_0/T$ with $\xi_0 = 8$).

of single-domain ferromagnetic particles, which involves the solution of differential recurrence relations resembling Eq. (35), and in the theory of quantum noise in ring laser gyroscopes [40].

Figure 6 shows a comparison of typical experimental data for the symmetric top molecule CH_3Cl with the theoretical spectrum predicted by Eqs. (37) and (51). It appears that the model provides a reasonable description of both the microwave and far-infrared spectra for this species with a fair description of the temperature dependence.

F. Conclusions

We have shown in this section how the equations of motion of the cage model of polar fluids originally proposed by Hill may be solved exactly using matrix continued fractions. Furthermore, these equations automatically lead to (a) a microwave absorption peak having its origin in the reorientational motion of the cage and (b) a pronounced far-infrared absorption peak having its origin in the oscillations of the dipole in the cage–dipole interaction potential. The equations of motion of dipole and cage may be justified in terms of a generalization of the Onsager model of the static susceptibility of polar fluids to include the dynamical behavior. A conclusion that may be drawn from this is that the origin of the Poley absorption peak is the interaction between the dipole and its surrounding cage of neighbors which causes librations of the dipole. The model appears to reproduce satisfactorily the main features of the microwave and far-infrared absorption of CH_3Cl. In addition, it is shown that the simple closed-form expression, Eq. (53), for the complex susceptibility which was originally derived in the context of the small oscillation version of the cage model (that is the itinerant oscillator model) can also provide a reasonable approximation to the susceptibility for large values of the reaction field parameter.

The result given here may also be extended to rotation in space. As far as the cage motion is concerned, the complex susceptibility will still be governed by the Rocard equation because the equations of motion factorize. However, the solution for the dipole correlation function is much more complicated because of the difficulty of handling differential recurrence relations pertaining to rotation in space in the presence of a potential.

We remark that activation processes that involve crossing of the encaged dipole over an internal potential barrier may also be incorporated into the present model by adding a $\cos 2\theta$ term to the potential in Eq. (15). This may give rise to a Debye-like relaxation process at very low frequencies with relaxation time governed by the Arrhenius law, the prefactor of which may be calculated precisely using the Kramers theory of escape of particles over potential barriers (see Section III). We remark that a $\cos 2\theta$ term in the potential has also been considered by Polimeno and Freed [41] in their discussion of a many-body stochastic approach to rotational motions in liquids. Their discussion is given in the context of a simplified description of a liquid in which only three bodies are retained, namely, a solute molecule (body 1), a slowly relaxing local structure or solvent cage (body 2), and a fast stochastic field as a source of fluctuating torque. In general, one may remark that this approach which is based on projection operators and numerical solution of many body Fokker–Planck–Kramers equations is likely to be of much interest in the context of the extension of the model described in this review to rotation in space.

Yet another direction in which the present model may be extended is fractional dynamics where one supposes that the surroundings of the dipole or cage obeys not the simple Langevin equation, Eq. (14), but instead a generalized Langevin equation [13,14] with a memory function given by the Riemann–Liouville fractional derivative [13] while Eq. (15) remains a conventional (or if one prefers also a fractional) Langevin equation. In this way it will be possible to incorporate slow relaxation effects into the cage model as we demonstrate in Section IV.

III. APPLICATION TO FERROFLUIDS

We have mentioned in the general introduction that a long-standing problem in the theory of magnetic relaxation of ferrofluids is how the solid-state or Néel mechanism of (longitudinal) relaxation of internal rotation of the magnetic dipole moment with respect to the crystalline axes inside the particle, the associated transverse modes [which may give rise to ferromagnetic resonance (FMR)], and the mechanical Brownian relaxation due to physical rotation of the ferrofluid particle in the carrier fluid may be treated in the context of a single model comprising both relaxation processes. (We recall that the slowest mode of the longitudinal relaxation process describes the reversal of the magnetization over the potential barrier created by the internal anisotropy of the particle, which of course will be modified by an external applied field. The time to cross the barrier is known as the Néel relaxation time and follows the Arrhenius law.)

We have mentioned that the question posed above was answered in part by Shliomis and Stepanov [9]. They showed that for uniaxial particles, for weak applied magnetic fields, and in the noninertial limit, the equations of motion of the ferrofluid particle incorporating both the internal and the Brownian relaxation processes decouple from each other. Thus the reciprocal of the greatest relaxation time is the sum of the reciprocals of the Néel and Brownian relaxation times of both processes considered *independently*—that is, those of a *frozen Néel* and a *frozen Brownian* mechanism! In this instance the joint probability of the orientations of the magnetic moment and the particle in the fluid (i.e., the crystallographic axes) is the product of the individual probability distributions of the orientations of the axes and the particle so that the underlying Fokker–Planck equation for the joint probability distribution also factorizes as do the statistical moments. Thus the internal and Debye processes are statistically independent. If the applied field is sufficiently strong, however, no such decoupling can take place.

The Shliomis–Stepanov approach [9] to the ferrofluid relaxation problem, which is based on the Fokker–Planck equation, has come to be known in the literature on magnetism as the egg model. Yet another treatment has recently been given by Scherer and Matuttis [42] using a generalized Lagrangian formalism; however, in the discussion of the applications of their method, they limited themselves to a frozen Néel and a frozen Brownian mechanism, respectively.

Here, we reexamine the egg model (which is of course a form [24] of the itinerant oscillator model) noting the ratio of the free Brownian diffusion (Debye) time to the free Néel diffusion time and discarding the assumption of a weak applied field [12]. The results will then be used to demonstrate how the ferrofluid magnetic relaxation problem in the noninertial or high mechanical friction limit is essentially similar to the Néel relaxation in a uniform magnetic field applied at an oblique angle to the easy axis of magnetization [44–48]. Unlike in the solid-state mechanism, however, the orientation of the field with respect to the easy axis is now a function of the time due to the physical rotation of the crystallographic axes [9] arising from the ferrofluid. The fact that the behavior is essentially similar in all other respects to the solid-state oblique field problem suggests that a strong *intrinsic* dependence of the greatest relaxation time on the damping (independent of that due to the free diffusion time) arising from the coupling between longitudinal and transverse modes occurs. Alternatively, the set of eigenvalues that characterize the longitudinal relaxation now depends strongly on the damping, unlike in axial symmetry. This precession-aided (so called because the influence of the precessional term is proportional to the inverse of the damping coefficient) longitudinal relaxation is absent in the weak field case [9]. Here the equations of motion decouple into those describing a frozen Néel (pure Debye or Brownian) and a frozen Brownian (pure Néel) mechanism of relaxation, respectively. Thus the Néel or longitudinal relaxation is governed by an axially symmetric potential. Hence no *intrinsic* dependence of the greatest relaxation time on the damping exists. Moreover, the longitudinal set of eigenvalues is independent of the damping, with the damping entering [48,49] via the free (Néel) diffusion time only. It follows that in the linear response to a weak applied field, the only effect of the fluid carrier is to further dampen, according to a Debye or Rocard (inertia-corrected Debye) [21] mechanism, both the longitudinal and transverse responses of the solid-state mechanism. Furthermore, it will be demonstrated that in this case *the transverse relaxation process in the ferrofluid, which may give rise to ferromagnetic resonance, may be accurately described by the solid-state transverse result for axial symmetry* because the Brownian relaxation time greatly exceeds the characteristic relaxation times of all the transverse modes. The latter conclusion is reinforced by the favorable agreement of the linear response result with experimental observations [12] of the complex susceptibility of four ferrofluid samples, which is presented in Section III.E.

In order to illustrate how precession-aided relaxation effects may manifest themselves in a ferrofluid, it will be useful to briefly summarize the differences in the relaxation behavior for axially symmetric and nonaxially symmetric potentials of the magnetocrystalline anisotropy and applied field, when the Brownian relaxation mode is frozen. Thus only the solid-state (Néel) mechanism is operative; that is, the magnetic moment of the single-domain particle may reorientate only with respect to the crystalline axes.

A. Langevin Equation Formalism

Our starting point is the Landau–Lifshitz or Gilbert (LLG) equation for the dynamics of the magnetization $\mathbf{M}$ of a single-domain ferromagnetic particle, namely [48–51],

$$2\,\tau_N \frac{d}{dt}\mathbf{M} = \beta(\alpha^{-1}M_s[\mathbf{M} \times \mathbf{H}] + [[\mathbf{M} \times \mathbf{H}] \times \mathbf{M}]) \tag{68}$$

where

$$\tau_N = \frac{\beta(1 + \alpha^2)M_s}{2\gamma\alpha} \tag{69}$$

is the free diffusion time (Néel diffusion time) of the magnetic moment (solid-state mechanism), α is the dimensionless damping (dissipation) constant, M_s is the saturation magnetization, γ is the gyromagnetic ratio, $\beta = v_m/(kT)$, v_m is the volume (domain volume) of the particle, and α^{-1} determines the magnitude of the precession term. The magnetic field $\mathbf{H}$ consists of applied fields (Zeeman term), the anisotropy field $\mathbf{H}_a$, and a random white noise field accounting for the thermal fluctuations of the magnetization of an individual particle.

Equation (68) is the Langevin equation of the solid-state orientation process [48–51]. The field $\mathbf{H}$ may be written as

$$\mathbf{H} = \mathbf{H}_{ef} + \mathbf{H}_n(t) \tag{70}$$

Here

$$\mathbf{H}_{ef} = -\frac{\partial V}{\partial \mathbf{M}} = -\frac{\partial U}{\partial \mathbf{m}} \tag{71}$$

is the conservative part of $\mathbf{H}$, which is determined from the free energy density V (U is the free energy, $m = M_s v_m$ is the magnitude of the magnetic moment $\mathbf{m}$ of the single domain particle.) The random field $\mathbf{H}_n(t)$ [50] has the properties [the angular braces denote the statistical average over the realizations of $\mathbf{H}_n(t)$]:

$$\langle H_n(t) \rangle = 0 \tag{72}$$

$$\langle H_n^{(i)}(t) H_n^{(j)}(t') \rangle = \frac{2kT\alpha}{\gamma M_s v_m} \delta_{ij}\delta(t - t') \tag{73}$$

Here δ_{ij} is Kronecker's delta; $i, j = 1,2,3$ correspond to Cartesian (the crystalline) axes, $\delta(t)$ is the Dirac delta-function. The random variable $\mathbf{H}_n(t)$ must also obey Isserlis's theorem [21]. By introducing [9] the dipole vector

$$\mathbf{e} = \frac{\mathbf{M}}{M_s} = \frac{\mathbf{m}}{m} \tag{74}$$

we find that Eq. (68) becomes

$$\frac{d\mathbf{e}}{dt} = \frac{\gamma}{1 + \alpha^2}(\mathbf{e} \times \mathbf{H}) + \frac{\alpha\gamma}{1 + \alpha^2}(\mathbf{e} \times \mathbf{H}) \times \mathbf{e} \tag{75}$$

Equation (75) has the form [21] of a kinematic relation involving the angular velocity $\boldsymbol{\omega}_e$ of the dipole vector $\mathbf{e}$:

$$\frac{d\mathbf{e}}{dt} = \boldsymbol{\omega}_e \times \mathbf{e} = (\boldsymbol{\omega}_L + \boldsymbol{\omega}_R) \times \mathbf{e} \tag{76}$$

Here

$$\boldsymbol{\omega}_L = -\frac{\gamma\mathbf{H}}{1 + \alpha^2} = -\frac{\gamma(\mathbf{H}_{ef} + \mathbf{H}_n(t))}{1 + \alpha^2} \tag{77}$$

is the angular velocity of free (Larmor) precession of $\mathbf{m}$ in the field $\mathbf{H}_{ef}$, superimposed on which is the rapidly fluctuating $\mathbf{H}_n(t)$ and

$$\boldsymbol{\omega}_R = \frac{\gamma\alpha}{1 + \alpha^2}(\mathbf{e} \times \mathbf{H}) \tag{78}$$

is the *relaxational* component of $\boldsymbol{\omega}_e$. Equations (77) and (78) differ from Eqs. (8) and (9) of Shliomis and Stepanov [9] because they contain the noise field [51] and the factor $(1 + \alpha^2)^{-1}$, since the Gilbert equation is used rather than the Landau–Lifshitz equation. The kinematic relation, Eq. (76), and the coupled Langevin equations, Eqs. (77) and (78), are stochastic differential equations describing the motion of the dipole vector **e** relative to the crystallographic axes that is the internal or solid state relaxation. Differential recurrence relations (equivalent to the Fokker–Planck equation) for the statistical moments governing the dynamical behavior of **e** may be deduced from Eqs. (76)–(78) as described in Refs. 21 and 34, 45–47. If we now, following Ref. 9 and allowing for the factor $(1 + \alpha^2)$, introduce the magnetic viscosity

$$\mu = \frac{M_s}{6\alpha\gamma}(1 + \alpha^2) \tag{79}$$

Eq. (78) becomes

$$6\mu v_m \omega_R = -\mathbf{m} \times \frac{\partial U}{\partial \mathbf{m}} + \mathbf{m} \times \mathbf{H}_n(t) \tag{80}$$

Equation (80) will be the key equation in our discussion of precession aided Néel relaxation.

B.   Precession-Aided Effects in Néel Relaxation

The discussion given above holds for arbitrary free energy U. We now specialize our discussion to a particle with uniaxial anisotropy, so that

$$v_m V(\mathbf{M}) = U = -mH_0(\mathbf{e} \cdot \mathbf{h}) - K v_m(\mathbf{e} \cdot \mathbf{n})^2 \tag{81}$$

$$\mathbf{h} = \frac{\mathbf{H}_0}{H_0} \tag{82}$$

H_0 is the amplitude of the external uniform magnetic field (the polarizing field), $K > 0$ is the constant of the effective magnetic anisotropy, and **n** is a unit vector along the easy magnetization axis. The fact that **h** is not necessarily parallel to **n** means that the axial symmetry characteristic of uniaxial anisotropy will be broken. Thus Eq. (80) becomes

$$6\mu v_m \omega_R - mH_0[\mathbf{e} \times \mathbf{h}] - 2K v_m(\mathbf{e} \cdot \mathbf{n})[\mathbf{e} \times \mathbf{n}] = \mathbf{m} \times \mathbf{H}_n(t) = \lambda_R(t) \tag{83}$$

say. Next we write U explicitly as

$$U = -mH_0 \cos \Theta + K v_m \sin^2 \vartheta \tag{84}$$

where

$$\cos \Theta = \cos \vartheta' \cos \vartheta + \sin \vartheta \sin \vartheta' \cos (\varphi - \varphi') \tag{85}$$

Here ϑ and φ are the polar angles of $\mathbf{e}$ with respect to the easy axis, which is the polar axis: ϑ' and φ' are the polar angles of the external field direction $\mathbf{h}$ (again with respect to the easy axis) which in the solid-state problem are constants independent of the time. Analytic expressions for the greatest relaxation time τ in the bistable potential given by Eq. (84) in the intermediate to high damping case (IHD) (where α is such that the energy loss per cycle of the motion of the magnetization at the saddle point energy trajectory, $\Delta E \gg kT$) may be obtained using Langer's theory [51,52] of the decay of metastable states applied to the two-degree-of-freedom system specified by ϑ and φ, since in the solid-state case ϑ' and φ' are fixed. Likewise, the Kramers energy controlled diffusion method [51,53] may be used to obtain τ in the very low damping (VLD) case where $\Delta E \ll kT$.

We summarize as follows [54]. The free energy $v_m V(\mathbf{M})$, Eq. (81), has a bistable structure with minima at $\mathbf{n}_1$ and $\mathbf{n}_2$ separated by a potential barrier containing a saddle point [54] at $\mathbf{n}_0$. If $(\alpha_1^{(i)}, \alpha_2^{(i)}, \alpha_3^{(i)})$ denote the direction cosines of $\mathbf{M}$ and $\mathbf{M}$ is close to a stationary point $\mathbf{n}_i$ of the free energy, then $V(\mathbf{M})$ can be approximated to second order in $\alpha^{(i)}$ as [51,54]

$$V = V_i + \frac{1}{2}[c_1^{(i)}(\alpha_1^{(i)})^2 + c_2^{(i)}(\alpha_2^{(i)})^2] \tag{86}$$

The relevant Fokker–Planck equation may then be solved near the saddle point, yielding [51,54]

$$\tau = \tau_{IHD} \sim \left\{ \frac{\Omega_0}{2\pi\omega_0}[\omega_1 e^{\beta(V_1 - V_0)} + \omega_2 e^{\beta(V_2 - V_0)}] \right\}^{-1} \tag{87}$$

(equations for the expansion coefficients $c_i^{(j)}$ and V_i for the potential given by Eq. (84) are given elsewhere [51,54]);

$$\omega_1^2 = \gamma^2 M_s^{-2} c_1^{(1)} c_2^{(1)}, \quad \omega_2^2 = \gamma^2 M_s^{-2} c_1^{(2)} c_2^{(2)}, \quad \omega_0^2 = -\gamma^2 M_s^{-2} c_1^{(0)} c_2^{(0)} \tag{88}$$

are the squares of the well and saddle angular frequencies, respectively, and

$$\Omega_0 = \frac{\beta}{4\tau_N}\left[-c_1^{(0)} - c_2^{(0)} + \sqrt{(c_2^{(0)} - c_1^{(0)})^2 - 4\alpha^{-2} c_1^{(0)} c_2^{(0)}} \right] \tag{89}$$

Equation (89) is effectively the smallest positive (unstable barrier crossing mode) eigenvalue of the noiseless Langevin equation [Eq. (75) omitting the $\mathbf{H}_n(t)$ term] linearized in terms of the direction cosines about the saddle point. We remark that the influence of the precessional term on the longitudinal relaxation is represented by the α^{-2} term in Eq. (89). Furthermore, the relative magnitudes of the precessional and aligning terms in the Langevin equations are determined by α^{-1}, so that Eqs. (87)–(89) and (90) below describe precession-aided longitudinal relaxation.

Equation (87) applies when $\Delta E \gg kT$ (IHD). If $\Delta E \ll kT$ (VLD), we have for the escape from a single well [48,51]

$$\tau = \tau_{LD} \sim \frac{\pi kT}{\omega_1 \Delta E} e^{\beta(V_0 - V_1)} \tag{90}$$

where $\Delta E \approx \alpha V_m |V_0|$. The IHD and VLD limits correspond to $\alpha \geq 1$ and $\alpha \leq 0.01$, respectively. However, for crossover values of α (about $\alpha \approx 0.1$), neither Eq. (87) nor Eq. (90) yields reliable quantitative estimates. Thus a more detailed analysis is necessary [51]. The most striking aspect of the precession-aided relaxation is the behavior of the complex susceptibility for a small alternating-current field superimposed on the strong field $\mathbf{H}_0$. This is particularly sensitive to the longitudinal and transverse mode coupling, exhibiting a strong dependence of the high-frequency ferromagnetic resonant modes (characterized by ω_L) on the aligning (Néel) mode characterized by ω_R and vice versa. Furthermore, suppression of the barrier crossing mode in favor of the fast relaxation modes in the wells of the bistable potential given by Eq. (84) occurs if the applied field is sufficiently strong. In terms of the differential recurrence relations generated by the Langevin or Fokker–Planck equations by means of a Fourier expansion in the spherical harmonics $Y_l^m(\vartheta, \varphi)$, the coupling effect manifests itself in recurrence relations *inextricably* mixed in the characteristic numbers l and m, unlike in axial symmetry.

Finally, we remark on the asymptote for axial symmetry which arises if $\mathbf{H}_0$ is reversed, or is applied parallel to the easy axis $\mathbf{n}$. In the axially symmetric case ($\mathbf{H}_0 = 0$), τ is given by Brown's asymptotic expression [50] for simple uniaxial anisotropy:

$$\tau \sim \frac{\tau_N \sqrt{\pi}}{2\sigma^{3/2}} \exp(\sigma), \qquad \sigma > 2 \tag{91}$$

where $\sigma = \beta K$. Thus, τ normalized by τ_N unlike Eqs. (87)–(90) is independent of α, so that the mode coupling effect completely disappears. Bridging formulas, which illustrate how the asymptotic Eqs. (87) and (90) join smoothly onto the asymptotic Eq. (91) in the limit of small $\mathbf{H}_0$ have been extensively discussed in Ref. 51.

C. Effect of the Fluid Carrier: (i) Response to a Weak Applied Field

Let the crystallographic axes now rotate with angular velocity ω_n corresponding to physical rotation of the ferrofluid particle due to the stochastic torques imposed by the liquid and the aligning action of **H** (we now have 5 degrees of freedom, viz., the dipole angles ϑ, φ as before and the Euler angles ϑ', φ', ψ', which instead of being constant are now functions of the time due to the physical rotation of the easy axis). The relative angular velocity of the dipole and easy axis is then $\boldsymbol{\omega}_R - \boldsymbol{\omega}_n$, so that Eq. (83), describing the motion of **m**, must be modified to

$$6\mu v_m(\boldsymbol{\omega}_R - \boldsymbol{\omega}_n) - mH_0[\mathbf{e} \times \mathbf{h}] - 2Kv_m(\mathbf{e} \cdot \mathbf{n})[\mathbf{e} \times \mathbf{n}] = \lambda_R(t) \qquad (92)$$

The corresponding mechanical equation of motion of the particle (the particle is treated as a rigid sphere, I is the moment of inertia of the sphere about a diameter) is by Newton's third law

$$I\dot{\boldsymbol{\omega}}_n + 6\mu v_m(\boldsymbol{\omega}_n - \boldsymbol{\omega}_R) + 6\eta v\boldsymbol{\omega}_n - 2Kv_m(\mathbf{e} \cdot \mathbf{n})[\mathbf{n} \times \mathbf{e}] = \lambda_n(t) - \lambda_R(t) \quad (93)$$

where

$$6\eta v = \zeta \qquad (94)$$

is the mechanical drag coefficient of the particle in the fluid, v is its hydrodynamic volume, and η is the viscosity of the fluid. Thus, by addition of Eqs. (92) and (93) we have the mechanical equation

$$I\dot{\boldsymbol{\omega}}_n + 6\eta v\boldsymbol{\omega}_n - mH_0[\mathbf{e} \times \mathbf{h}] = \lambda_n(t) \qquad (95)$$

where the white noise $\lambda_n(t)$ torque arising from the fluid carrier obeys

$$\langle \lambda_n(t) \rangle = 0 \qquad (96)$$

$$\langle \lambda_n^{(\alpha)}(t)\lambda_n^{(\beta)}(t') \rangle = 2kT\zeta\delta_{\alpha\beta}\delta(t - t') \qquad (97)$$

α, $\beta = 1$, 2, 3 refer to the orientation of the crystallographic axes relative to Cartesian axes fixed in the liquid. The $\lambda_n(t)$ again obey Isserlis's theorem [21], and we shall suppose that $\lambda_n(t)$ and $\lambda_R(t)$ are uncorrelated.

Equations (92) and (95) in general are coupled to each other inextricably by the external field term $\mathbf{e} \times \mathbf{h}$. If that vanishes, however, so that U depends only on $(\mathbf{e} \cdot \mathbf{n})^2$, they become

$$6\mu v_m(\boldsymbol{\omega}_R - \boldsymbol{\omega}_n) - 2Kv_m(\mathbf{e} \cdot \mathbf{n})[\mathbf{e} \times \mathbf{n}] = \lambda_R(t) \qquad (98)$$

$$I\dot{\boldsymbol{\omega}}_n + 6\eta v\boldsymbol{\omega}_n = \lambda_n(t) \qquad (99)$$

The equations thus separate into the equation of motion of **m** relative to the easy axis, Eq. (98), and the equation of motion of the easy axis itself, Eq. (99). The mechanical equation, Eq. (99), is governed by two characteristic times [51]: the Brownian diffusion or Debye relaxation time

$$\tau_B = \frac{3\eta v}{kT} \tag{100}$$

and the frictional time

$$\tau_\eta = \frac{I}{6\eta v} \tag{101}$$

Thus the dynamical behavior of Eq. (99) is governed by the inertial parameter (corresponding to the γ of the polar fluid case of the model in Section II)

$$a = \frac{\tau_\eta}{\tau_B} = \frac{kT}{3\eta v} \cdot \frac{I}{6\eta v} \tag{102}$$

If $a \to 0$, we have the noninertial response. This is treated by Shliomis and Stepanov [9], who were able to factorize the joint distribution of the dipole and easy axis orientations in the Fokker–Planck equation into the product of the two separate distributions. Thus as far as the internal relaxation process is concerned, the axially symmetric treatment of Brown [50] applies. Hence no intrinsic coupling between the transverse and longitudinal modes exists; that is, the eigenvalues of the longitudinal relaxation process are independent of α. The distribution function of the easy axis orientations $\mathbf{n}$ is simply that of a free Brownian rotator excluding inertial effects.

The picture in terms of the decoupled Langevin equations (98) and (99) (omitting the inertial term $I\dot{\omega}_n$ in Eq. (98)) is that the orientational correlation functions of the longitudinal and transverse components of the magnetization in the axially symmetric potential, $Kv_m \sin^2 \vartheta$, are simply multiplied by the liquid state factor, $\exp(-t/\tau_B)$, of the Brownian (Debye) relaxation of the ferrofluid stemming from Eq. (99). As far as the ferromagnetic resonance is concerned, we shall presently demonstrate that this factor is irrelevant.

We summarize as follows: The longitudinal and transverse magnetic susceptibilities characterizing the solid-state process are approximately described by [55]

$$\chi_{\|}(\omega) = \frac{\chi'_{\|}(0)}{1 + i\omega\tau} \tag{103}$$

$$\chi_{\perp}(\omega) = \frac{\chi'_{\perp}(0)[(1 + i\omega\tau_2) + \Delta]}{(1 + i\omega\tau_2)(1 + i\omega\tau_{\perp}) + \Delta} \tag{104}$$

$$\Delta = \frac{\sigma\tau_2}{\alpha^2 \tau_N^2}(\tau_N - \tau_{\perp}) \tag{105}$$

In Eqs. (102) and (103), $\chi'_{||}(0)$ and $\chi'_{\perp}(0)$ are the static susceptibilities $||$ and $\perp$ to the easy axis and τ is rendered reasonably accurately [56] by Brown's uniaxial anisotropy asymptote, Eq. (91), if $\sigma \geq 2$. Furthermore, the transverse suscept-ibility, Eq. (104), which is derived by truncating the infinite hierarchy of differential recurrence relations for the correlation functions (with $U = Kv_m \sin^2 \vartheta$) at the quadrupole term using the effective eigenvalue method, yields an accurate result for the transverse response, provided that $\sigma \geq 5$ [39]. The effective eigenvalue solution, Eq. (104), fails [39] for $\sigma < 5$ [where Eq. (91) and Eq. (103) also cease to be entirely reliable] since at small to moderate barrier heights a spread of the precession frequencies of the magnetization in the anisotropy field exists. Thus the hierarchy must be solved exactly using matrix continued fractions. In Eq. (104), $\tau_{\perp}$ is the effective relaxation time of the autocorrelation functions of the components $\sin\vartheta\cos\varphi$, $\sin\vartheta\sin\varphi$ [which are linear combinations of the autocorrelation functions of $Y_1^1(\vartheta,\varphi)$ and $Y_1^{-1}(\vartheta,\varphi)$] of the dipole moment relaxation mode, while τ_2 is the effective relaxation time of the quadrupole moment relaxation mode, which is a linear combination of the autocorrelation functions of $Y_2^1(\vartheta,\varphi)$ and $Y_2^{-1}(\vartheta,\varphi)$. The effective relaxation times τ_2 and $\tau_{\perp}$ may both be expressed [55] in terms of Dawson's integral and decrease monotonically with σ, each having asymptotes

$$\frac{\tau_N}{\sigma}, \qquad \sigma \gg 1 \tag{106}$$

Now according to linear response theory

$$\chi_{||,\perp} = f_{||,\perp}(0) - i\omega \int_0^\infty f_{||,\perp} e^{i\omega t} dt = \tilde{f}_{||,\perp}(0) - i\omega \tilde{f}_{||,\perp}(\omega) \tag{107}$$

where $f_{||,\perp}(t)$ are the longitudinal and transverse aftereffect functions, for example,

$$f_{||}(t) \cong \chi'_{||}(0)\exp(-t/\tau) \tag{108}$$

Moreover, in the frequency domain (the tilde denoting the one-sided Fourier transform)

$$\tilde{f}_{||}(\omega) = \frac{\chi'_{||}(0)}{i\omega + 1/\tau} \tag{109}$$

$$\tilde{f}_{\perp}(\omega) = \frac{\chi'_{\perp}(0)\left(i\omega + \frac{1}{\tau_2}\right)}{\left(i\omega + \frac{1}{\tau_2}\right)\left(i\omega + \frac{1}{\tau_{\perp}}\right) + \frac{\sigma}{\tau_N\alpha^2}\left(\frac{1}{\tau_{\perp}} - \frac{1}{\tau_N}\right)} \tag{110}$$

We note that both the effective relaxation times and the zero-frequency susceptibilities may be written in terms of Dawson's integral. The detailed expressions are given in Ref. 55.

The complex susceptibility of a ferrofluid in a weak applied field may be written directly from Eqs. (109) and (110) and the Langevin equations (98) and (99) [taking note of Eq. (102)] using the shift theorem for one-sided Fourier transforms, Eq. (30). Thus

$$\chi_{||}(\omega) = \frac{\chi'_{\perp}(0)}{1 + i\omega T_{||}} \tag{111}$$

with the Néel relaxation time modified to

$$T_{||} = \frac{\tau\tau_B}{\tau + \tau_B} \tag{112}$$

Moreover [55], provided that σ is not small so that the effective eigenvalue truncation of the hierarchy of recurrence relations for the statistical moments is valid, we obtain

$$\chi_{\perp}(\omega) = \frac{\chi'_{\perp}(0)\left[i\omega T_2 + 1 + \frac{\Delta T_{\perp}T_2}{\tau_{\perp}\tau_2}\right]}{(i\omega T_2 + 1)(i\omega T_1 + 1) + \frac{\Delta T_{\perp}T_2}{\tau_{\perp}\tau_2}} \tag{113}$$

Thus the effective relaxation time of the dipole mode is modified to

$$T_{\perp} = \frac{\tau_{\perp}\tau_B}{\tau_{\perp} + \tau_B} \tag{114}$$

and that of the quadrupole mode becomes

$$T_2 = \frac{\tau_2\tau_B}{\tau_2 + \tau_B} \tag{115}$$

In a weak measuring field the particle anisotropy axes are oriented in a random fashion. Hence [9,57] the susceptibility (averaged over particle orientations) is given by

$$\chi = \frac{1}{3}(\chi_{||} + 2\chi_{\perp}) \tag{116}$$

Now [58] in ferrofluids where the Néel mechanism is blocked we have $\sigma \geq 8$, and so we will have in Eq. (112) $\tau \gg \tau_B$; thus

$$T_{||} \cong \tau_B \tag{117}$$

Furthermore, in Eqs. (114) and (115) we may use the fact [55] that $\tau_\perp$ and τ_2 are monotonic decreasing functions of σ and also that usually [57], for the ratio of the free diffusion times,

$$\frac{\tau_N}{\tau_B} \sim 10^{-2} \tag{118}$$

in order to ascertain which times may be neglected in Eq. (113). Thus we deduce that in Eq. (113) for all σ

$$T_\perp \cong \tau_\perp \tag{119}$$

$$T_2 \cong \tau_2 \tag{120}$$

Hence we may conclude, recalling the exact [39] transverse relaxation solution for $\chi_\perp$ from the Fokker–Planck equation, that the *solid-state effective eigenvalue solution embodied in Eq. (104) can also accurately describe the ferromagnetic resonance in ferrofluids* except in the range of small σ (<5). Then the exact solid-state solution based on matrix continued fractions as in Section II must be used. The conclusion appears to be in agreement with that of Fannin [59] and Fannin et al. [60], who have extensively analyzed experimental data on ferrofluids using Eq. (104). By way of illustration, a detailed comparison with experimental data for four colloidal suspensions is given in Section III.E. We also remark that the very large σ (high barrier) limit of Eq. (113) (Landau–Lifshitz limit) agrees with the result of Scaife [58], who analyzed the problem using an entirely different method. In the limit of high damping where $\alpha \gg 1$, Eq. (113) reduces to a pure relaxation equation in complete agreement with Shliomis and Stepanov [9], namely their Eqs. (31) and (32).

Having justified how the FMR effect in ferrofluids consisting of uniaxial particles, subjected to a weak external field, may be accurately described by the solid-state response [that is, the factor $\exp(-t/\tau_B)$ arising from the mechanical motion may be discounted in the transverse response], we shall now very briefly describe inertial effects arising from the term $I\dot{\omega}_n$ in Eq. (99). Furthermore, we shall justify the neglect of this term as far as ferrofluid relaxation is concerned.

If inertial effects are included, the correlation functions pertaining to longitudinal and transverse motions will still be the product of the correlation functions of the free Brownian motion of a sphere [9] and the solid-state correlation functions $\langle \cos \vartheta(0) \cos \vartheta(t) \rangle_0$, and so on; however, the composite expressions will be much more complicated for an arbitrary inertial parameter a, Eq. (102). The reason is that the orientational correlation functions for the Brownian motion of a sphere may only be expressed exactly [18,19] as the inverse Laplace transform of an infinite continued fraction in the frequency domain. The parameter a has, however, been evaluated by Raikher and Shliomis [57] for typical values of ferrofluid parameters and is of the order 10^{-5}. Hence, one may entirely neglect inertial effects unlike [21] in polar dielectrics, where the inertial effects become progressively more important at high frequencies. We remark that our treatment will apply not only to uniaxial anisotropy but to an arbitrary nonaxially symmetric potential $U(\mathbf{e} \cdot \mathbf{n}, t)$ of the magnetocrystalline anisotropy. Since the potential is a function of $\mathbf{e} \cdot \mathbf{n}$ only, the autocorrelation functions of the overall process (e.g., cubic anisotropy) will be the product of the individual orientational autocorrelation functions of the freely rotating sphere and the solid-state mechanism. However, unlike uniaxial anisotropy even though the correlation functions still factorize, substantial coupling between the transverse and longitudinal modes (which now have α dependent eigenvalues) [cf. Eqs. (87) and (90)] will exist. The reason is that the nonaxial symmetry is now an intrinsic property of the particles. This phenomenon should be observable in measurements of the complex susceptibility of such particles.

D. Effect of the Fluid Carrier: (ii) Mode–Mode Coupling Effects for Particles with Uniaxial Anisotropy in a Strong Applied Field

We commence by recalling Eqs. (92) and (95). In the noninertial limit, Eq. (95) becomes

$$6\eta v\omega_n - mH_0[\mathbf{e} \times \mathbf{h}] = \lambda_n(t) \tag{121}$$

and on eliminating ω_n in Eq. (92) with the aid of Eq. (121) we obtain

$$\omega_R - mH_0[\mathbf{e} \times \mathbf{h}]\left(\frac{1}{6\mu v_m} + \frac{1}{6\eta v}\right) - \frac{K}{3\mu}(\mathbf{e} \cdot \mathbf{n})[\mathbf{e} \times \mathbf{n}] = \frac{\lambda_R(t)}{6\mu v_m} + \frac{\lambda_n(t)}{6\eta v} \tag{122}$$

Equation (122) is of the same form as Eq. (83) describing the motion of the dipole moment with a frozen ($\eta \to \infty$) Brownian mechanism in the presence of a field $\mathbf{H}_0$ at an oblique angle to the easy axis. It immediately follows that the effect of the fluid on the solid-state or internal mechanism of relaxation is to alter the magnetic drag coefficient ζ_m of the solid-state process so that

$$\frac{1}{\zeta_m} = \frac{1}{6\mu v_m} + \frac{1}{6\eta v} \tag{123}$$

The corresponding change in the dimensionless damping coefficient α of the solid-state process is

$$\alpha' = \alpha\left(1 + \frac{\mu v_m}{\eta v}\right) = \alpha(1 + \tau_N/\tau_B) \tag{124}$$

Here the ratio τ_N/τ_B represents the coupling between the magnetic and mechanical motions arising from the nonseparable nature of the Langevin equations, Eqs. (121) and (122). Thus the correction to the solid-state result imposed by the fluid is once again of the order 10^{-2}. Hence we may conclude, *despite the nonseparability of the equations of motion, that the Néel relaxation time of the ferrofluid particle should still be accurately represented in the IHD and VLD limits by the solid-state relaxation time formulae, Eqs. (87) and (90). Furthermore, Eq. (122) should be closely approximated by the solid-state relaxation equation*

$$\boldsymbol{\omega}_R - mH_0(\mathbf{e} \times \mathbf{h})\frac{1}{6\mu v_m} - \frac{K}{3\mu}(\mathbf{e} \cdot \mathbf{n})(\mathbf{e} \times \mathbf{n}) = \frac{\lambda_R(t)}{6\mu v_m} \tag{125}$$

Hence, just as in the solid-state problem, one would also expect the following effects to occur [48] in a ferrofluid for a large direct-current bias field superimposed on which is a small alternating-current field:

(a) A strong dependence of τ/τ_N on α (i.e., a frictional dependence of the smallest nonvanishing eigenvalue) unlike in the weak field case (axial symmetry), which is a signature of the coupling between the longitudinal and transverse modes.

(b) Suppression of the Néel and barrier crossing modes in favor of the fast relaxation modes in the deep well of the bistable potential created by Eq. (84), if the reduced [48] bias field h_c exceeds a certain critical value. This gives rise to a high-frequency Debye-like relaxation band [43,48].

(c) A very-high-frequency FMR peak due to excitation of the transverse modes having frequencies close to the precession frequency.

In the longitudinal response, however, guided by the relaxation time Eq. (112), it will be necessary to have particles with $\tau < \tau_B$ so that the characteristic α dependence of the response is not masked by the Brownian process due to the damping imposed by the fluid.

We now return to the mechanical equation, Eq. (95), of motion of the particles, which for small damping (if the small inertial term is retained) predicts a damped oscillation of fundamental frequency [57]

$$\omega_0 = \sqrt{mH_0/I} \tag{126}$$

which would appear in the spectrum as a high-frequency resonant absorption, as has been verified [25] in the theory of dielectric relaxation. In ferrofluids, however, it has been estimated [57] that fields of order of magnitude 10^7 Oe are needed for oscillatory effects, which is higher than any that may be obtained under terrestrial conditions. Hence one may rule out this resonant mode of the motion.

It remains to discuss the influence of $\mathbf{H}_0$ on the mechanical relaxation modes. An estimate of this may be made by recalling that Eq. (95) is basically the equation of motion of a rigid dipole in a strong constant external field. Moreover, if inertial effects are neglected, it has been shown in Refs. 21 and 61 that the longitudinal and transverse effective relaxation times decrease monotonically with field strength from τ_B, having asymptotic behavior

$$\tau_{\parallel} \sim \frac{\tau_B}{\xi}, \qquad \tau_{\perp} \sim \frac{2\tau_B}{\xi} \tag{127}$$

where

$$\xi = \frac{v_m M_s H_0}{kT} \gg 1 \tag{128}$$

Thus, the principal effect of the external field in Eq. (121) is to reduce the Brownian relaxation time.

Since detailed experimental data for ferrofluid susceptibilities in a strong oblique polarizing field (of intensity 10 T and higher)—superimposed on which is a weak alternating-current field—are not yet readily available, we shall in Section III.E confine ourselves to an illustration of how the weak alternating-current field susceptibilities [Eqs. (111), (113), and (116), which incorporate the effect of the fluid carrier] compare favorably with experiment. We shall also demonstrate the effect that a weak polarizing field $\mathbf{H}_0$ (0–100 kA/m) has on the susceptibility profiles (Figs. 7, 8, and 9) below. Furthermore, we shall show how

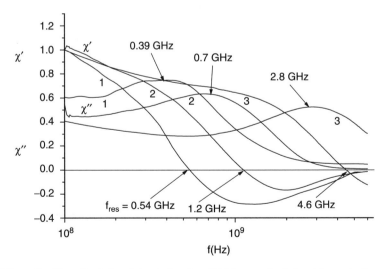

Figure 7. Normalized plot of χ' and χ'' against f(Hz) for samples 1, 2 and 3 over the frequency range 100 MHz to 6 GHz.

to determine the average value of the internal field of a particle, the anisotropy constant and the gyromagnetic ratio.

E. Comparison with Experimental Observations of the Complex Susceptibility Data

In order to support our theoretical discussions, we now present [12] room temperature complex susceptibility data for four colloidal suspensions, samples 1, 2, 3, and 4, respectively. The samples are as follows (in all of the samples the surfactant is oleic acid):

Sample 1 is a 150-Gauss (0.015-T) fluid consisting of $Ni_{0.5}Zn_{0.5}Fe_2O_4$ particles suspended in a low-vapor-pressure hydrocarbon (isopar M). The particles have a median diameter of 9 nm and a bulk saturation magnetization of 0.15 T.

Sample 2 is a 300-Gauss (0.03-T) fluid consisting of $Mn_{0.66}Zn_{0.34}Fe_2O_4$ particles suspended in isopar M. The particles have a median diameter of 9 nm and a bulk saturation magnetization of 0.31 T.

Sample 3 is a 400-Gauss (0.04-T) fluid consisting of cobalt particles suspended in a diester carrier. The particles have a median diameter of 7.8 nm and a bulk saturation magnetization of 1 T.

Sample 4 is a 760-Gauss (0.076-T) fluid consisting of magnetite particles suspended in isopar M. The particles have a median diameter of 9 nm and a bulk saturation magnetization of 0.4 T.

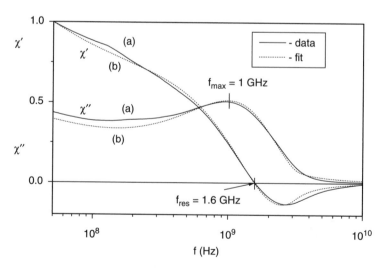

Figure 8.   Normalized plot of (a) χ' and χ'', and (b) corresponding fits using Eqs. (111), (113), and (116) for sample 4.

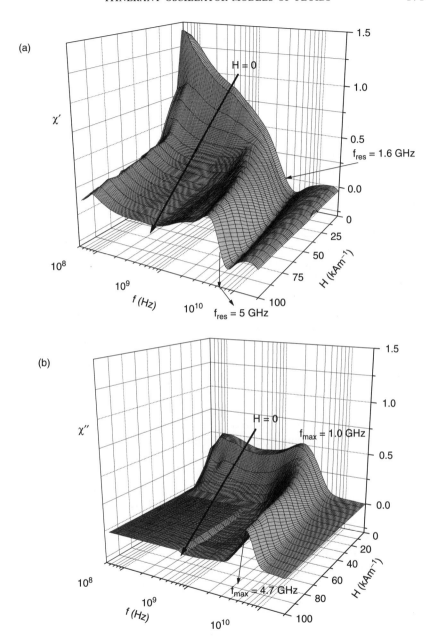

Figure 9. (a) 3-D plot of χ' against f (Hz) for sample 4 over the frequency range 100 MHz to 15 GHz for 17 values of polarizing field, $\mathbf{H}_0$, over the range 0–100 kAm^{-1}. (b) 3-D plot of χ'' against f(Hz) for sample 4 over the frequency range 100 MHz to 15 GHz for 17 values of polarizing field, H_0, over the range 0–100 kAm^{-1}.

Referring to samples, 1, 2, and 3 [12], Fig. 7 shows the results obtained for the real, χ', and imaginary, χ'', susceptibility components over the frequency range 100 MHz to 6 GHz. It is apparent that a resonant-like profile, indicated by the χ' component changing sign, at a frequency f_{res}, is characteristic of all the samples. f_{res} is seen to vary from 0.54 GHz for sample 1 to 4.4 GHz for sample 3. As $2\pi f_{res} = \gamma \bar{H}_A = \gamma 2\bar{K}/M_s$ (where $\bar{H}_A$ and $\bar{K}$ are average values of the particle internal field and anisotropy constant, respectively), this result is a signature of the difference in internal field and anisotropy constant of the particles. The susceptibility profiles have the same resonant form as that obtained for sample 4 (see Fig. 8); these profiles are shown to have the form predicted by Eqs. (111), (113), and (116). Furthermore, the actual values of $\bar{H}_A$ and $\bar{K}$ can be determined by means of polarized measurements as will be demonstrated now for sample 4.

Plot (a) of Fig. 8 shows the susceptibility components obtained for sample 4 over the wider frequency range of 50 MHz to 10 GHz, while plot (b) shows the fit obtained using Eqs. (111), (113), and (116). As magnetic fluids have a distribution of particle shape and size, these parameters are accounted for by modifying the above equations to include a normal distribution of anisotropy constant, $\bar{K}$, and a Nakagami distribution [59,60] of radii, r. Here the fit was obtained for a mean $\bar{K} = 1 \times 10^4 \, \text{J/m}^3$ with a standard deviation of 6×10^3 and a Nakagami distribution of radii, r, with a width factor $\beta = 4$ and a mean particle radius $\bar{r} \approx 4.5$ nm and a saturation magnetization of 0.4 T. The value used for α, the damping parameter, was 0.1, a figure within the range of values normally quoted [60] for α.

As far as measurements with a weak polarizing field are concerned, we remark that variation of the polarizing field, $\mathbf{H}_0$, over the range 0–100 kAm^{-1} will result in f_{res} increasing from 1.6 GHz to 5.0 GHz. However, a much clearer understanding of the effect of $\mathbf{H}_0$ on χ' and χ'' can be gleaned from a 3-D representation of the spectra as illustrated in Fig. 9a for the χ' component and in Fig. 9b for the χ'' component. Initially, both components reduce with increasing $\mathbf{H}_0$; this effect is a manifestation of the contribution of the relaxational components to the susceptibility. Beyond approximately 400 MHz a relaxation to resonance transition occurs with the χ' component going through zero at the resonant frequency f_{res}.

A plot of f_{res} against $\mathbf{H}_0$ for the sample is shown in Fig. 10; and because $\omega_{res} = 2\pi f_{res} = \gamma(H_0 + \bar{H}_A)$, the value of $\bar{H}_A$ is determined from the intercept of Fig. 10 and is found to be 41 kAm^{-1}, corresponding to a mean value of anisotropy constant, $\bar{K}$, at room temperature and bulk M_S of 0.4 T, or $8.2 \times 10^3 \, \text{J/m}^3$. The gyromagnetic ratio, γ, is found to be $2.26 \times 10^5 \, \text{S}^{-1} \, \text{A}^{-1}$ m from the slope of Fig. 10 and is in close agreement with the theoretical value of $2.21 \times 10^5 \, \text{S}^{-1} \, \text{A}^{-1}$ m. Similar polarized studies undertaken [62,63] on samples 1, 2, and 3 produced approximate $\bar{H}_A$ values of 1.6, 27, and

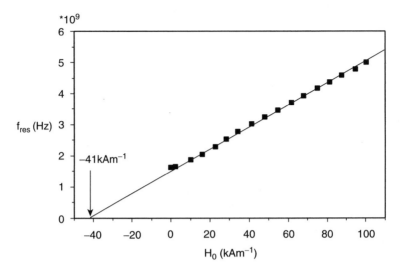

Figure 10. Plot of f_{res} against $\mathbf{H}_0$ for sample 4 in determining the average value of internal field, $\bar{H}_A = 41$ kAm^{-1}.

114 kAm^{-1}, respectively. The corresponding values of $\bar{K}$ were 120, 4×10^3, and 6×10^4 J/m^3.

F. Conclusions

The approach to the combined mechanical and magnetic motions of a ferrofluid particle and their mutual behavior, which is based on rearrangement of the Langevin equations and a consideration of the various characteristic time scales, indicates how the physical effects of the fluid carrier on the magnetic relaxation may be explained without elaborate and detailed solution (which will always involve supermatrix continued fractions) of the various equations describing the system. The relative orders of magnitude of the time scales involved determine which of the existing independent internal and Brownian mode solutions may be applied to the ferrofluid particle in any given situation. These considerations hold even in the strong field case, where the variables cannot be separated in the underlying Langevin equations. In particular, for zero or very weak external fields, we have shown that the high-frequency behavior may be accurately modeled by the solid-state result, Eq. (104), because the Brownian relaxation time τ_B simply cancels out of Eq. (113) due to the relative orders of magnitude of the various time scales. These are dictated by the ratio of the free diffusion times and the monotonic decrease of the effective relaxation times with barrier height that is a consequence of Dawson's integral. This appears to be the explanation of the success of Eq. (104) in explaining the experimental results of Section III.E and of Fannin et al. [60,62]. We further remark that no assumptions beyond that of the effective eigenvalue truncation of the set of differential recurrence relations have been made to obtain this result, since for zero or weak field the Langevin equations will always decouple. Furthermore, for $\sigma < 5$, where the effective eigenvalue solution is not an accurate representation of the exact transverse susceptibility solution [39], that solution may always be found from the underlying set of differential recurrence relations by using matrix continued fractions.

The experimental data on the linear response—that is, the weak a.c. field susceptibility and the effect of a weak polarizing field, which we have presented in Section III.E—strongly support our conjectures concerning the application of the solid-state transverse response result, Eq. (104), to magnetic fluids.

Finally, although the equations of motion do not separate in a strong bias field, it has been shown by considering the ratio of the free diffusion times that the relaxation behavior is essentially similar to the Néel relaxation in an oblique field. Thus, because all the eigenvalues now depend strongly on the damping, one would expect various precession-aided relaxation and resonant effects to appear in ferrofluids just as in the solid state. The precession-aided relaxation is also of interest in connection with the stochastic resonance phenomenon [64,65]. This phenomenon automatically appears in bistable potentials such as Eq. (84) and should be acutely sensitive to the magnitude and direction of the bias field because of the depletion effect produced by such a field in the shallower of the two wells of the potential [66–69]. The stochastic resonance effect should also be acutely sensitive to the weak alternating-current field if the bias field is near to the critical value at which the switchover of the greatest relaxation time from Arrhenius-like to non-Arrhenius-like behavior takes place.

In conclusion, we remark that a detailed review of matrix continued fraction methods for the solution of differential recurrence relations is available in Ref. 70, while a detailed account of the rotational Brownian motion of the sphere is available in Ref. 33.

IV. ANOMALOUS DIFFUSION

The theory of the Brownian motion on which the itinerant oscillator model is based is distinguished by a characteristic feature, namely, the concept of a *collision rate* which is the inverse of the time interval between successive collision events of the Brownian particle with its surroundings. In the words of Einstein [71]:

> We introduce a time interval τ in our discussion, which is to be very small compared with the observed interval of time, but, nevertheless of such a magnitude that the movements executed by a particle in two consecutive intervals of time τ are to be considered as mutually independent phenomena.

This concept, which is based on a random walk with a well-defined characteristic time and which applies when collisions are frequent but weak [13], leads to the Smoluchowski equation for the evolution of the concentration of Brownian particles in configuration space. If inertial effects are included (see Note 8 of Ref. 71, due to Fürth), we obtain the Fokker–Planck equation for the evolution of the distribution function in phase space which describes normal diffusion.

On the other hand, in continuous-time random walks [13], on which the theory of anomalous diffusion is based, the concept of a collision rate does not hold and so the mean of the time intervals between successive collision events *diverges*. Furthemore, since no characteristic time scale exists for such problems [13], the number of collisions in a time internal $(0, t)$ is not proportional to t for large times. In other words the central limit theorem [21] (law of large numbers) is no longer valid, leading to non-normal behavior. In order to describe such anomalous diffusion processes, Barkai and Silbey [13] consider a "Brownian" test particle that moves freely in one dimension and that collides elastically at random times with particles of the heat bath which are assumed to move much more rapidly than the last particle. The times between collision events are assumed to be independent identically distributed random variables implying that the number of collisions is a time interval $(0, t)$ is a renewal process. This is reasonable according to Barkai and Silbey when the bath particles thermalize rapidly and when the test particle is slow. Thus the time intervals between collision events $\{\tau_i\}$ are described by a probability density $\Psi(\tau)$, which is independent of the mechanical state of the test particle. Hence [13] the process is characterized by free motion with constant velocity for a time τ_1, then an elastic collision with a bath particle, then a free motion evolution for a time $\{\tau_2\}$, then again a collision. The most important assumption of the model is that $\Psi(\tau)$ decays as a power law for long times, namely that

$$\Psi(\tau) \sim \tau^{-1-\alpha} \qquad (129)$$

with $0 < \alpha < 1$.

Now the mean time between collisions diverges; that is,

$$\int_0^\infty \tau\Psi(\tau)\, d\tau \to \infty \qquad (130)$$

The above stochastic collision model then leads to a generalization of the Fokker–Planck equation for the evolution of the phase distribution function for mechanical particles [51] known as the fractional Klein–Kramers equation where the velocities acquire a fractional character [13]. We shall now describe how this equation may be used in the context of rotational Brownian motion. As an example we will consider the dielectric relaxation of a rotator without an external potential which would describe, for example, the motion of the surroundings in the itinerant oscillator model. Following Ref. 72, we shall now spend a little time to describe the application of continuous time random walks in rotational relaxation.

A. Application of Continuous-Time Random Walks in Rotational Relaxation

Relaxation functions for fractal random walks are fundamental in the kinetics of complex systems such as liquid crystals, amorphous semiconductors and polymers, glass forming liquids, and so on [73]. Relaxation in these systems may deviate considerably from the exponential (Debye) pattern. An important task in dielectric relaxation of complex systems is to extend [74,75] the Debye theory of relaxation of polar molecules to fractional dynamics, so that empirical decay functions—for example, the stretched exponential of Williams and Watts [76]—may be justified in terms of continuous-time random walks.

We have remarked earlier that a marked deficiency of the Debye theory is that it cannot describe the dielectric relaxation process at short times [of the order of the characteristic decay time of the dipole angular velocity correlation function] or at high frequencies (in the FIR region) because, like the Einstein theory of the translational Brownian motion [21], it is based on the Smoluchowski equation. The approximate equation describing the evolution of the probability density function (p.d.f.) of the orientations of the dipoles in configuration space [51] is only valid when dipole inertial effects are negligible because it assumes instantaneous equilibrium of the angular velocities. Thus the Debye theory predicts *infinite integral absorption*.

We have seen that the incorporation of such inertial effects in the Debye theory, ensuring a return to optical transparency at high frequencies, was first achieved by Rocard [77] and a completely rigorous treatment was given by Gross [18] and Sack [19]. They used the Klein–Kramers equation applied to rotational Brownian motion, describing the evolution of the p.d.f. in configuration-angular velocity space. Hence they obtained exact expressions for the complex susceptibility of noninteracting rigid dipoles as infinite continued fractions. The simple relaxation equation of Rocard [77] then appears for small inertial effects—that is, where the ratio of the characteristic relaxation time of the angular velocity correlation function τ_ω to the (Debye) relaxation time τ of the dipole relaxation function is small.

B. The Fractional Klein–Kramers Equation

In the introduction we have seen that Barkai and Silbey [13] have proposed recently a generalization of the Klein–Kramers equation for the inertia corrected translational Brownian motion to fractional dynamics. Here we demonstrate how a similar fractional Klein–Kramers equation (FKKE) for the anomalous rotational diffusion may be solved using the generalized integration theorem [73] of Laplace transformation yielding the exact complex dielectric susceptibility in continued fraction form as in the normal Brownian motion. The method may also be obviously extended to all problems where solutions of the Klein–Kramers, or more generally the Fokker–Planck equation, may be expressed as recurrence relations involving ordinary or matrix continued fractions. Thus the itinerant oscillator model may also be solved in fractional dynamics.

We illustrate by considering the simplest microscopic model of dielectric relaxation, namely, an assembly of noninteracting rigid dipoles of moment μ each rotating about a fixed axis through its center. This system would then represent the motion of the cage in the itinerant oscillator model. As usual, a dipole has moment of inertia I and is specified by the angular coordinate ϕ so that it constitutes a system of 1 (rotational) degree of freedom. The FKKE for the p.d.f. $W(\phi, \dot{\phi}, t)$ in the phase space $(\phi, \dot{\phi})$ is identical to that of the one-dimensional translational Brownian motion of a particle [13]; however, rotational quantities (angle ϕ, moment of inertia I, etc.) replace translational ones (position x, mass m, etc.) so that [72]

$$\frac{dW}{dt} = \frac{\partial W}{\partial t} + \dot{\phi}\frac{\partial W}{\partial \phi} - \frac{\mu E \sin \phi}{I}\frac{\partial W}{\partial \dot{\phi}} = {}_0D_t^{1-\alpha}L_{FP}W \qquad (131)$$

where

$${}_0D_t^{1-\alpha}L_{FP}W = {}_0D_t^{1-\alpha}\vartheta\beta\left(\frac{\partial}{\partial\dot{\phi}}(\dot{\phi}W) + \frac{kT}{I}\frac{\partial^2 W}{\partial\dot{\phi}^2}\right)$$

is the fractional Fokker–Planck operator, k is the Boltzmann constant, T is the temperature, $\beta = \zeta/I$, ζ is the viscous damping coefficient of a dipole, $\vartheta = \tau^{1-\alpha}$, α is the anomalous exponent [73] characterizing the fractal time process, and τ is the intertrapping time scale identified with the Debye relaxation time ζ/kT (at ambient temperatures, τ is of the order 10^{-11} sec for molecular liquids and solutions). Thus the fractional dynamics emerges from the competition of Brownian motion events of average duration τ interrupted by trapping events whose duration is broadly distributed in accordance with our introduction. In writing Eq. (131) we assume that a *weak* uniform electric field **E** applied along the initial line is suddenly switched off at time $t = 0$ so that linear response theory [21] may be used to describe the ensuing response. In the absence of **E**, Eq. (131) describes anomalous diffusion; the value $\alpha = 1$ corresponds to normal diffusion [13]. The operator ${}_0D_t^{1-\alpha} \equiv \frac{\partial}{\partial t}{}_0D_t^{-\alpha}$ in

Eq. (131) is defined in terms of the convolution (the Riemann–Liouville fractional integral definition) [73]

$$_0D_t^{-\alpha}W(\phi, \dot{\phi}, t) = \frac{1}{\Gamma(\alpha)} \int_0^t \frac{W(\phi, \dot{\phi}, t')\, dt'}{(t - t')^{1-\alpha}} \qquad (132)$$

so that the fractional derivative is a type of memory function [73] or *stosszahlansatz* for the Boltzmann equation underlying the Fokker–Planck equation. We remark that a slowly decaying power-law kernel in the Riemann–Liouville operator, Eq. (132), is typical of memory effects in complex systems. In this context we remark that the operator $_0D_t^{1-\alpha}$ in Eq. (131) does not act on $\dot{\phi}\frac{\partial W}{\partial \phi}$ and $-\mu EI^{-1}\sin\phi\frac{\partial W}{\partial\dot{\phi}}$ (i.e., the Liouville or conservative terms) in the convective derivative $\dot{W}$ so that the conventional form [51] of the Boltzmann equation with Fokker–Planck *stosszahlansatz* as modified by $_0D_t^{1-\alpha}$ is preserved. If the fractional derivative acts on the convective term [i.e., the conservative part of Eq. (131)], then nonphysical behavior (i.e., infinite integral absorption) of the dielectric absorption coefficient occurs as in the Debye theory of dielectric relaxation.

C. Solution of the Fractional Klein–Kramers Equation

We may seek a solution of Eq. (131) as in normal diffusion as

$$W(\phi, \dot{\phi}, t) = e^{-\eta^2\dot{\phi}^2} \sum_{p=-\infty}^{\infty} \sum_{n=0}^{\infty} c_{p,n}(t)e^{ip\phi}H_n(\eta\dot{\phi}) \qquad (133)$$

where $\eta = \sqrt{I/2kT}$ and the $H_n(x)$ are the Hermite polynomials. The linearized initial (at $t = 0$) p.d.f. is

$$W(\phi, \dot{\phi}, 0) \approx \frac{1}{2\pi^{3/2}} \eta e^{-\eta^2\dot{\phi}^2}\left[1 + \frac{\mu E}{kT}\cos\phi\right] \qquad (134)$$

Straightforward manipulation [72] of the recurrence relations of the H_n leads, by orthogonality (as in Section II), to a differential recurrence relation for the coefficients $c_{p,n}(t)$, which may be written for the only case of interest (since the linear response is considered), namely, $p = 1$:

$$\frac{d}{dt}c_{1,n}(t) + \frac{i}{2\eta}[2(n + 1)c_{1,n+1}(t) + c_{1,n-1}(t)] = -_0D_t^{1-\alpha}\vartheta n\beta c_{1,n}(t) \qquad (135)$$

On using the integration theorem of Laplace transformation as generalized to fractal calculus [73], namely,

$$L\{_0D_t^{1-\alpha}f(t)\} = \begin{cases} s^{1-\alpha}\tilde{f}(s) - _0D_t^{-\alpha}f(t)|_{t=0} & (0 < \alpha < 1) \\ s^{1-\alpha}\tilde{f}(s) & (1 \le \alpha < 2) \end{cases}$$

where $\tilde{f}(s) = L\{f(t)\} = \int_0^\infty e^{-st}f(t)\,dt$, we have from Eq. (135)

$$[2\tau s + n\gamma'^2(\tau s)^{1-\alpha}]\tilde{c}_{1,n}(s) + i\gamma'[2(n+1)\tilde{c}_{1,n+1}(s) + \tilde{c}_{1,n-1}(s)] = c_{1,n}(0) \quad (136)$$

Here $\gamma' = \tau/\eta == \zeta\sqrt{2/IkT}$ is chosen as the inertial effects parameter ($\gamma' = \sqrt{2/\gamma}$ is effectively the inverse square root of the parameter γ used earlier in Section I). Noting the initial condition, Eq. (134), all the $c_{1,n}(0)$ in Eq. (136) will vanish with the exception of $n = 0$. *Furthermore, Eq. (136) is an example of how, using the Laplace integration theorem above, all recurrence relations associated with the Brownian motion may be generalized to fractional dynamics.* The normalized complex susceptibility $\hat{\chi}(\omega) = \hat{\chi}'(\omega) - i\hat{\chi}''(\omega)$ is given by linear response theory as

$$\hat{\chi}(\omega) = \frac{\chi(\omega)}{\chi'(0)} = 1 - i\omega\frac{\tilde{c}_{1,0}(i\omega)}{c_{1,0}(0)} \quad (137)$$

where $\chi'(0) = \mu^2 N/(2kT)$ and N is the number of dipoles per unit volume. [We remark that $\tilde{c}_{1,0}(s)$ also yields the Laplace transform of the characteristic function of the configuration space p.d.f.]

Equation (136), which is a three-term algebraic recurrence relation for the $\tilde{c}_{1,n}(s)$, can be solved in terms of continued fractions [21], thereby yielding the generalization of the familiar Gross–Sack result [18,19] for a fixed axis rotator to fractional relaxation, namely,

$$\hat{\chi}(\omega) = 1 - \cfrac{B(i\omega\tau)^\sigma}{B(i\omega\tau)^\sigma + \cfrac{B}{1 + B(i\omega\tau)^\sigma + \cfrac{2B}{2 + B(i\omega\tau)^\sigma + \cfrac{3B}{3 + \cdots}}}} \quad (138)$$

where $\sigma = 2 - \alpha$ and $B = 2\gamma'^{-2}(i\omega\tau)^{2(\alpha-1)}$. Equation (138), in turn, can be expressed in terms of the confluent hypergeometric (Kummer) function $M(a, b, z)$ [78], namely,

$$\hat{\chi}(\omega) = 1 - \frac{(i\tau\omega)^\sigma}{1 + (i\tau\omega)^\sigma}M(1, 1 + B[1 + (i\tau\omega)^\sigma], B) \quad (139)$$

Equation (139) can readily be derived by comparing Eq. (138) with the continued fraction

$$\frac{M(a,b,z)}{(b-1)M(a-1,b-1,z)} = \cfrac{1}{b-1-z+\cfrac{az}{b-z+\cfrac{(a+1)z}{b+1-z+\cdots}}} \qquad (140)$$

with $a = 1$, $z = B$, and $b = 1 + B[1 + (i\tau\omega)^\sigma]$ and by noting that $M(0, b-1, z) = 1$ [78]. The continued fraction, Eq. (140), can be obtained from the known recurrence relation [78]

$$b(1 - b + z)M(a,b,z) - azM(a+1,b+1,z) + b(b-1)M(a-1,b-1,z) = 0$$

We remark that $M(1, 1 + b, z) = bz^{-b}e^z\gamma(b,z)$, where $\gamma(b,z) = \int_0^z e^{-t}t^{b-1}\, dt$ is the incomplete gamma function [78].

For $\sigma = 1$, Eq. (139) can be reduced to Sack's result [Eq. (3.19c) of Ref. 19]. In the high damping limit ($\gamma' \gg 1$), Eq. (139) can be simplified, yielding the generalization to fractional dynamics of the Rocard [21,77] equation, namely,

$$\hat{\chi}(\omega) = \frac{1}{1 + (i\omega\tau)^\sigma - 2(\omega\tau/\gamma')^2} \qquad (141)$$

On neglecting inertial effects ($\gamma' \to \infty$), Eq. (139) becomes

$$\hat{\chi}(\omega) = \frac{1}{1 + (i\omega\tau)^\sigma} \qquad (142)$$

that is, the result previously proposed from empirical considerations [75]. For $\sigma = 1$, Eq. (142) reduces to the Debye equation.

D. Dielectric Loss Spectra

Dielectric loss spectra $\hat{\chi}''(\omega)$ versus $\log_{10}(\omega\eta)$, for various values of α and γ', are shown in Figs. 11–13. It is apparent that the spectral parameters (the characteristic frequency, the half-width, the shape) strongly depend on both α (which pertains to the velocity space) and γ'. Moreover, the high-frequency behavior of $\hat{\chi}''(\omega)$ is *entirely determined* by the inertia of system. It is apparent, just as in Brownian dynamics, that inertial effects produce a much more rapid falloff of $\hat{\chi}''(\omega)$ at high frequencies. Thus the Gordon sum rule for the dipole integral absorption of one-dimensional rotators [14,28], namely

$$\int_0^\infty \omega\chi''(\omega)\, d\omega = \frac{\pi N\mu^2}{4I} \qquad (143)$$

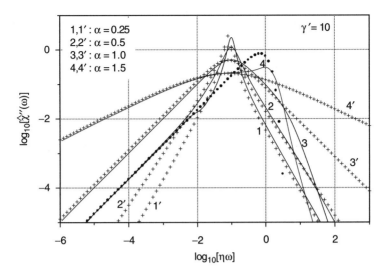

Figure 11. Dielectric loss spectra $\hat{\chi}''(\omega)$ for $\gamma' = 10$ and various values of α: $\alpha = 0.25$ (curves 1 and 1′), $\alpha = 0.5$ (2, 2′), $\alpha = 1$ (3, 3′), and $\alpha = 1.5$ (4, 4′). Solid lines (exact solution) (1, 2, 3, and 4): Eq. (139); crosses (1′, 2′, 3′, and 4′): Eq. (142); filled circles: Eq. (141).

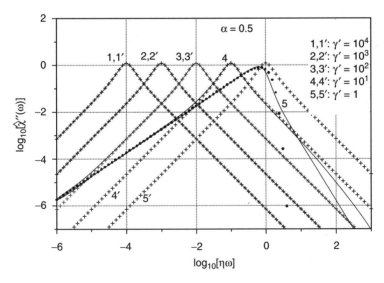

Figure 12. Dielectric loss spectra $\hat{\chi}''(\omega)$ for $\alpha = 0.5$ (enhanced diffusion) and various values of γ': $\gamma' = 10^4$ (curves 1 and 1′), $\gamma' = 10^3$ (2, 2′), $\gamma' = 10^2$ (3, 3′), $\gamma' = 10$ (4, 4′), and $\gamma' = 1$ (5, 5′). Solid lines (exact solution) (1, 2, 3, 4, and 5): Eq.(139); crosses (1′, 2′, 3′, 4′, and 5′): Eq. (142); filled circles: Eq. (141).

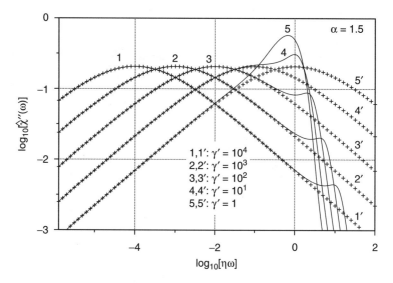

Figure 13. The same as in Fig. 12 for $\alpha = 1.5$ (subdiffusion).

is satisfied. It is significant that the right-hand side of Eq. (143) is determined by molecular parameters only and does not depend on temperature. Such behavior is quite unlike that resulting from the hypothesis that the fractional derivative acts on the convective term leading always [28] to infinite integral absorption as in the Debye (excluding inertial effects) and Van Vleck–Weisskopf kinetic models [27,28,79]. Moreover, it is apparent from Figs. 11 and 13, assuming that Eq. (131) also describes subdiffusion in configuration space (that is, $\sigma < 1$ or $2 > \alpha > 1$); that Eq. (139) also provides a physically acceptable description of the loss spectrum. Here, in the high damping limit ($\gamma' \gg 1$), the low-frequency part of $\hat{\chi}''(\omega)$ may be approximated by the modified Debye equation, Eq. (142). For $\alpha < 1$ corresponding to $\sigma > 1$ (enhanced diffusion in configuration space), the low-frequency behavior of $\hat{\chi}''(\omega)$ is similar (see Figs. 11 and 13) to that of the dielectric loss $\hat{\chi}''_{FR}(\omega)$ in the free rotation limit ($\zeta = 0$), which is given by [18,19]

$$\hat{\chi}''_{FR}(\omega) = \sqrt{\pi}\eta\omega e^{-\eta^2\omega^2} \tag{144}$$

E. Angular Velocity Correlation Function

The high-frequency dielectric absorption in the FIR region is proportional to the spectrum of the equilibrium angular velocity correlation function $\langle \dot{\phi}(0)\dot{\phi}(t)\rangle_0$ and can be used to determine that function experimentally [14]. In order to calculate the $\langle\dot{\phi}(0)\dot{\phi}(t)\rangle_0$ in the fractional dynamics, one can simply adapt the result of Barkai and Silbey [13] for the translational velocity correlation function $\langle\dot{x}(0)\dot{x}(t)\rangle_0$ replacing translational quantities by rotational ones. We have

$$\langle\dot{\phi}(0)\dot{\phi}(t)\rangle_0 = \frac{kT}{I}E_\alpha\left(-\frac{\gamma'^2}{2}\left(\frac{t}{\tau}\right)^\alpha\right) \tag{145}$$

where $E_\alpha(z)$ is the Mittag–Leffler function [72] defined by

$$E_\alpha(z) = \sum_{n=0}^{\infty}\frac{z^n}{\Gamma(1+\alpha n)} \tag{146}$$

Equation (145) represents the generalization of the a.v.c.f. of the Ornstein–Uhlenbeck [21] (inertia-corrected Einstein) theory of the Brownian motion to fractional dynamics. The long-time tail due to the asymptotic $(t\gg\tau)$ $t^{-\alpha}$-like dependence [72] of the $\langle\dot{\phi}(0)\dot{\phi}(t)\rangle_0$ is apparent, as is the stretched exponential behavior at short times $(t\ll\tau)$. For $\alpha>1$, $\langle\dot{\phi}(0)\dot{\phi}(t)\rangle_0$ exhibits oscillations (see Fig. 14) which is consistent with the large excess absorption occurring at high frequencies.

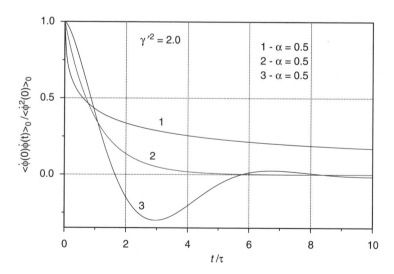

Figure 14. The normalized angular velocity correlation function $\langle\dot{\phi}(0)\dot{\phi}(t)\rangle_0/\langle\dot{\phi}^2(0)\rangle_0$ for $\gamma'^2=2$ and $\alpha=0.5$ (curve 1), $\alpha=1.0$ (curve 2), and $\alpha=1.5$ (curve 3).

F. Conclusions

We have demonstrated how existing Brownian motion solutions for dielectric relaxation may be generalized to fractional dynamics using continued fractions [21] and the Laplace transform of the Riemann–Liouville definition of the fractional derivative. The result is of particular interest in dielectric relaxation because it demonstrates how the unphysical high-frequency divergence of the absorption coefficient due to the neglect of inertia may be removed in fractional relaxation as in inertia-corrected Debye relaxation (see Figs. 11–13). The methods we have outlined are obviously applicable on account of Eq. (132) to other relaxation models such as the itinerant oscillator, which attempt to incorporate both resonance and relaxation behavior in a single model with the purpose of simultaneously explaining the Debye (low-frequency) and FIR absorption spectra of complex dipole systems (recent applications of fractional models to anomalous dielectric relaxation are given, e.g., in Refs. 79–83).

In passing, we finally note that possible new applications of the methods described in this review are the fluctuations of a rubidium atom on the surface of a fullereme cage [84] and orientation moments of spheroids in simple shear flow [85].

Acknowledgments

The support of this work by the Enterprise Ireland Research Collaboration Fund 2001, the Trinity College Dublin Trust, USAF, EOARD (contract F61773-01-WE407), and INTAS (project no. 01-2341) is gratefully acknowledged. W. T. C. is the holder of a Berkeley Fellowship 2001–2002 at Trinity College Dublin for the academic year 2001–2002.

References

1. N. E. Hill, *Proc. Phys. Soc. Lond.* **82**, 723 (1963).

2. V. F. Sears, *Proc. Phys. Soc. Lond.* **86**, 953 (1965).

3. G. A. P. Wyllie, *J. Phys. C* **4**, 564 (1971).

4. J. H. Calderwood and W. T. Coffey, *Proc. R. Soc. Lond.* **A365**, 269 (1977).

5. W. T. Coffey, P. M. Corcoran, and M. W. Evans, *Proc. R. Soc. Lond.* **A410**, 61 (1987).

6. W. T. Coffey, P. M. Corcoran, and J. K. Vij, *Proc. R. Soc. Lond.* **A414**, 339 (1987).

7. W. T. Coffey, P. M. Corcoran, and J. K. Vij, *Proc. R. Soc. Lond.* **A425**, 169 (1989).

8. M. I. Shliomis, *Sov. Phys. Usp.* **17**, 153 (1974).

9. M. I. Shliomis and V. I. Stepanov, in *Relaxation Phenomena in Condensed Matter, Advances in Chemical Physics*, Vol. 87, W. T. Coffey, ed., Wiley, New York, 1994, p. 1.

10. P. S. Damle, A. Sjölander, and K. S. Singwi, *Phys. Rev.* **165**, 277 (1968).

11. N. G. van Kampen, *Stochastic Processes in Physics and Chemistry*, 2nd ed., North-Holland, Amsterdam, 1992.

12. W. T. Coffey and P. C. Fannin, *J. Phys. Cond. Matter* **14**, 1(2002).

13. E. Barkai and R. S. Silbey, *J. Phys. Chem. B.* **104**, 3866 (2000).

14. M. W. Evans, G. J Evans, W. T. Coffey, and P. Grigolini, *Molecular Dynamics*, Wiley, New York, 1982.

15. W. T. Coffey, M. W. Evans, and P. Grigolini, *Molecular Diffusion and Spectra*, Wiley, New York 1984; Russian translation: Mir, Moscow, 1987.

16. J. P. Poley, *J. Appl. Sci.* **B4**, 337 (1955).

17. G. P. Johari, *J. Non-Crystalline Solids* **307**, 114 (2002).

18. E. P. Gross, *J. Chem. Phys.* **23**, 1415 (1955).

19. R. A. Sack, *Proc. Phys. Soc.* B **70** 402, 414 (1957).

20. *Dielectric Properties and Molecular Behavior*, N. E. Hill, W. E. Vaughan, A. H. Price, and M. Davies, eds., D. Van Nostrand, London, 1969, p. 90.

21. W. T. Coffey, Y. P. Kalmykov, and J. T. Waldron, *The Langevin Equation*, World Scientific, Singapore, 1996, reprinted 1998.

22. P. Debye, *Polar Molecules*, Chemical Catalog, New York 1929, reprinted by Dover Publications, New York, 1954.

23. G. A. P. Wyllie, in *Dielectric and Related Molecular Processes*, vol. 1, Specialist Periodical Reports, Senior Reporter, M. Davies, The Chemical Society, London, 1972, p. 21.

24. W. T. Coffey, *J. Chem. Phys.* **107**, 4960 (1997).

25. W. T. Coffey, Y. P. Kalmykov, and S. V. Titov, *J. Chem. Phys.* **115**, 9895 (2001).

26. W. T. Coffey, *J. Chem. Phys.* **93**, 724 (1990).

27. B. K. P. Scaife, *Principles of Dielectrics*, Oxford University Press, London, 1989, revised edition 1998.

28. B. K. P. Scaife, *Complex Permittivity*, The English Universities Press, London, 1971.

29. W. T. Coffey, C. Rybarsch, and W. Schröer, *Chem. Phys. Lett.* **99**, 31 (1983).

30. H. Fröhlich, *Theory of Dielectrics*, Oxford University Press, London (1958).

31. H. Fröhlich, personal communication to W. T. Coffey, 1987.

32. W. T. Coffey and M. E. Walsh, *J. Chem. Phys.* **106**, 7625 (1997).

33. J. R. McConnell, *Rotational Brownian Motion and Dielectric Theory*, Academic, London, 1980.

34. Yu. P. Kalmykov, *J. Mol. Liq.* **69**, 53 (1996).

35. J. T. Waldron, Yu. P. Kalmykov, and W. T. Coffey, *Phys. Rev. E* **49**, 3976 (1994).

36. M. E. Walsh and P. M. Déjardin, *J. Phys. B* **32**, 1 (1998).

37. H. Mori, *Prog. Theor. Phys.* **33**, 423 (1965).

38. E. C. Titchmarsh, *An Introduction to the Theory of Fourier Integrals*, Oxford University Press, London, 1937.

39. Yu. P. Kalmykov and W. T. Coffey, *Phys. Rev. B* **56**, 3325 (1997).

40. W. T. Coffey, Y. P. Kalmykov, and E. S. Massawe, *Phys. Rev. E* **48**, 699 (1993).

41. A. Polimeno and J. H. Freed, *Adv. Chem. Phys.* **83**, 89 (1993).

42. C. Scherer and H. G. Matuttis, *Phys. Rev. E* **63**, 1 (2001).

43. W. T. Coffey, D. S. F. Crothers, J. L. Dormann, L. J. Geoghegan, and E.C. Kennedy, *Phys. Rev. B* **58**, 3249 (1998).

44. W. T. Coffey, D. S. F. Crothers, J. L. Dormann, Y. P. Kalmykov, E. C. Kennedy, and W. Wernsdorfer, *Phys. Rev. Lett.* **80**, 5655 (1998).

45. Y. P. Kalmykov and S. V. Titov, *Phys. Rev. Lett.* **82**, 2967 (1999).

46. Y. P. Kalmykov and S. V. Titov, *Phys. Solid State* (St. Petersburg) **40**, 1492 (1998).

47. W. T. Coffey, P. J. Cregg, and Y. P. Kalmykov, *Adv. Chem. Phys.* **83** 263 (1993).

48. W. T. Coffey, D. S. F. Crothers, Y. P. Kalmykov, and S.V. Titov, *Phys. Rev. B* **64** (012411) (2001).

49. W. F. Brown, Jr., *IEEE Trans. Magn.* **MAG-15**, 196 (1979).

50. W. F. Brown, Jr., *Phys. Rev.* **130**, 1677 (1963).

51. W. T. Coffey, D. A. Garanin, and D. J. McCarthy, *Adv. Chem. Phys.* **117**, 483 (2001).

52. J. S. Langer, *Ann. Phys. (N.Y.)* **54**, 258, 1969.

53. H. A. Kramers, *Physica (Utrecht)* **7**, 284 (1940).

54. L. J. Geoghegan, W. T. Coffey, and B. Mulligan, *Adv. Chem. Phys.* **100**, 475 (1997).

55. W. T. Coffey, Y. P. Kalmykov, and E. S. Massawe, in *Modern Nonlinear Optics*, Part 2, M. W. Evans and S. Kielich, eds., *Advances in Chemical Physics*, Vol. 85, Wiley, New York, 1993, p.667.

56. W. T. Coffey, D. S. F. Crothers, Y. P. Kalmykov, E. S. Massawe, and J. T. Waldron, *Phys. Rev. E* **49**, 1869 (1994).

57. Y. L. Raikher and M. I. Shliomis, in *Relaxation Phenomena in Condensed Matter*, W. T. Coffey, ed., *Advances in Chemical Physics*, Vol. 87, Wiley, New York, 1994, p. 595.

58. B. K. P. Scaife, *Adv. Chem. Phys.* **109**, 1 (1999).

59. P. C. Fannin, *Adv. Chem. Phys.* **104**, 181 (1998).

60. P. C. Fannin, T. Relihan, and S. W Charles, *Phys. Rev B* **55**, 21 (1997).

61. J. T. Waldron, Y. P. Kalmykov, and W. T. Coffey, *Phys. Rev. E* **49**, 3976 (1994).

62. P. C. Fannin, T. Relihan, and S. W. Charles, *J. Magn. Magn. Mater.* **162**, 319 (1996).

63. P. C. Fannin, S. W. Charles, and J. L. Dormann, *J. Magn. Magn. Mater.* **201**, 98 (1999).

64. Y. L. Raikher and V. I. Stepanov, *J. Phys. Cond. Matter* **6**, 4137 (1994).

65. L. Gammaitoni, P. Hänggi, P. Jung, and F. Marchesoni, *Rev. Mod. Phys.* **70**, 223 (1998).

66. W. T. Coffey, D. S. F. Crothers, and Y. P. Kalmykov, *Phys. Rev. E* **55**, 4812 (1997).

67. D. A. Garanin, *Phys. Rev. E* **54**, 3250 (1996).

68. W. T. Coffey, *Adv. Chem. Phys.* **103**, 259 (1998).

69. W. T. Coffey, D. S. F. Crothers, Y. P. Kalmykov, and J. T. Waldron, *Phys. Rev. B* **51**, 15947 (1995).

70. J. L. Déjardin, Y. P. Kalmykov, and P. M. Déjardin, *Adv. Chem. Phys.* **117**, 275 (2001).

71. A. Einstein. *Investigations on the Theory of the Brownian Movement*, R. Fürth, ed., Dover Publications, New York, 1956.

72. W. T. Coffey, Y. P. Kalmykov, and S. V. Titov, *Phys. Rev. E* **65**, 032102 (2002).

73. R. Metzler and J. Klafter, *Adv. Chem. Phys.* **116**, 223 (2001).

74. W. G. Glöckle and T. F. Nonnenmacker, *J. Stat. Phys.* **71**, 741 (1993).

75. K. Weron and M. Kotulski, *Physica A* **232**, 180 (1996).

76. G. Williams and D. C. Watts, *Trans. Faraday Soc.* **66**, 80 (1970).

77. M. Y. Rocard, *J. Phys. Radium* **4**, 247 (1933).

78. M. Abramowitz and I. Stegun, eds., *Handbook of Mathematical Functions*, Dover, New York, 1964.

79. J. H. Van Vleck and V. F. Weisskopf, *Rev. Mod. Phys.* **17**, 227 (1945).

80. R. R. Nigmatullin and Y. A. Ryabov, *Fizika Tverdogo Tela (St. Petersburg)* **39**, 101 (1997) [*Phys. Solid State* **39**, 87 (1997)].

81. V. V. Novikov and V. P. Privalko, *Phys. Rev. E* **64**, 031504 (2001).

82. W. T. Coffey, Y. P. Kalmykov, and S. V. Titov, *Phys. Rev. E* **65**, 032102 (2002).

83. W. T. Coffey, Y. P. Kalmykov, and S. V. Titov, *J. Chem. Phys.* **116**, 6422 (2002).

84. P. Dugourd, R. Antoine, D. Rayane, E. Benichou, and M. Broyer, *Phys. Rev. A* **62**, 011201 (2000).

85. K. Asokan, T. R. Ramamohan, and V. Kumaran, *Phys. Fluids* **14**, 75 (2002).

86. M. C. Wang and G. E. Uhlenbeck, *Rev. Mod. Phys.* **17**, 323 (1945).

STATISTICAL MECHANICS OF STATIC AND LOW-VELOCITY KINETIC FRICTION

MARTIN H. MÜSER[*]

Institut für Physik, Johannes Gutenberg-Universität, Mainz, Germany

MICHAEL URBAKH

School of Chemistry, Tel Aviv University, Tel Aviv, Israel

MARK O. ROBBINS

Department of Physics and Astronomy, The Johns Hopkins University, Baltimore, Maryland, U.S.A.

CONTENTS

[*] *Current address*: Department of Applied Mathematics, University of Western Ontario, London, Ontario, Canada

Advances in Chemical Physics, Volume 126, Edited by I. Prigogine and Stuart A. Rice.
ISBN 0-471-23582-2. © 2003 John Wiley & Sons, Inc.

I.   PRELIMINARY REMARKS

A.   Introduction

Since prehistoric times, major technological advances have gone hand-in-hand with fundamental advances in tribology, the science of friction, lubrication, and wear. In each era, new mechanical technology has resulted in improved experiments that drive new theoretical understanding, and new understanding has led to further improvements in technology. D. Dowson's book on the *History of Tribology* [1] gives an impressive overlook from the invention of fire rods by the Neanderthals (friction-induced chemical reactions are now called *tribochemistry*) to the attempts of eminent scientists such as Leonardo da Vinci, Coulomb, and Euler to formulate and to understand the basic laws of solid friction. We are currently in the midst of a new revolution in tribology, driven by (a) the advent of experimental techniques that allow controlled friction measurements at atomic scales and (b) computers that allow the complex dynamics in atomic scale contacts to be analyzed. This new line of study, dubbed *nanotribology*, is playing a central role in the quest to build robust machines with nanometer-scale moving parts and is poised in turn to benefit from the resulting advances in nanotechnology.

The three major new atomic-scale experimental methods developed in the last decade are the quartz crystal microbalance (QCM) [2–4], atomic and friction force microscopes (AFM/FFM) [5,6], and the surface force apparatus (SFA) [7,7a,8]. These new tools reveal complementary information about tribology at the nanometer scale. The QCM measures dissipation as an adsorbed film of submonolayer to several monolayer thickness slides over a substrate. AFM and FFM explore the interactions between a surface and a tip whose radius of curvature is 10–100 nm [9]. The number of atoms in the contact ranges from a few to a few thousand. Larger radii of curvature and contacts have been examined by gluing spheres to an AFM cantilever [10,11]. SFA experiments measure shear forces in even larger-diameter ($\sim 10\,\mu m$) contacts, but with angstrom-scale control of the thickness of lubricating films.

Pioneering experiments with these new techniques have reinvigorated theoretical studies of tribology, which had languished for the previous half century. The difficulty in testing hypotheses was a major factor in this theoretical hiatus. A tremendous number of parameters affect the lateral forces between two solids, yet the primary output of a typical set of macroscopic experiments is a single number: the dimensionless ratio of lateral and normal forces, or friction coefficient μ. Thus verification or falsification of theories was basically impossible. The advent of controlled atomic scale measurements changed this situation, and new computational techniques allowed theorists to study models that included more of the internal degrees of freedom of the contacting surfaces [12,13].

The goal of this chapter is to review statistical mechanical approaches to the molecular origins of friction. A brief outline of macroscopic experimental results and the fundamental theoretical issues they raise is provided in the remainder of this section. Then progressively more detailed models of the contacting solids are described. Our goal is to illustrate basic phenomena with the simplest model that captures them, and then to generalize the lessons learned to more complicated models. We begin by presenting results for completely rigid solids with no internal degrees of freedom (Section II). Then elastic and nonelastic deformations within the solids are considered in Sections III and IV. In most practical cases the solids are separated by an intervening layer of lubricant, physisorbed contaminants, dust, and so on. The effect of such intervening material is described in Section V. The next section considers the origin of the complex oscillatory dynamics observed in some sliding systems. The interface between macroscopic objects generally contains many micron-scale contacts, and we discuss the interplay between these contacts in Section VII. Section VIII presents brief concluding remarks.

B. Some Fundamental Definitions and Questions

Two types of friction are commonly measured and calculated. The static friction F_s is defined as the minimum lateral force needed to initiate sliding of one object over a second, while the kinetic friction $F_k(v)$ is the force needed to maintain sliding at a steady velocity v. Observation of static friction implies that the contacting solids have locked into a local free-energy minimum, and F_s represents the force needed to lift them out of this minimum. It is a threshold rather than an actual force acting on the system, and it limits lateral motion in any direction. No work is done by the static friction, since no motion occurs. The kinetic friction is intrinsically related to dissipation mechanisms, and it equals the work done on the system by external forces divided by the distance moved.

One central issue in tribology is why static friction is so universally observed between solid objects. How does any pair of macroscopic objects, placed in contact at any position and orientation, manage to lock together in a local free energy minimum? A second issue is why experimental values of F_s and F_k tend to be closely correlated. The two reflect fundamentally different processes and their behavior is qualitatively different in many of the simple models described below.

A third central issue concerns the relationship between friction and the normal force or load L that pushes the two objects together. The macroscopic laws of friction found in textbooks were first published by the French engineer Amontons about 300 years ago [14], albeit the first recorded studies go back even further to the Italian genius da Vinci. Both found that the friction F_s between two solid bodies is (i) independent of the (apparent) area of contact and (ii) proportional to L. These laws can be summarized in the equation

$$F = \mu L \tag{1}$$

and different friction coefficients μ_s and μ_k are obtained for static and kinetic friction.

These simple laws apply fairly well to an abundance of materials, but also have many exceptions. For example, most materials exhibit friction at zero load; and many materials, such as tape, exhibit friction at negative loads. These forces tend to be small for the relatively rough surfaces that were available to

Amontons and da Vinci, but researchers going back to Coulomb have accounted for them by adding a term proportional to area to Eq. (1). This is often motivated as an additional load coming from the adhesive interactions between the surfaces and is discussed further in Section I.D.

Coulomb contributed what is often called the third law of friction, i.e. that μ_k is relatively independent of sliding velocity. The experiments discussed in Section I.D show that the actual dependence is logarithmic in many experimental systems and that μ_k often *increases* with decreasing velocity. Thus there is a fundamental difference between kinetic friction and viscous or drag forces that decrease to zero linearly with v. A nearly constant kinetic friction implies that motion does not become adiabatic even as the center-of-mass velocity decreases to zero, and the system is never in the linear response regime described by the fluctuation–dissipation theorem. Why and how this behavior occurs is closely related to the second issue raised above.

C. Basic Scenarios for Friction

The earliest attempts to explain Amontons's laws were based on the idea that macroscopic peaks or asperities on one surface interlocked with valleys on the opposing surface [1]. As illustrated in Fig. 1a, the bottom surface then forms a ramp that the top surface must be lifted up over in order to slide. If the typical angle of the ramp is θ, and there is no microscopic friction between the surfaces,

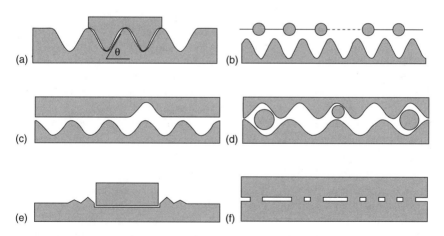

Figure 1. Schematic representation of various possible friction mechanisms: (a) Geometric interlocking of asperities with typical angle θ, (b) elastic deformation (stretched dashed bonds) to interlock atoms and/or macroscopic peaks, resulting in multiple metastable states, (c) defect pinning (circles), (d) pinning by an intervening layer of weakly bound material, (e) plastic deformation or plowing, and (f) material mixing or cold welding.

then the lateral force needed to produce motion is just $F_s = L \tan \theta$. Thus this simple model predicts a load independent coefficient of friction $\mu_s = \tan \theta$ that can be consistent with any experimental value between zero and infinity. However, there are both practical and fundamental difficulties with this picture for static friction.

One obvious difficulty is that it is hard to imagine how the roughness on two independent surfaces can always manage to interlock. Section II examines to what degree the corrugations of a rigid slider can interlock with those of a rigid substrate. A more subtle issue is that this picture cannot explain a constant kinetic friction. Once the top surface climbs past a peak on the bottom surface, the lateral force changes sign as the surface slides to the next interlocked state. Since the energy of successive interlocked states is statistically equivalent, the net force and average work vanish. Adding damping terms to the model only produces a kinetic friction that vanishes linearly with velocity. Leslie [15] noted that the observation of a constant kinetic friction at low velocities suggests that, like Sisyphus, the top wall always climbs upward without ever getting any higher.

To our knowledge, Brillouin was the first to recognize that a nonvanishing dissipation at low velocity can only occur if some *quasi-discontinuous* jumps from one state to another become unavoidable [16]. During these jumps, some part of the system becomes linearly unstable and the instantaneous motion remains far from equilibrium even as the average velocity goes to zero. This requires that some part of the system be compliant enough to allow for different mechanically stable states at the same center-of-mass position. The simplest model illustrating such instabilities and their relation to hysteresis is generally attributed to Tomlinson [17], but had been introduced earlier by Prandtl [18]. In the Prandtl–Tomlinson model (Section III.A), a simple mass point is pulled over a corrugated surface by a harmonic spring, which models the elastic coupling within the slider. If the spring is soft enough, the mass advances in a sequence of discontinuous pops and there is a constant kinetic friction at low velocity. Also other single-particle models such as one lubricant atom embedded between two plates [17a,17b] successfully predict various tribological phenomena. Another type of rapid instability leading to friction has been suggested recently [19,20]. In this model, the top surface advances by propagation of a small crack across the surface at a velocity that remains comparable to the speed of sound for any sliding velocity.

Elastic deformations within the solid can also produce static friction between surfaces that would not otherwise interlock. Figure 1b illustrates a system where the spacing between peaks on the top surface is stretched to conform to the bottom surface. This could occur at the scale of either macroscopic asperities [21] (Section VII) or individual surface atoms (Section III). The elastic energy required to displace each peak into an opposing valley must be compensated by the gain in potential energy due to interactions between the surfaces. This requires that the interactions between surfaces be strong compared to the interactions within each surface, and theoretical studies indicate that this mechanism is unlikely to allow locking of each peak [22–25]. However, the elastic energy cost can be lowered by deforming the system more gradually. The energy required for interlocking at rare defects (Fig. 1c) vanishes as the separation between them grows [21,26,27], although the friction force also vanishes in this limit (Sections III.C.5 and VII.A). While this mechanism is important for pinning of charge-density waves and flux lattices in superconductors [28,29], it is unlikely to be able to explain most friction measurements (Section VII).

One recently proposed scenario (Section V) is that unstable jumps and plastic deformation occur within a layer of weakly bound material between the two surfaces (Fig. 1d) [30]. This layer could consist of physisorbed molecules, dust, grit, wear debris, and so on. If interactions within the layer are weak, it is free to rearrange into a metastable state that conforms to any geometry of the bounding surfaces. Once in this state, it will resist any change in geometry due to translation, leading to static friction. When forced to slide, the constituent elements advance independently in a series of rapid pops, leading to a constant kinetic friction.

All of the above mechanisms can be characterized as wearless and therefore are consistent with steady-state sliding. During most of the last century, the predominant explanation of the origin of static and small-velocity kinetic friction was based on irreversible plastic deformation within the two contacting solids (Section IV) [31]. If stresses are strong enough to permanently deform the solids, their surfaces will naturally interlock to produce static friction, and the work needed to deform material during sliding will produce a constant kinetic friction. Obvious examples where this mechanism is important include scratching or "plowing" of a substrate (Fig. 1e) and large-scale removal of material as in machining (Section IV.D). The plastic deformation may result from large stresses in the contact, or be driven by thermodynamics as in cold welding or material mixing (Fig. 1f). Plastic deformation and the concurrent generation of wear and debris make it impossible to define a steady-state value for the kinetic friction force. However, one may not infer that time variation in F_k implies that plastic deformation is the main source of friction.

All the above-mentioned scenarios may be relevant in specific systems, and more than one may be active at the same time in any real macrotribological system. Our goal below is to examine the implications and limits of each mechanism and identify the cases where they are most likely. Before embarking on this quest, we briefly review important experimental findings.

D. Phenomenology of Static and Low-Velocity Kinetic Friction

Experimental measurements show that rough surfaces do not interpenetrate as envisioned in Fig. 1a, at least at the most macroscopic scales. Electrical [31],

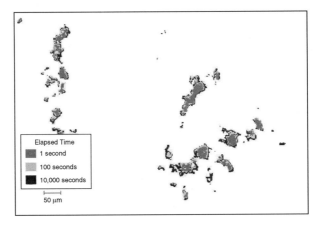

Figure 2. Images of the contacts between two acrylic surfaces as a function of the time since sliding stopped. Even at large loads, the contacts cover a small fraction of the interfacial area. With permission from Ref. 33. *Pure Appl. Geophys.* **143**, 238 (1994).

mechanical [32], and optical [33,34] measurements show that contact only occurs where the peaks of asperities on opposing surfaces coincide (Fig. 2). As a result, the real area of molecular contact A_{real} is much smaller than the apparent geometrical area of the surfaces A_{app}. It was the latter that da Vinci and Amontons found did not affect the friction force in most cases.

Elastic and plastic deformations flatten the contact regions into micrometer-scale patches, which are many times larger than molecular scales. The static friction corresponds to an average yield stress within these contacts of $\tau_s = F_s/A_{real}$, and a similar relation gives the local shear stress τ_k corresponding to the kinetic friction. Experimental studies of a wide range of materials indicate that τ rises linearly with the local pressure P [35–40]:

$$\tau = \tau_0 + \alpha P \tag{2}$$

This provides a simple phenomenological explanation for Amontons's laws as well as some exceptions to them. Summing over A_{real}, one finds

$$F_s = \tau_0 A_{real} + \alpha L \tag{3}$$

or

$$\mu_s = \alpha + \tau_0/p \tag{4}$$

where $p = L/A_{real}$ is the average pressure in the contact. Note that Amontons's laws that μ_s is independent of load and macroscopic contact area are obeyed if

p is constant or large compared to τ_0. The former holds if the surfaces deform completely plastically [31] so that the local pressure is pinned at the hardness of the material. The pressure can also be constant if the asperities have a random distribution of heights and deform elastically [41–43]; however, the constant is not a material property like the hardness. Since τ_0 is generally positive, Eq. (3) gives a finite friction force at zero or even negative loads, thus avoiding the most glaring failure of Amontons's laws. A large number of experiments [31,35,44] show an apparent divergence of μ at small loads that is consistent with the second term in Eq. (4).

Experiments also show that the friction is very sensitive to the last layer of atoms on the contacting surfaces [31,44]. For example, graphite is a good lubricant in ambient air because of water adsorbed on the surface. It is not a useful lubricant in space or vacuum applications. Another dramatic example is that a single molecular layer of lubricant can decrease the friction between two metallic surfaces by an order of magnitude [45]. This is clearly inconsistent with the mechanical interlocking picture (Fig. 1a), since the macroscopic slope of the surfaces remains unaffected. Other evidence against macroscopic interlocking comes from the fact that the friction generally begins to rise when surfaces are made sufficiently smooth. This observation can be understood from Eq. (3) and the fact that A_{real} increases as surfaces are made smoother. Note that the above observations do not imply that shear and dissipation are always confined to the interface. Plastic deformation can extend to a great depth beneath the surface, particularly when the hardnesses of the two materials in contact are very different [31].

Experiments clearly show that the frictional force is history-dependent. The static friction generally increases with the time of contact. The kinetic friction fluctuates in magnitude and direction, and its average value depends on past sliding velocities. Rabinowicz has shown that the friction fluctuations in some systems result from the presence of a limited number of large asperities, and he has used the force autocorrelation function to extract the characteristic size of these asperities [44]. Memory of past sliding velocities indicates that sliding produces a gradual evolution in the area, structure, or composition of the contacts between surfaces. Phenomenological "rate-state" models have been very successful in describing such memory effects in macroscopic friction measurements.

Rate-state models assume that the friction depends on the rate and a small number of "state variables" that describe the properties of the interface. Different physical interpretations of the microscopic properties that these state variables describe have been proposed, such as the amount of dilation at the interface [46] or the degree of crystallinity of an intervening film [47,48]. Most approaches leave the nature of the state variable unspecified and merely assume that it depends on some average of recent velocities [49–51]. The coefficients of a simple dynamical equation are fit to experiment and can then be used to describe a wide range of dynamic sliding conditions, particularly stick–slip motion (Section VI).

A simple rate-state model that describes dry sliding friction between rocks, plastics, wood, and many other rough materials [33,52,53] was proposed by Dieterich [50] and developed by Ruina [51]. The instantaneous friction coefficient is written as

$$\mu = \mu_0 + A \ln(v/v_0) + B \ln(\theta/\theta_0) \qquad (5)$$

where v and θ are the current velocity and state, and μ_0 is the friction coefficient at some reference state where $v = v_0$ and $\theta = \theta_0$. The equation of motion for the state variable is

$$\partial_t \theta = 1 - \left(\frac{\theta}{D_c}\right) v - \left(\frac{C\theta}{B\sigma}\right)\partial_t \sigma \qquad (6)$$

where θ has units of time and can be thought of as the age of the contact, $\partial_t \theta$ is its time derivative, σ is the applied normal stress, and D_c is a characteristic sliding distance over which the contact evolves. Experimental values of D_c vary from 2 to 100 μm, which is comparable to the contact diameters determined by Rabinowicz from force fluctuations [44] and seen in Fig. 2. There are also four positive dimensionless fit parameters in this theory: μ_0, A, B, and C. For typical rocks, $\mu_0 \sim 0.6$, A and B are from 0.005 to 0.015, and C is between 0.25 and 0.5.

Under static conditions, θ rises linearly with time and the static friction coefficient rises as $\ln t$. The steady-state value of θ during sliding can be found by setting the time derivatives in Eq. (6) to zero, yielding $\theta_{ss} = D_c/v$. Inserting this in Eq. (5) gives $\mu_k = \mu_0 + (A - B)\ln(v/v_0)$. Thus the steady-state kinetic friction may increase or decrease with increasing velocity depending on the relative size of A and B. However, an instantaneous change in velocity always increases μ because A is positive.

Figure 3 shows the predicted and observed changes in friction after a jump in velocity for a variety of systems. In each case the friction jumps down after a sudden decrease in velocity and then rises after a sliding distance of order D_c. Increasing the velocity to its initial value causes a sharp rise in μ followed by a gradual decrease. It is important to note that the friction relaxes over the same sliding distance after the two changes, even though the sliding times are different by an order of magnitude. This is the motivation for expressing memory in terms of a characteristic distance in Eq. (6).

In a beautiful set of experiments, Dieterich and Kilgore [33,34] were able to show that for many systems the state variable is directly related to the contact geometry. Using transmitted light, they could observe each asperity contact and

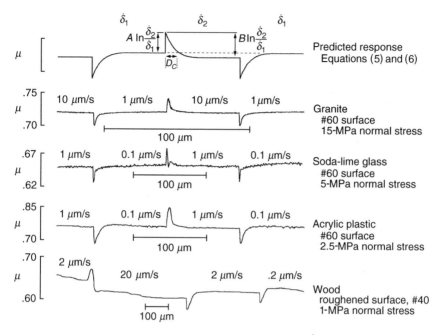

Figure 3. Comparison of the change in friction with velocity $\dot{\delta}$ predicted by Eq. (5) and observations on a number of materials. With permission from Ref. 33. *Pure Appl. Geophys.* **143**, 238 (1994).

its evolution with time (Fig. 2). Changes in the state variable correlated directly with changes in the real area of contact obtained by summing over all asperities. Moreover, the change in friction over the length scale D_c could be quantitatively described by changes in the distribution of contacts. After a given sliding distance d, contacts that are smaller than d have disappeared and been replaced by new contacts whose area reflects the current sliding velocity. The distance to reach steady state is thus directly related to the distribution of contact sizes. Dieterich and Kilgore observed a wide range of contact diameters, from their resolution of about 1 μm up to ∼50 μm. The value of D_c lies in this range. Dieterich and Kilgore also found a logarithmic increase with time in the contact area between two static surfaces that could be correlated with the increase in static friction. They concluded that for the rough surfaces and high loads in their experiments the contacts are always near the plastic limit, and that A_{real} increases with time through a creep process.

Dieterich and Kilgore's observations suggest that for their systems the friction force can be written as the product of a real contact area A_{real} that depends on past history and a shear stress τ that depends on the instantaneous velocity. They find that $A_{\text{real}}(\theta) = A_{\text{real}}(\theta_0)(1 + (B/\mu_0)\ln(\theta/\theta_0)$. If the shear stress rises logarithmically with velocity as $\tau(v) = \tau_0(1 + A/\mu_0\ln(v/v_0))$, one recovers Eq. (5) up to linear order in the small parameters A and B. Baumberger and collaborators have also observed increases in contact area with increasing resting time or decreasing velocity [54].

Logarithmic dependences of shear stress on shear rate are seen in many bulk materials such as amorphous polymers. The logarithm follows naturally from a simple model proposed by Eyring [55] that assumes shear is thermally activated and that the activation barrier decreases linearly with the applied stress. If the equilibrium barrier U is much larger than $k_B T$, one can ignore backward motion and write that the velocity $v = v_0 \exp(-(U - \tau V^*)/k_B T)$ where the coefficient V^* has units of volume and is called the *activation volume*. Solving for τ, one finds

$$\tau = \tau_0 + (k_B T/V^*)\ln(v/v_0) \tag{7}$$

where $\tau_0 = U/V^*$. If V^* is independent of T, then the coefficient of the $\ln(v)$ term is proportional to T. This behavior is observed both in the bulk shear stress of polymers and in their frictional force. Dieterich and Kilgore also find that A and B rise roughly linearly with temperature in the much harder materials that they study. The variation of B can be understood if the creep that causes A_{real} to increase is thermally activated, and recent work on polymers has shown that there is a correlation between the distance from the glass transition temperature and the rate of area growth [56]. However, the above work does not address the type of processes that control activation at the interface and thus A. Recent work on this topic is discussed in Section V.D.

To summarize this section, the total area of real contact between macroscopic surfaces is usually much smaller than the apparent area, and consists of many local contacts with a broad range of diameters from ~ 1 to 100 μm. The friction can be expressed as a product of A_{real} and the average yield or shear stress τ within the contacts. The number of atoms within the contacts is large enough that statistical mechanical methods can be used to calculate τ. The friction depends on past history due to two types of creep process. Creep normal to the interface leads to an (almost) ever-increasing A_{real}. As a result, F_s increases with resting time and F_k increases with decreasing velocity. Lateral creep leads to a reduction of friction as v is decreased. Ultimately, at *astronomically* small sliding velocities, A_{real} should saturate at A_{app}. The effect of lateral creep would then dominate, and one can even expect a linear relation between F_k and v as discussed by Estrin and Bréchet [57] from a material science point of view and by Chauve and co-workers [58,59] based on a finite-temperature renormalization group analysis. However, this limit is not generally reached in practical experiments.

II.   RIGID WALLS

An important step in developing an understanding of friction is to start with as few explicitly treated degrees of freedom as possible and to gradually increase the level of complexity. It is thus quite natural to consider solids as completely rigid and impenetrable in a first approximation. Lubricant atoms—if present at all—are in complete thermal equilibrium so that the slider sees a well-defined (free) energy profile. The first historic attempts by Amontons, Coulomb, Euler, Bélidor, and others [1] to understand friction as a purely geometric interlocking of rough but essentially rigid surfaces may be referred to as a *rigid wall model*. The underlying assumptions of this model are often far from realistic. Incorporating elastic or plastic deformations, cold welding, and so on, will qualitatively change the tribological behavior of the two solids in contact. It is yet important to understand the properties of the rigid-wall model quantitatively, since most models are in one way or another generalizations of it; for example, the Prandtl–Tomlinson and the Frenkel–Kontorova model correspond to a rigid-wall model in the limit of infinitely strong springs. It is, moreover, conceivable that the picture of rigid walls is rather accurate for nanoscale objects.

The simplest and most illustrative starting point is to consider an infinitely large, single-crystalline substrate and a rigid slider. For reasons of symmetry, the interactions V between the two rigid objects must then share the periodicity of the substrate. For example, motion along direction with period b yields

$$V = V_0 \cos(2\pi x/b) + V_1 \cos(4\pi x/b) + \cdots \tag{8}$$

where x denotes the relative, lateral displacement between slider and substrate at given normal load L and where the V_n's are expansion coefficients.

There are two important issues addressed in the literature. One is the size of the prefactors V_n. These prefactors determine the maximum possible lateral force between the two objects, which provides a meaningful upper bound for the static friction force. The other issue is how the friction is altered through thermal fluctuations/activation that stem from the not explicitly treated internal degrees of freedom. These two questions will be discussed separately in Sections II.A and II.B before some applications are presented in Section II.C.

We wish to make a final general comment on the properties of the rigid-wall model. As pointed out by Leslie [15] nonzero static friction does not imply nonzero kinetic friction due to energy recovery. The energy required to lift the slider to the top of the barrier can be regained, in principle, by moving it downhill in a controlled, adiabatic way. A compliant driving device, however, will not allow us to move the tip in a controlled way. New effects emerge leading to energy dissipation as most simply described by the Prandtl–Tomlinson model (Section III.A).

A. Geometric Interlocking

The reason why a rigid-wall model is appealing can be seen from Fig. 1a: In order to initiate sliding through the application of a lateral force F, the upper solid has to be lifted up a maximum slope θ. This slope is independent of the normal load L if we assume simple hard-disk interactions; that is, as soon as the two solids overlap, the interaction energy increases from zero to infinity. The maximum lateral force F_s between the two solids would then simply be

$$F_s = L \tan \theta \qquad (9)$$

if we assume that there is no friction on smaller scales. As noted in the previous section, this equation is consistent with Amontons's law, Eq. (1). In 1737, Bélidor [1] modeled rough surfaces by means of such hard-disk asperities of spherical shape arranged to make commensurate crystalline walls (in these days, we call those asperities *atoms*!). Two surfaces are called *commensurate* if they share the same periodicities within the interface. Bélidor found $\mu_s \approx 0.35$, which is a "typical" value for measured friction coefficients. An automatic consequence of commensurability is the proportionality between F_s, or alternatively V_n in Eq. (8), with L irrespective of the area of contact, provided that adhesive interactions are negligible.

One problem with Bélidor's explanation is the assumption of commensurate walls. Surface corrugations of two macroscopic bodies usually do not match as well as sketched in Fig. 1a, unless they are specifically designed to interlock. Even in those SFA experiments in which two chemically identical surfaces are slid with respect to each other, special care has to be taken in order to orient the two surfaces perfectly such that commensurability is obtained.

In general, one may expect that the surface corrugations of two solid bodies do not match at any length scale—or to be more precise, they only match in a stochastic way. In the following paragraph, the effect of the correlation of surface roughness on the shear force will be discussed for smooth surfaces with a roughness exponent $H = 0$. Extremely rough and self-affine surfaces for which $H > 0$ will be considered later in Section VII. For $H = 0$, it is relatively easy to make general quantitative predictions for lateral forces between the surfaces as a function of the symmetry of the surfaces. In order to calculate the lateral forces between two rigid solid bodies from an atomistic point of view, it is necessary to make assumptions about the interaction between the solids. However, it turns out that the details of the interactions do not influence important, qualitative results. Two different models [60,61] yield similar properties for the scaling of the maximum shear forces F_s with increasing size of the interface (for instance, contact area A or rod length L) at a fixed normal pressure $p_\perp$. Panyukov and Rabin [60] considered randomly, irreversibly charged rods and plates, where the net charge vanished on each object. Müser et al. [61] analyzed two rigid solids that experience forces that grow exponentially fast with the overlap between them. The latter model roughly mimics the exponential repulsion due to the Pauli principle as electronic orbitals belonging to different solids begin to overlap.

For both models, it is possible to calculate the energy landscape generated by relative translation analytically. Both times it is found that (i) $\tau_s = F_s/A$ is independent of A if the two periods of the two surfaces match, (ii) τ_s decreases as $A^{-1/2}$ if the two surfaces are random, and (iii) τ_s is zero if the surfaces are incommensurate. Contributions from the circumference of finite contacts between incommensurate surfaces yield contributions to τ_s that vanish with a higher power law than $A^{-1/2}$ (see also Fig. 6).

Further results of the overlap model [61] are as follows: (iv) The prefactor that determines the strength of the exponential repulsion has no effect on F_s at fixed normal load L, (v) the lateral force scales linearly with L for any fixed lateral displacement between slider and substrate, (vi) allowing for moderate elastic interactions within the bulk does not necessarily increase F_s, because the roughness may decrease as the surfaces become more compliant, and (vii) the prefactor of F_s for nonidentical commensurate surfaces decreases exponentially with the length of the common period $\mathcal{L}_c$. This last result had already been found by Lee and Rice [62] for a yet different model system. We note that the derivation of properties (iv) and (v) relied strongly on the assumption of exponential repulsion or hard disk interactions. Therefore one must expect charged objects to behave differently concerning these two points.

Assuming constant normal pressure in the contacts, which is equivalent to a linearity between L and A, the static friction coefficient obeys the following general scaling laws for rigid objects where roughness exponent $H = 0$ applies:

$$\mu_s \propto \begin{cases} \exp(-\mathcal{L}_c) & \text{commensurate surfaces} \\ 1/\sqrt{A} & \text{random surfaces} \\ 0 & \text{incommensurate surfaces} \end{cases} \tag{10}$$

where in all cases finite-size corrections apply that fall faster than $A^{-1/2}$. The experimental test of Eq. (10) and the above mentioned hypotheses is a challenging task, since it requires the use of flat and chemically inert surfaces. However, atomistic computer simulations, in which the necessary, idealized conditions can be more easily implemented, confirm the conclusions, in particular those concerning commensurate and incommensurate interfaces [22–24,30,61,63]. The scaling predictions (i)–(vi) were tested systematically by Müser et al. [61] via molecular dynamics simulations, in which the two opposing walls contained discrete atoms. The intrasurface interactions consisted of an elastic coupling of atoms to their lattice site, while intersurface interactions consisted of Lennard-Jones (LJ) interactions. Despite allowing the walls some degree of elasticity and despite the interactions being different from the simple exponential repulsion model, agreement with the theoretical conclusions (i)–(vi) was found. Only (iv) and (v) had to be slightly modified by adding an adhesive offset load, when the adhesive part of the LJ interactions was taken into account.

B. Thermal Effects

Before considering the coupling of the wall to a driving device, it is instructive to study the motion of the slider under a constant force in the presence of thermal fluctuations. This constant force can be exerted by inclining the substrate or by putting a test charge on the slider in an electric field.

The substrate is still assumed to be crystalline, and for the sake of simplicity the interaction V between the slider and the substrate is described by one single harmonic; thus

$$V = f_0 b' \cos (x/b') \qquad (11)$$

Here x is the position of the slider relative to the substrate, $b'/2\pi$ is the substrate's period, and f_0 is the (zero-temperature) static friction force, whose scaling with the area of contact and normal load we just discussed. In order to incorporate the effects of thermal fluctuations on the motion of the slider, one can exploit the isomorphism to the motion of a Brownian particle moving on a substrate. A nice description of that problem is given by Risken in Chapter 11 of Ref. 64. Here we will discuss some of the aspects that we believe to be important for friction.

Thermal effects can be incorporated into the model by adding a thermal random force $\Gamma(t)$ and a damping term $m\gamma v$ to the conservative force between slider and substrate:

$$m\ddot{x} + m\gamma\dot{x} = f_0 \sin(x/b') + \Gamma(t) \qquad (12)$$

where m is the slider's mass. The random force should satisfy the fluctuation–dissipation theorem; that is, it must have zero mean $\langle \Gamma(t) \rangle = 0$ and be δ-correlated [65,66]:

$$\langle \Gamma(t)\Gamma(t') \rangle = 2m\gamma k_B T \delta(t - t') \qquad (13)$$

where k_B denotes the Boltzmann constant and T denotes the temperature. The random forces and the damping term $-m\gamma v$ arise from interactions with phonons and/or other excitations that are not treated explicitly. Typically, the time scales associated with these excitations are short compared to the motion of the slider from one minimum to another. This justifies the assumption of δ correlation in the random forces.

In the small-velocity limit, one can usually neglect the inertial term $m\ddot{x}$ in Eq. (12) for the calculation of the friction force. For this limit, Risken and Vollmer [66] derived a matrix-continued-fraction method that allows one to calculate the nonlinear response to a constant external force F as an expansion in terms of inverse damping constants. The expansion also converges for very small (but finite) damping constants. It is worth discussing certain limits. The central quantity is the mobility μ_{mob}, which is defined as

$$\langle v \rangle = \mu_{mob} F \tag{14}$$

where $\langle v \rangle$ is the average sliding velocity. From a phenomenological point of view, one may interpret μ_{mob}^{-1} as an effective viscosity η_{eff}.

In the limit of large damping and small F (linear response regime), Risken and Vollmer find the following for the mobility of a particle moving in a simple cosine potential:

$$\mu_{mob} = \frac{\tilde{T}}{\gamma} \left\{ I_0(\tilde{T}^{-1}) \right\}^{-2} + \mathcal{O}\left(\frac{1}{\gamma^3}\right) \tag{15}$$

$$\approx \frac{1}{\gamma} \times \begin{cases} \left(1 - \frac{1}{2}\tilde{T}^{-2}\right) & \text{for } \tilde{T} \gg 1 \\ \tilde{T}^{-1} \exp(-2\tilde{T}^{-1}) & \text{for } \tilde{T} \ll 1 \end{cases} \tag{16}$$

where I_0 is the modified Bessel function and $\tilde{T} = k_B T / f_0 b'$ a reduced temperature. The two limiting approximations given in Eq. (16) are valid at large and small thermal energies, respectively. At large thermal energies, the corrugation of the substrate potential barely influences the motion and phononic dissipation dominates the friction. At small thermal energies, the motion of the slider becomes activated. The exponential decrease of the mobility, or the exponential increase of η_{eff}, is prototypical for creep. Note that the mobilities (or η_{eff}^{-1}) given in Eq. (15) are only valid for sufficiently small external forces F—that is, when the system is sufficiently close to thermal equilibrium.

When F increases, μ_{mob} starts to deviate from its equilibrium values as shown in Fig. 4. At large F and/or at large temperatures, the distribution of the variable x becomes flat and the influence of the corrugation does not play an important role any longer. One must thus distinguish between three different concepts: (i) dissipation, which is reflected in the phenomenological damping constant γ, (ii) equilibrium mobility $\mu_{mob}(F \to 0)$ or η_{eff}^{-1}, which arises from an interplay of dissipation, surface corrugation, and thermal fluctuations, and

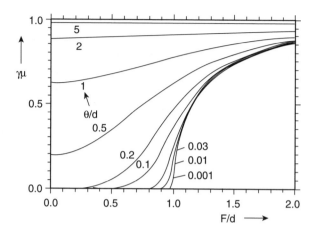

Figure 4. Effective mobility as a function of reduced forces for various reduced temperatures θ/d. With permission from Ref. 64. *The Fokker Planck Equation*, Springer, Heidelberg, 1984.

(iii) nonequilibrium mobility μ_{mob} at finite, sufficiently large values of F. In the present case, the increase of the mobility could be interpreted as an apparent shear thinning, despite the fact that the rigid wall has no internal degrees of freedom. On the other hand, one may interpret x as an internal degree of freedom of the combined system and say that due to the externally imposed shear, x spends different fractions of time at a given point of phase space at different forces. This argument would be consistent with the effect of shear thinning.

Additional information conveyed by Fig. 4 is that at finite temperatures, the mobility remains finite, although it may become exponentially small as T tends to zero. Hence strictly speaking, the zero-velocity kinetic friction F_k is always zero as long as temperature T is finite; thus

$$\lim_{T \to 0^+} \lim_{v \to 0} F_k(v) = 0 \tag{17}$$

On the other hand, if we first send temperature to zero, then F_k will tend to the value f_0; thus

$$\lim_{v \to 0} \lim_{T \to 0^+} F_k(v) = f_0 \tag{18}$$

The fact that two limits cannot be exchanged as shown in Eqs. (17) and (18) for the limits $v \to 0$ and $T \to 0^+$ is frequently encountered in tribological contexts. This concerns, for example, the determination of static friction F_s as a function of (a) the speed $\dot{\sigma}_{ext}$ with which the externally applied shear force is ramped up and (b) the time we are willing to wait to see the slider move by one lattice constant. It also concerns the way in which the thermodynamic limit is realized as discussed by Müser and Robbins [24]: For a finite system, static friction will always tend to zero in the limit $\dot{\sigma}_{ext} \rightarrow 0$, even if the walls are commensurate. Thus, the meaningful determination of static friction forces requires a sufficiently large time-scale separation between the creep motion of the junction and the vibronic degrees of freedom.

C. Applications

1. Nanotubes

AFM manipulations of carbon nanotubes (CNT) present a unique opportunity to study the friction between rigid objects as a function of commensurability. One important case is sliding of a CNT on a highly oriented graphite substrate [67,68]. The local structure of the two surfaces consists of nearly identical hexagonal rings, and the commensurability between them can be changed by rotating the axis of the CNT with respect to the crystalline substrate. The second case is relative sliding of concentric multiwalled CNTs [69]. Here the commensurability is determined by the intrinsic structure of the two tubes.

Figure 5 illustrates the behavior of a CNT on a graphite substrate as an AFM tip begins to displace it [67,68]. The CNT is initially in an incommensurate state and slides smoothly over the graphite. The AFM tip exerts a torque that rotates the CNT about its center of mass. However, this motion is interrupted at discrete in-plane orientations where the CNT "locks" into a low-energy state [68,70] indicated by an increase in the force required to move the CNT. Figure 5 shows a lateral force trace illustrating the pronounced change in the force on going from the commensurate to incommensurate state and vice versa [68]. As the nanotube is rotated in-plane, several of these discrete commensurate orientations are observed, each separated by 60 ± 1 degrees (Fig. 5, upper inset). The contrast of the friction in the commensurate and incommensurate states is dramatic (order of magnitude) and abrupt (a discrete change within experimental uncertainty). It was also found [67,68] that for this system, the transition from the incommensurate to commensurate state is accompanied by a transition from sliding motion in which the CNT rotates in-plane, to gearlike rolling motion.

Molecular statics calculations by Buldum and Ciraci [71] support the hypothesis that the observed lock-in orientations are directly related to commensurate registry, and the particular set of commensurate orientations is determined by the CNT chirality (the wrapping orientation of the outer graphene sheet of the CNT). Thus the friction experiments provide a novel method for measuring the nanotube chirality. Large multiwall CNTs of different

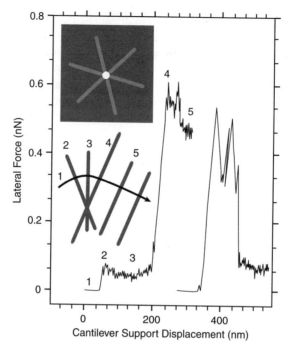

Figure 5. Lateral force trace as a CNT is rotated into (left trace) and out of (right trace) commensurate contact [68]. The lower inset shows a top view (schematically) of the process for the left trace. The AFM tip is moving along in contact with the graphite substrate (1), the CNT is contacted (2) and begins rotating in-plane (3), the commensurate state is reached and the lateral force rises dramatically (4) before rolling motion begins (5). The right trace begins with the tip on the substrate; the tip then contacts the CNT in the commensurate state, begins rolling and then pops out of commensurate contact, and begins rotating in plane with a corresponding drop in lateral force. The upper inset is a composite of three AFM images of a CNT in its three commensurate orientations. Note that the nanotube was not simply rotated about its center as the composite implies. The images were translated laterally in order to emphasize the 60-degree intervals. With permission from Ref. 68. *Phys. Rev. B* **62**, 10665 (2000).

diameters are expected to have different chiralities and should show different commensurate orientations [71].

Sliding between concentric multiwalled carbon nanotubes presents a simple geometry, which restricts interlayer motion to a single (axial) direction with a fixed interlayer orientation of stiff, smooth layers. Each layer in a concentric multiwalled carbon nanotube is indexed by two integers (n, m) that give the circumference in graphitic lattice coordinates. The difference in radii between successive layers frustrates the circumferential interlayer registry. The axial period of a single layer is $3\sqrt{n^2 + nm + m^2}/GCF(2n + m, n + 2m)$ bond lengths, where GCF indicates the greatest common factor. Two layers with $n_1/m_1 \neq n_2/m_2$ are typically axially incommensurate [72].

Potential energy calculations by Kolmogorov and Crespi [73] demonstrated that the (7,7)/(12,12) commensurate system shows a relatively large barrier to sliding which is defined as the maximal variation in total energy as the outer layer slides through the full unit cell of the inner layer. The barrier increases linearly with system size. In contrast, the incommensurate (14,0)/(16,10) nanotube has a potential corrugation from 0.0 eV to 0.3 eV which, excepting fluctuations, is independent of system size. Thus the energetic barrier to interlayer sliding in defect-free nanotubes containing thousands of atoms is comparable to that for a single unit cell of crystalline graphite. Efficient cancellation of registration-dependent interactions in incommensurate tubes can induce extremely small and nonextensive shear strengths. For finite (and therefore imperfectly) incommensurate systems, a barrier to sliding arises from fluctuations in the finite-sampled registry-averaged interlayer binding energy. The size of the fluctuations depends on the number and distribution of the sample points for the different local registries. The unusual nonextensive behavior observed here is in fact a generic behavior of (unlubricated) incommensurate systems.

Structural defects (Fig. 1c) can play a very important role in the response to interlayer sliding. The commensurate nanotubes with uniform-wrapping angle have the expected behavior under shear: Disruptions in translational symmetry due to, for example, on-wall defects should typically lower the corrugation below that expected for a perfect crystal. However, for incommensurate layers, the defects play the opposite role: The perfect system has an extremely low corrugation, and the introduction of a small concentration of defects increases the shear strength, through a contribution that scales linearly with length at constant defect density (Section III.C.5).

2. Curved, Nanoscale Contacts

Wenning and Müser [74] extended the considerations made above for athermal, flat walls to the interaction between a curved tip and a flat substrate by including Hertzian contact mechanics. Since the Hertzian contact area A increases proportionally to $L^{2/3}$, they concluded that for a dry, nonadhesive, *commensurate* tip–substrate system, F_s should scale linearly with L, since μ_s is independent of A. This has now been confirmed experimentally by Miura and Kamiya for MoS_2 flakes on MoS_2 surfaces [74a]. For a dry, nonadhesive, *disordered* tip pressed on a crystalline substrate, they obtained $F_s \propto L^{2/3}$, which was obtained by inserting $A \propto L^{2/3}$ into $F_s \propto L/\sqrt{A}$. The predictions were confirmed by molecular dynamics simulations, in which special care was taken to obtain the proper contact mechanics. The results of the friction–force curve are shown in Fig. 6.

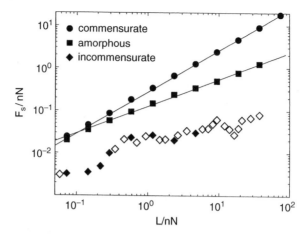

Figure 6. Static friction force F_s versus normal load L for different tip geometries. In all cases, the radius of curvature was $R_c = 70$ Å and contacts were nonadhesive. Straight lines are fits according to $F_s \propto L^\beta$ with the results $\beta = 0.97 \pm 0.005$ (commensurate) and $\beta = 0.63 \pm 0.01$ (amorphous). [With permission from Ref. 74. *Europhys. Lett.* **54**, 663 (2001); additional data provided by L. Wenning (open diamonds).]

The power law found in their simulations for the interaction of an amorphous tip with a crystalline substrate is similar to that observed in AFM experiments in argon atmosphere [75,76]. Also AFM studies of friction on layered materials [77] showed striking similarity with Fig. 6 in the wearless regime, although the experiments were conducted in atmosphere. In the latter case, there is even an order of magnitude agreement between the experimentally measured friction and simulated friction. The prefactors observed in another set of experiments [75,76] are, however, distinctly larger than those in the simulations. Moreover, the experimental results by Schwarz et al. [76] obtained with tips of different radii of curvature seem consistent with a picture in which the friction force is proportional to the area of contact and independent of the normal pressure. It thus remains an open question what friction mechanism is predominant in these experiments.

III. DRY, ELASTIC FRICTION

A. Properties of the Prandtl–Tomlinson Model

1. Definition of the Model and the Low Velocity Limit

The Prandtl–Tomlinson (PT) model [17] (usually only referred to as the Tomlinson model) is the simplest model that allows for elastic instability and hence for pinning between two incommensurate solids. In its original version (see Fig. 7), atoms in the upper wall are coupled harmonically to their ideal lattice sites. The substrate is assumed to be rigid with a fixed center of mass. Thus the surface atoms from the upper wall experience a force that is periodic in the substrate's lattice constant b. Dissipation due to phononic or electronic damping is assumed to be linear in the sliding velocity. The equation of motion for individual atoms then reads

$$m\ddot{x}_n + m\gamma\dot{x}_n = -k(x_n - x_{0,n}) + f_0 \sin(x_n/b') \qquad (19)$$

Here, x_n and $x_{0,n}$ denote, respectively, the current position and the ideal lattice site of atom n with mass m; γ is a damping rate, k is the coupling strength to the ideal lattice site, f_0 is the substrate's corrugation strength and $b' = b/2\pi$. Neighboring lattice sites are assumed to be separated by a distance a; hence $x_{0,n+1} - x_{0,n} = a$. Despite its seeming simplicity, the analytical treatment of the PT model or simple generalizations thereof can be rather complex. It is still used widely in order to interpret tribological properties of systems in relative lateral motion [13,78–82].

A paper by Prandtl [18] on the kinetic theory of solid bodies, which was published in 1928, one year prior to Tomlinson's paper [17], never achieved the recognition in the tribology community that it deserves. Prandtl's model is similar to the Tomlinson model and likewise focused on elastic hysteresis effects within the bulk. Nevertheless, Prandtl did emphasize the relevance of his work to dry friction between solid bodies. In particular, he formulated the condition that can be considered the Holy Grail of dry, elastic friction: "*If the elastic coupling of the mass points is chosen such that at every instance of time a fraction of the mass points possesses several stable equilibrium positions, then the system shows hysteresis. . . .*" In the context of friction, hysteresis translates to finite static friction or to a finite kinetic friction that does not vanish in the limit of small sliding velocities. Note that the dissipative term that is introduced ad hoc in Eq. (19) does vanish linearly with small velocities.

Prandtl's condition is satisfied if the elastic coupling to a lattice site is sufficiently small, namely $k < k^*$ with

$$k^* = f_0/b \qquad (20)$$

Suppose for a moment that $k \geq k^*$. In this case there is a unique static solution for every x_n in Eq. (19), irrespective of the value of $x_{0,n}$. When the upper solid is moved at a constant (small) velocity v_0 relative to the substrate, each atom is always close to its unique equilibrium position. This equilibrium position moves with a velocity that is of the order of v_0. Hence the friction force is of the order of $m\gamma v_0$, and consequently F_k vanishes linearly with v_0 as v_0 tends to zero. The situation becomes different for $k < k^*$. Atoms with more than one stable equilibrium position will now pop from one stable position to another one when

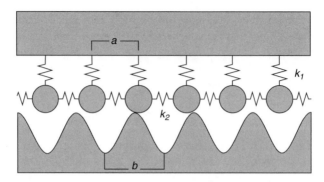

Figure 7. Schematic representation of the one-dimensional Frenkel–Kontorova–Tomlinson model. a and b denote the lattice constant of the upper solid and the substrate, respectively. The substrate is considered rigid, and its center of mass is kept fixed. In the slider, each atom is coupled with a spring of lateral stiffness k_1 to its ideal lattice site and with a spring of stiffness k_2 to its neighbor. The PT model is obtained for $k_2 = 0$, while the Frenkel–Kontorova model corresponds to $k_1 = 0$. We will drop the subscripts for these two cases since a single spring is relevant.

the slider is moved laterally. For small pulling velocities v_0, this occurs when there is no mechanically stable position at time $t + \delta t$ that is near an atom's position at time t.

Figure 8 illustrates the unstable motion for $k < k^*$ by showing the evolution of the time-dependent potential energy $V(x)$ associated with the conservative forces, namely,

$$V(x) = f_0 b' \cos(x/b') + \frac{1}{2} k(x - x_0)^2 \qquad (21)$$

with $\dot{x}_0 = v_0 > 0$. In the "popping" processes (indicated by the thick solid line in Fig. 8), the velocities $\dot{x}_n$ will exceed v_0 by orders of magnitudes for $v_0 \to 0$. At small v_0, the dynamics along most of the thick solid line is rather independent of the precise value of v_0 and the dissipated energy $\int dx \gamma m v$ has a well-defined positive limit $F_k(v_0 = 0)$. Hence in the absence of thermal fluctuations that have been disregarded so far, F_k remains finite even in the limit of infinitely small v_0. We refer to Fig. 9 in Ref. 82 for a discussion in similar terms. We note in passing that a detailed analysis of the motion along the thick solid line, which is emphasized in Fig. 8, lasts a time $t \propto v_0^{-1/3}$ resulting in $F_k(v_0) - F_k(0) \propto v_0^{2/3}$ [29,83].

The number of free parameters that define the athermal Prandtl–Tomlinson model can be reduced to three by a convenient choice of units. b' can be used to define the unit of the length scale, $f_0 b'$ is the unit of the energy scale, and

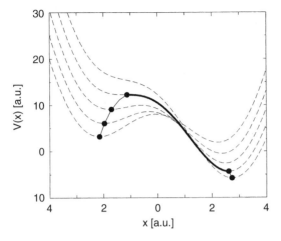

Figure 8. Schematic representation of the time evolution of the potential energy in the Prandtl–Tomlinson model (dashed lines); see Eq. (21). All curves are equidistant in time, separated by a time interval Δt. The circles denote mechanically stable positions, and the solid line indicates the motion of an overdamped point particle from left to right. Motion is linearly unstable on the thick portion of the line.

$\sqrt{f_0/mb'}$ is the unit of the frequency. Eq. (19) then reads in reduced units indicated by the tilde:

$$\ddot{\tilde{x}}_n + \tilde{\gamma}\dot{\tilde{x}}_n = -\tilde{k}(\tilde{x}_n - \tilde{x}_{0,n}) + \sin(\tilde{x}) \tag{22}$$

with $\tilde{x}_{0,n+1} - \tilde{x}_{0,n} = \tilde{a}$. Hence, the free parameters are $\tilde{k}$, $\tilde{\gamma}$, and $\tilde{a}$. It is usually assumed that the size or the length of the interface is infinitely large. Depending on the values of the reduced parameters, one distinguishes the following regimes: The (strongly) pinned regime where to $\tilde{k} < \tilde{k}^* = 1$ in which $\tilde{F}_k$ remains finite even at infinitesimally small $\tilde{v}_0$ and the unpinned (or weakly pinned) regime where $\tilde{k} \geq \tilde{k}^*$ for which $\tilde{F}_k(0) = 0$. One also distinguishes between underdamped motion $\tilde{\gamma} \ll 1$ and overdamped motion $\tilde{\gamma} \gg 1$. In the case of overdamped motion, inertial effects can be neglected. If $\tilde{a} = 1$ the two solids are called perfectly commensurate. In this case one finds $\tilde{F}_s = 1$ independent of the values for $\tilde{k}$ and $\tilde{\gamma}$. If $\tilde{a}$ corresponds to a rational number, the two solids are called commensurate and $\tilde{F}_s$ is finite even in the weakly pinned regime. However, the values for $\tilde{F}_s$ are typically rather small unless $\tilde{a} = 1$. The consideration of higher harmonics in Eq. (19) can result in larger F_s for other rational values of $\tilde{a}$; however, for most practical purposes, higher-order contributions are typically small and may be neglected [13]. If $\tilde{a}$ is irrational, the surfaces are called incommensurate.

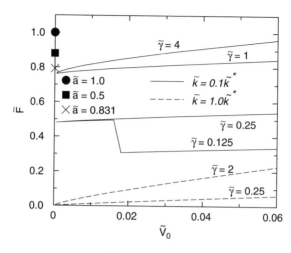

Figure 9. Average kinetic friction $\tilde{F}$ (independent of a) in the athermal Prandtl–Tomlinson model at low velocities $\tilde{v}$ for two different spring strengths $\tilde{k}$ and various damping coefficients $\tilde{\gamma}$. The symbols at $\tilde{v}_0 = 0$ indicate the static friction force for $\tilde{k} = 0.1\tilde{k}^*$. All units are reduced units.

Concerning the low-velocity kinetic friction in the pinned, overdamped regime, we want to stress that the kinetic friction $\tilde{F}_k(v_0)$ barely depends on the precise value of $\tilde{\gamma}$ in the limit $\tilde{v}_0 \to 0$ since the dissipated energy is proportional to $\int dx m \tilde{\gamma} \tilde{v}$ and the product of $\tilde{\gamma}$ and $\tilde{v}$ is relatively independent of $\tilde{\gamma}$ during most of the pop; that is, small values of $\tilde{\gamma}$ invoke large $\tilde{v}$ and vice versa. As argued by Fisher [29], the precise value of m in Eq. (19) remains irrelevant below a certain threshold value for the energy dissipation during sliding. The low-velocity limit of the friction force is shown in Fig. 9. One can see that the low-velocity $\tilde{F}_k$ can be distinctly reduced when the motion becomes underdamped. Atoms can pick up momentum at small γ that can carry them past metastable configurations.

2. The Static Friction Force

So far we have only investigated the average value of $\tilde{F}_k$ in the PT model for small $\tilde{v}_0$ under the assumption that the slider's center-of-mass moves smoothly with velocity $\tilde{v}_0$. Smooth sliding can be achieved by driving the system in an appropriate way, which will be discussed later. For the calculation of the smooth sliding $\langle F_k \rangle$, it is indeed sufficient to investigate the coupling of a *single* atom to its ideal lattice site and to the substrate, since the motion of all atoms relative to their preferred positions is identical up to temporal shifts. $\langle \tilde{F}_k(\tilde{v}_0 \to 0) \rangle$ therefore only depends on $\tilde{k}$ and $\tilde{\gamma}$ but not on $\tilde{a}$. The discussion of the Prandtl–Tomlinson model [82,84] is often reduced to this particular calculation of $\tilde{F}_k$ which is, as argued above, a one-particle property. For the *instantaneous* values of $\tilde{F}_k$ and also for $\tilde{F}_s$, however, it does matter how the contributions from different atoms add up (interfere) to yield a net force. This, in turn, depends on the spacing between neighboring atoms. The summation to the net force is relatively straightforward for the two limiting cases of perfectly commensurate walls and for incommensurate walls.

If $\tilde{a} = 1$, every atom in the slider has the same velocity at every instant of time, once steady state (not necessarily smooth sliding) has been reached. Hence the problem is reduced to the motion of a single particle, for which one obtains $\tilde{F}_s = 1$. This provides an upper bound of $\tilde{F}_s$ for arbitrary $\tilde{a}$. If the walls are incommensurate or disordered, one can again make use of the argument that the motion of all atoms relative to their preferred positions is the same up to temporal shifts once steady state has been reached. Owing to the incommensurability, the distribution of these temporal shifts with respect to a reference trajectory cannot change with time in the thermodynamic limit, and the instantaneous value of F_k is identical to $\langle \tilde{F}_k \rangle$ at all times. This gives a lower bound for $\tilde{F}_s$ for arbitrary $\tilde{a}$. The static friction for arbitrary commensurability and/or finite systems lies in between the upper and the lower bound.

As argued by Fisher, pinned and sliding solutions can only coexist in some range of the externally applied force if the inertial term exceeds a certain threshold value [29]. This can lead to stick-slip motion as described in Section VI.A. For sufficiently small inertial terms, Middleton [85] has shown for a wide class of models, which includes the PT model as a special case, that the transition between pinned and sliding states is nonhysteretic and that there is a unique average value of F_k which does depend on v_0 but not on the initial microstate. The instantaneous value of F_k can nevertheless fluctuate, and the maximum of F_k can be used as a lower bound for the static friction force F_s. The measured values of F_s can also fluctuate, because unlike F_k they may depend on the initial microstate of the system [85].

3. Thermal Effects

Temperature can be incorporated into the PT model by including random forces in the equation of motion, Eq. (22), similar to those introduced in the rigid-wall model, Eqs. (12) and (13). The additional complication in the PT model is the elastic coupling of the central particle to a (moving) equilibrium site, which makes it difficult to apply the rigorous concepts developed for the Brownian motion of a particle in a periodic potential to the PT model.

The first discussion of the effect of thermal fluctuations on friction forces in the Prandtl–Tomlinson model was given by Prandtl in 1928 [18]. He considered a mass point attached to a single spring in a situation where the spring k_1 in Fig. 7 was compliant enough to exhibit elastic instabilities, but yet sufficiently strong to allow at most two mechanically stable positions; see also Fig. 8, in which this scenario is shown. Prandtl argued that at finite temperatures, the atom would not follow the thick line shown in Fig. 8, because thermal fluctuations would assist the atom in hopping from one minimum to another one. He then wrote down what we now call a master equation or a transition state theory for the conversion between the mechanically stable states and concluded that the low-temperature, low-velocity *kinetic* friction force should have a logarithmic dependence on sliding velocity:

$$F(v) = F(v_{\mathrm{ref}}) + \chi \ln v/v_{\mathrm{ref}} \qquad (23)$$

where v_{ref} is a reference velocity and χ a constant. This equation is applicable for velocities small enough such that the (average) kinetic energy is distinctly smaller than the energy barriers, but still large enough in order for the system to be out of thermodynamic equilibrium.

The same equation can be derived within a simple Kramers picture [55,80,86] for the escape from a well (locked state), assuming that the pulling force produces a small constant potential bias that reduces a height of a potential barrier. The progressive increase of the force results in a corresponding increase of the escape rate that leads to a creep motion of the atom. However, this behavior is different from what occurs when the atom (or an AFM tip) is driven across the surface and the potential bias is continuously ramped up as the support is moved [87,88]. The consequences of this effect will be discussed in more detail in Section III.B.2.

Corrections similar to those in Eq. (23) apply for the static friction, which depends on the rate of loading $\dot{F}'$:

$$F_s = F_s(\dot{F}'_{\mathrm{ref}}) + \chi \ln(\dot{F}'/\dot{F}'_{\mathrm{ref}}) \qquad (24)$$

This has been discussed extensively by Evans and Ritchie in the context of the dynamic adhesive strength of molecular bonds [89]. They also consider regimes in which F' is extremely large and extremely small (approaching thermal equilibrium).

B. Applications of the Prandtl–Tomlinson Model

1. General Remarks

The PT model is frequently used as a minimalistic approximation for more complex models. For instance, it is the mean-field version of the Frenkel–Kontorova (FK) model as stressed by D. S. Fisher [29,83] in the context of the motion of charge-density waves. The (mean-field) description of driven, coupled Josephson junctions is also mathematically equivalent to the PT model. This equivalence has been exploited by Baumberger and Caroli for a model that, however, was termed the lumped junction model [84] and that attempts to relate the parameters of the PT model to properties of the system embedded between two sliding surfaces. This analogy will be discussed in further detail in Section VI.B.

The PT model [90] and a generalized PT model [91] were also used to interpret molecular dynamics (MD) simulations [90] of a sliding, commensurate interface between ordered, organic monolayers. Glosli and McClelland [90] identified instabilities ("plucking motion") that lead to a constant low-velocity kinetic friction between the two walls. Harrison and coworkers [91a,91b] have identified a variety of instabilities that lead to finite kinetic friction between diamond surfaces with different terminations, including a novel turnstile like rotation of methyl groups. The commensurability in these simulations certainly facilitated the description in terms of a model that is as simple as the PT model, because the relevant degrees of freedom all acted in a rather coherent way. However, in order to understand the experimentally relevant instabilities, it would be necessary to also consider incommensurate alignments of the two surfaces, which has not yet been reported to the best of our knowledge. In the following discussion, we focus on the interpretation of AFM experiments.

2. Interpretation of Friction on the Nanometer Scale

Experimentally, the static friction force is determined as a maximal force needed to initiate sliding motion. The question arises whether the static friction obtained in this way is a unique (inherent) property of the system, or whether it depends also on the conditions of the measurement. Section II.C.2 only discussed the maximum possible shear force between a tip and a substrate, but the effect of thermal fluctuations and the effect of the driving device—for instance, the stiffness of the driving device or the tip itself—was not included.

Recent experimental studies on atomic force microscopy [80,92] rederived the logarithmic velocity dependence of kinetic friction and argued that, due to thermal fluctuations, the maximal value of the friction force (the static friction) has a logarithmic dependence on loading rate as described in Eq. (24). That equation has also been suggested for interpretation of single molecular mechanical probe experiments [86,93–96] in which the molecule under study is connected to a driven "spring," an AFM cantilever or a laser trap. Investigated processes include specific binding of ligand receptor pairs [93], protein unfolding [94,95], and mechanical properties of single polymer molecules such as DNA [96]. In all these experiments, one probes forces along a reaction coordinate and their maximal values are ascribed to rupture or unfolding forces.

In the following, we focus on the application of the Prandtl–Tomlinson model to the interpretation of AFM experiments. As mentioned in Chapter III.A.3, the potential bias is continuously ramped up as the support of an AFM tip is moved. This results in a different friction–velocity relationship than Eq. (23). Thermal fluctuations contribute to the response of the tip in two opposite directions: (a) They help in getting out of locked states at the minima of the total potential $V(x, t)$

$$V(x, t) = V_0 \cos\left(\frac{2\pi}{b}x\right) + \frac{k}{2}(x - vt)^2 \tag{25}$$

(see also dashed lines in Fig. 8), and (b) they can make a sliding tip return back to a locked state. The fluctuation-assisted motion is expected to cause a decrease in the frictional force at low velocities (stick–slip regime), when the activation dominates over the potential. At high velocities the second effect becomes relevant and causes an enhancement of the friction in the sliding regime.

An analytical expression for the velocity dependence of the static friction that includes thermal activation of the tip out of the locked state has been given recently by Sang et al. [87] and Dudko et al. [88]. In the absence of noise the driven tip leaves a locked state [a minimum of the total potential $V(x,t)$] only when the potential barrier vanishes—that is, at the instability point where $d^2V/dx^2 = 0$. At this point the spring force reaches the maximum value of $2\pi V_0/b$. In the presence of noise the transition to sliding occurs earlier, and the probability $P_0(t)$ to find the tip in the locked state is given by the following kinetic equation:

$$\frac{dP_o(t)}{dt} = -\omega_0 P_0(t) \exp(-E(t)/k_B T) \tag{26}$$

where $E(t)$ is the energy distance between a neighboring minimum and maximum of the potential $V(x, t)$ and ω_0 is a characteristic frequency of the order of $(2\pi/b)\sqrt{V_0/m}$. In a locked state, the time dependence of the tip position is given by the equilibrium condition $dV(x,t)/dx = 0$. The value of the activation energy $E(t)$ changes between $2V_0$, at zero driving force, and zero at the instability point. For a weak spring near the instability point, $E(t)$ can be written in the form

$$E(t) = V_0 \left(2V_0 - \frac{kvtb}{\pi V_0}\right)^{3/2} \tag{27}$$

This leads to the following equation for the observed maximal force:

$$F_{\max} = \text{const} - \frac{\pi}{b} V_0^{1/3} \left[k_B T \, \ln\left(v\frac{Kb}{2\pi V_0 \omega_0}\right)\right]^{2/3} \tag{28}$$

This result describes well the numerical calculations according using a modified Tomlinson model [88] and, moreover, is in a good agreement with experiment

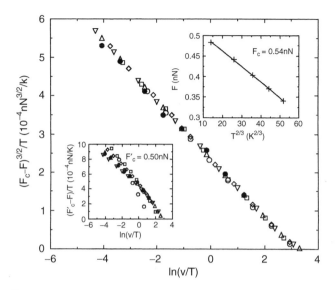

Figure 10. Test of scaling prediction made implicitly in Eq. (28). Experimental data for different temperatures between 53 and 373 K were taken from Ref. 80. The upper inset shows the maximum friction force as a function of temperature. The lower inset shows significantly worse scaling for linear creep, using $\ln(v/T)$. The units of velocity v are nanometers per second, and temperature is in degrees kelvin. With permission from Ref. 87. *Phys. Rev. Lett.* **87**, 174301 (2001).

[87]. This is shown in Fig. 10, where the predicted velocity and temperature dependence agrees in an impressive way with high-precision experimental results by Gnecco et al. [80].

Equation (28) includes explicitly the dependence of the force on the parameters of the microscopic underlying potential and the macroscopic spring constant k. Thus the measured static friction depends not only on the properties of the system under study but also on the driving velocity and mechanical setup (k). For weak spring constants the $(\ln v)^{3/2}$ behavior holds over a wider range of temperatures and driving velocities. Equation (28) differs essentially from the phenomenological estimations [18,53,80,86] that predict a logarithmic variation of $F(v)$ as suggested in Eq. (23).

The leading correction to zero-velocity, zero-temperature kinetic friction of the form $(T \ln v)^{2/3}$ as described in Eq. (28) apparently also applies to more complicated elastic manifolds. Charitat and Joanny [97] investigated a polymer that was dragged over a surface containing sparsely distributed, trapping sites. They analyzed the competition between the soft elastic, intramolecular interactions, thermal noise, and the tendency of some monomers in the chain to remain pinned to a trapping site. Charitat and Joanny found Coulomb friction with a leading correction term equivalent to Eq. (28).

C. Properties of the Frenkel–Kontorova Model

1. Definition of the Model and Concepts

Another simple model that includes elasticity is the Frenkel–Kontorova (FK) model [98], which plays an important role in understanding various aspects of solid friction. The simplest form of the FK model [98] consists of a one-dimensional chain of N harmonically coupled atoms that interact with a periodic substrate potential (see Fig. 7). The potential energy is

$$V = \sum_{n=1}^{N} \frac{1}{2} k(x_{n+1} - x_n - a)^2 + V_0 \cos(x_n/b') \tag{29}$$

where a is the separation between two neighboring atoms in the free chain, k is the strength of the harmonic coupling between two atoms, $b = 2\pi b'$ is the substrate's periodicity, and V_0 is the interaction strength between atoms in the chain and the substrate.

Frenkel and Kontorova were not the first ones to use the model that is now associated with their names. However, unlike Dehlinger, who suggested the model [99], they succeeded in solving some aspects of the continuum approximation. Like the PT model, the FK model was first used to describe dislocations in crystals. Many of the recent applications are concerned with the motion of an elastic object over (or in) ordered [100] or disordered [101] structures. The FK model and generalizations thereof are also increasingly used to understand the friction between two solid bodies.

Unlike the PT model, the FK model allows for long-range elastic deformations (LRED), which are believed to play an important role in various aspects of solid friction. Also, new types of static and dynamic excitations such as kinks and solitons are possible in the FK model. As a consequence of the LRED, the lattice constant of the slider is allowed to relax in response to the slider's interaction with the substrate. The resulting average distance $\tilde{a}$ (sometimes imposed through periodic boundary conditions) may differ from the free chain's period a. The ratio $\tilde{\Omega} = \tilde{a}/b$ is sometimes referred to as the winding number. Similar to the PT model, the crucial parameters of the FK model are the ratios $\Omega = a/b$ ($\tilde{\Omega} = \tilde{a}/b$) and $\tilde{k} = kb'^2/V_0$. One novelty of the FK model with respect to the PT model is that the low-velocity limit of the (average) kinetic friction depends sensitively on the ratio $\Omega = a/b$. Also, the dependence of the static friction on the precise value of Ω is much stronger in the FK model than in the PT model.

For any given irrational value of Ω, there is an Ω-dependent (dimensionless) threshold spring constant $\tilde{k}_c$. Static friction vanishes for sufficiently strong springs $\tilde{k} > \tilde{k}_c$. Below the critical value $\tilde{k}_c$, metastability occurs and the static friction is finite. The transition from finite F_s to zero F_s is accompanied by a phase transition from a commensurate structure to an incommensurate structure [102,103]. For most values of Ω and fixed, nonzero k and V_0, the winding number $\tilde{\Omega}$ is a rational number near Ω. Two neighboring intervals for which $\tilde{\Omega}(\Omega)$ is a rational constant are separated by a point for which $\tilde{\Omega}$ and Ω are both irrational. The transition from commensurate to incommensurate is classified as a second-order transition [102,103], and consequently many properties are power laws as a function of $\kappa = \tilde{k} - \tilde{k}_c$; that is, for small κ we have

$$F_s \propto F_0 \kappa^{\psi'} \tag{30}$$

Critical exponents can also be introduced for other quantities such as the phonon gap and elastic coherence length. The various exponents can be explained in terms of renormalization theories (see, for instance, Ref. 104). We will summarize some properties of the (one-dimensional) FK model qualitatively before focusing on more quantitative studies. Reference 105 gives a pedagogical introduction into the FK model, and Ref. 100 provides an excellent overview of the rich dynamics of the FK model.

2. Continuum Approximation and Beyond

If the winding number is close to unity and k is large, then the displacement of an atom with respect to an ideal substrate site will be a slowly varying function of the index n. This implies that if one writes the position of atom n as

$$x_n/b = 2\pi n + \phi_n \tag{31}$$

then one may treat the index n in ϕ_n as a continuous variable [106]. This allows one to transform the discrete equation of motion:

$$m\ddot{x}_n = k(x_{n+1} + x_{n-1} - 2x_n) + \frac{V_0}{b'}\sin(x_n/b') \tag{32}$$

into the sine-Gordon equation (SGE)

$$m'\partial_t^2 \phi_{tt} = k'\partial_n^2 \phi_{nn} + \sin\phi \tag{33}$$

with $m' = mb'^2/V_0$ and $k' = kb'^2/V_0$. For many purposes, one also considers damped dynamics, in which case the inertial term on the left-hand side of Eq. (33) is replaced with the first derivative of time, namely with $\gamma\partial_t\phi$ where γ is a damping coefficient.

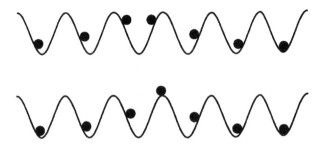

Figure 11. Stationary configurations of particles in the FK model representing a single kink. *Top*: Ground state. *Bottom*: Unstable configuration, corresponding to a saddle point. With permission from Ref. 100. *Phys. Rep.* **306**, 1 (1998).

In the present discussion, we assume periodic boundary conditions and moreover a situation where there is one more (or one less) atom in the chain than minima in the substrate potential. In the SGE, this translates into a boundary condition where φ makes a phase shift of ±2π as one moves through the periodically repeated sequence. For the energetically most stable solutions, atoms will mostly try to sit at the bottom of the potential. The required phase shift of ±2π will occur in a relatively narrow region. Such phase shifts are called walls, discommensurations, or kinks. A representation of a kink is given in Fig. 11. The motion of isolated (and also interacting) kinks under an externally applied force or field is, of course, crucial to friction in the FK model.

In the continuum model, the motion of a kink occurs without energy barriers if $\tilde{\Omega}$ is irrational. For rational $\tilde{\Omega}$ the distribution of ϕ_n in the ground state must be discrete, which makes it impossible to transform the phases continuously without extra energy. The dynamic solutions of Eq. (33) exploit the isomorphism with nonlinear relativistic wave equations [107,108] and a moving kink (soliton) can be interpreted as an elementary excitation with energy $E_k(v)$

$$E_k = m_k c^2 + \frac{1}{2} m_k v^2 + \mathcal{O}(m_k v^4 / c^2) \tag{34}$$

where $c = b\sqrt{k/m}$ is the sound velocity in the free chain. Furthermore, the effective kink mass m_k is given by

$$m_k = 8m\sqrt{V_0/kb^2} \tag{35}$$

where we have expressed all quantities in the dimensions of the discrete description.

While the system is invariant to any translation of the kink along the chain in the continuum limit (hence the kink energy E_k resembles the energy of a free particle), discrete models only need to be invariant with respect to a translation by a lattice constant b. If the chain is commensurate with the substrate and/or k' is sufficiently small, then a suitably defined kink coordinate X [109] will experience a potential periodic in b; thus

$$V_{PN} = \frac{1}{2} E_{PN} \cos(2\pi X/b) \qquad (36)$$

E_{PN} is the so-called Peierls–Nabarro barrier [110,111], which can be estimated analytically from extensions of the continuum approximation (see Ref. 100 for an overview). It is important to point out that in the thermodynamic limit of such a quasi-continuum treatment, there is a *finite* energy barrier for the initiation of kink motion (and ultimately sliding). This is different from what one finds in the PT model in the strong pinning limit, namely a well-defined shear pressure and consequently an energy barrier that increases linearly with system size.

3.   Thermal and Quantum Effects

Owing to the discreteness of the FK chain, even the small-velocity motion of a kink does not obey Newtonian dynamics in the sense that the equation of motion $m_k \ddot{X} = E_{PN} \sin(X/b')/2b'$ is satisfied. The kink can exchange energy with other internal degrees of freedom of the chain, which can effectively be described as a coupling of the kink to an external heat bath [112]. This suggests that on long time scales in thermal equilibrium, the motion of a kink (which is necessary for mass transport) is equivalent to the thermal motion of an isolated particle on a substrate (see Section III.A.3). Stressing this analogy implies that a kink in the one-dimensional FK model has finite mobility at finite temperature, provided that kinks are present. Hence the true static friction can be zero even if slider and substrate are perfectly commensurate. However, if the dimensionality of the interface is sufficiently large can the energy barrier for the creation of a kink diverge in the thermodynamic limit, as can be the case in two-dimensional FK models [113].

The discussion above anticipates the presence of kinks. For an open, one-dimensional chain that is not subject to periodic boundary conditions, kinks will always be present even for a slider commensurate with the substrate. The kinks can be thermally induced, because the energy to create a kink is finite. Although their number becomes exponentially small with decreasing temperature T, it remains finite at any finite value of T. Thus the concept of broken analyticity (no atom sits on top of an energy barrier) is strictly applicable only at absolute zero temperature. Bak and Fukuyama [114] found that quantum fluctuations also automatically destroy broken analyticity even at absolute zero. This finding has been challenged recently by Hu et al. [115] and remains an issue to be settled. For instance, a semiclassical treatment of charge density waves by Miyake and Matsukawa [116] suggests that quantum creep should be present at small temperatures, although it may be very small.

Another interesting quantum effect was suggested by Popov [117] and concerns the phononic friction between two incommensurate, weakly interacting solids rather than the friction due to elastic instabilities. He argues that an efficient energy transfer between the solids ceases to exist at small T, owing to the quantum mechanical nature of phonons; that is, the umklapp processes are frozen out at small T. The phononic drag force between the solids would then decrease with a power T^4 for temperatures small compared to the Debye temperature. Popov, however, did not include damping due to energy transfer between internal vibrations.

4.   Generalized Frenkel–Kontorova Models

Various generalizations of the Frenkel–Kontorova model have been proposed in the literature. An important generalization of the FK model with dramatic consequences for the resulting physical properties (critical exponents, scaling of friction force with system size, etc.) is to change the dimensionality of the interface and/or the dimensionality of the sliding elastic medium. Increasing the dimensionality of the sliding object makes the interactions between deformations at the interface effectively more long-ranged; and above a critical dimension, the critical behavior of the FK model and that of the PT model become identical. While many studies focus on the tribology of one-dimensional bodies and one-dimensional interfaces, the most relevant case for friction between solid bodies is of course a two-dimensional interface between two three-dimensional objects (Section III.C.5).

In many cases, however, only the interactions are altered with respect to Eq. (29). One may classify the alterations into three rubrics and combinations of those. (i) Changing the interactions of the atoms within the sliding chain: One possibility is to allow for anharmonicity in the next-neighbor interactions [118], another one is to include direct elastic long-range coupling [119]. These alterations can affect the universal behavior of the FK model. One special case of combining local elasticity of the slider as reflected in the FK model and mean-field elasticity such as in the PT model will be discussed in more detail in Section III.C.6. (ii) Changing the substrate slider potential: As long as the substrate is still crystalline, this can be done by allowing higher harmonics in the interaction potential. Another possibility, which facilitates analytical calculations, is to assume pairwise interactions modeled via Gaussians between slider and substrate atoms. Another possibility is the inclusion of disorder, which is important enough to be discussed further in Section III.C.5. A new type of behavior is also observed if the substrate is quasi-periodic as discussed by Vanossi et al. [120]. (iii) Allowing for elastic deformations within the substrate [121].

5. Role of Dimensionality and Disorder

Most real solids, and hence surfaces, are not perfectly periodic but contain a certain degree of disorder. This can change the tribological properties of a system qualitatively. There are many ways to introduce the effects of disorder into the FK model. One possibility is to assume that the substrate potential contains random elements, while another possibility is to assume that the bond lengths or the spring stiffnesses fluctuate around a mean value.

To discuss the effects of disorder and dimensionality qualitatively, let us consider a d_{int}-dimensional substrate surface in interaction with a d_{obj}-dimensional elastic solid, which we can envision as a generalization of the FK chain to higher dimensions. In such a situation, there will be a competition between the random substrate–slider interactions and the elastic coupling within the solids. An important question to ask is how the interactions change when we change the scale of the system; for example, how strong are the random and the elastic interactions on a scale $2L$ if we know their respective strengths on a scale L? Here L gives the linear dimension of the solids parallel and normal to the interface.

As discussed in Section II, the random forces between substrate and slider will scale with the square root of the interface's size; hence the random forces scale with $L^{d_{int}/2}$. The elastic forces, on the other hand, scale[1] with $L^{d_{obj}-2}$. In the thermodynamic limit $L \to \infty$, the effect of disorder will always dominate the elastic interactions or vice versa, *unless*

$$L^{d_{int}/2} = L^{d_{obj}-2} \tag{37}$$

If $\lim_{L\to\infty} L^{d_{int}/2}/L^{d_{obj}-2} \gg 1$, the random interactions will dominate; hence pinning via elastic instabilities cannot be avoided. This disorder-induced elastic pinning is similar to pinning of compliant, ordered systems as discussed above within the Prandtl–Tomlinson and the Frenkel–Kontorova model. If $\lim_{L\to\infty} L^{d_{int}/2}/L^{d_{obj}-2} \ll 1$, the long-range elastic forces dominate the random forces. The slider's motion can only be opposed by elastic instabilities if the elastic coupling is sufficiently weak at finite L in order to make local pinning possible, again akin to the case $\lambda > 1$ in the Prandtl–Tomlinson model.

---

[1]A linear chain can be more easily compressed if we replace one spring by two springs coupled in series. In two dimensions, springs are not only coupled in series but also in parallel, so that the elastic coupling remains invariant to a "block transformation." Each additional dimension strengthens the effect of "parallel" coupling.

The so-called *marginal* situation, in which both contributions scale with the same exponent, $d_{int}/2 = d_{obj} - 2$, occurs in the important case of three-dimensional solid bodies with two-dimensional surfaces. This case will be discussed in Section VII.A in the context of multiasperity contacts. In the marginal situation, the friction force can stay finite; however, one may expect the friction force per unit area and hence the friction coefficient to be exponentially small. For surfaces with roughness exponent zero and typical spacing of a few micrometers between individual microcontacts, Sokoloff predicted static friction coefficients of the order of 10^{-5}, provided that no additional local instabilities occurred in individual microcontacts [122]. The *marginal* dimension in the case of $d_{obj} = d_{int}$ (adsorbed monoatomic layers, charge density waves, etc.) is $d_{mar} = 4$ [29].

Above the marginal dimension, the system will be pinned if the systems are sufficiently compliant or "superlubric" if the systems are sufficiently stiff. Below the marginal dimension the disorder always wins over elasticity and a finite threshold force per unit area is obtained, provided that the force acts homogeneously on the particles. We note in passing that there is surprisingly little discussion of systems that are driven via a "hook," which might reflect some tribological situations better than constant force driving.

6. Frenkel–Kontorova–Tomlinson Model

As discussed above, there are many generalization of the PT model and the FK model One particularly interesting case is where the interactions between the atoms within the slider are altered, in an attempt to include both the possibility of long-range elastic deformations (LRED) and the counteracting long-range elastic interactions (LREI). In the PT model, LRED are suppressed and LREI are overestimated. In the FK model, LREI are neglected and LRED are overestimated if the slider is a three-dimensional solid. The Frenkel–Kontorova–Tomlinson (FKT) model [78,123] introduced by Weiss and Elmer is a one-dimensional model that includes both LREI and LRED. It mimics the motion of a soft solid on a hard plate and is shown in Fig. 7. Interactions between slider atom and substrate are described by the leading term of the potential in Eq. (8). The frictional properties predicted by this model strongly depend on the strength of the interaction between the sliding surfaces (amplitude of the periodic potential V_0) and the commensurability of the surface lattices. There are three threshold values of V_0 at which the behavior changes qualitatively.

Below the first threshold, $V_0^{(1)}$, the static friction is zero, whereas above $V_0^{(1)}$ it increases like a power law and approaches the value $F_s^{max} = 2\pi V_0 N/b$ in the asymptotic limit. In the commensurate case, where the ratio a/b of the lattice constants of both sliding surfaces is rational, $V_0^{(1)}$ is zero, because then the two surfaces automatically interlock geometrically. The exponent of the power law is given by the denominator of the fraction a/b. In the incommensurate case, $V_0^{(1)}$

is finite and increases with the ratio $\kappa = k_1/k_2$ of the stiffnesses of the particle–upper-plate and particle–particle springs. When κ equals the golden mean, the exponent is roughly 2. This behavior of F_s is similar to what is found for the depinning force of the ground state of the FK model [103], except that in the latter case the value of the exponent is close to 3.

For V_0 below the second threshold denoted by $V_0^{(2)}$, the kinetic friction is zero in the limit of quasi-static sliding—that is, for sliding velocity $v \to 0$. That is, for $V_0 < V_0^{(2)}$ the kinetic friction behaves like a viscous drag. For $V_0 > V_0^{(2)}$ the dynamics is determined by the Prandtl–Tomlinson-like mechanism of elastic instability, which leads to a finite kinetic friction. The threshold amplitude $V_0^{(2)}$ increases with κ and is always larger than zero. Therefore, in the commensurate case, vanishing kinetic friction does not imply vanishing static friction just like in the PT model. The FKT model for $V_0^{(1)} < V_0 < V_0^{(2)}$ is an example of a dry-friction system that behaves dynamically like a viscous fluid under shear even though the static friction is not zero.

The third threshold amplitude, $V_0^{(3)}$, is important for the precise meaning of the static friction F_s. Below this threshold the ground state of the undriven FKT model is the only mechanically stable state. For $V_0 > V_0^{(3)}$ additional metastable states appear. The first metastable state is not very different from the ground state. It can be described as a ground state plus a defect, which separates two equally sized domains. The motion of this defect allows sliding to occur.

Since the static friction is the force that is necessary to start sliding, it depends on the state of the system. Therefore static friction is not uniquely defined for $V_0 > V_0^{(3)}$. However, one can introduce the force that gives an upper limit for the actual static friction. It is defined as the smallest force above which there exists no stable state.

The relation between the thresholds introduced above depends on the commensurability of the surface lattices of the upper and lower bodies. In the incommensurate case all three thresholds are identical, $0 < V_0^{(1)} = V_0^{(2)} = V_0^{(3)}$, while for the commensurate case there is the following relation between them, $0 = V_0^{(1)} < V_0^{(2)} < V_0^{(3)}$. Although the FKT model is a very simplified model even for atomically flat surfaces, it is often believed to mimic some of the qualitative behavior of dry friction.

D. Applications of the Frenkel–Kontorova Model

1. Incommensurate Crystals

Hirano and Shinjo [22] calculated the condition for static friction between high-symmetry surfaces of fcc and bcc metals. Many of their results are consistent with the conclusion that the elastic interactions within the bulk dominate the interactions between the crystals in contact. This would correspond to $k > k^*$ in the PT model [see Eq. (20)] or to $V_0 < V_0^{(1)}$ in the FKT model (discussed in Section III.C.6). Hirano and Shinjo [22] tested contacts between various surface orientations of the same metal [i.e., (111) and (100) or (110) and (111)]. In all cases the interactions were too weak to produce static friction.

Lancon et al. [124] came to similar conclusions for grain boundaries in gold. Their simulations suggested zero static friction along the incommensurate direction, whereas finite barriers resisted sliding parallel to the commensurate axis. In later work, Lancon [125] varied the normal pressure p and hence the relative strength between interfacial and intrabulk interactions. At $p \approx 4$ GPa, he found an elastic instability and finite resistance to sliding also parallel to the incommensurate direction. While this is certainly an exciting result that shows that elastic instabilities are possible, it would nevertheless be difficult to prevent plastic flow or other nonelastic instabilities at the circumference of the contact if no periodic boundary conditions were present.

There have been relatively few experimental tests of the Frenkel–Kontorova model because of the difficulty in making sufficiently flat surfaces and of removing chemical contamination from surfaces. In one of the earliest experiments, Hirano et al. [126] examined the orientational dependence of the friction between atomically flat mica surfaces. They found as much as an order of magnitude decrease in friction when the mica was rotated to become incommensurate. Although this experiment was done in vacuum, the residual friction in the incommensurate case may have been due to surface contamination. When the surfaces were contaminated by exposure to air, there was no significant variation in friction with the orientation of the surfaces.

Similar results were obtained in studies of surfaces lubricated with MoS_2 [127]. The MoS_2 forms plate-like crystals that slide over each other within the contact. They appear to be randomly oriented and thus incommensurate. In ultrahigh vacuum the measured friction coefficient was always lower than 0.002 and in many cases dropped below the experimental noise. The authors infer that this low friction is due to sliding between incommensurate plates. When the films are exposed to air, the friction coefficient rises to 0.01–0.05. This suggests, as in the mica experiments, that a thin film between incommensurate surfaces produces a nonzero static friction. However, the rise in friction may also be related to chemical changes in the MoS_2 on exposure to oxygen [128].

In later work, Hirano et al. [129] studied friction as a function of the angle between a flat crystalline AFM tip and a crystalline substrate. Static friction was observed when the crystals were aligned, and it decreased below the threshold of detection when the crystals were rotated out of alignment. Unfortunately, the crystals were orders of magnitude smaller than typical contacts. As discussed in Section IV.A, edges can lead to significant friction in small contacts.

2. Adsorbed Layers

The most extensive studies of incommensurate systems have used a quartz-crystal microbalance (QCM) to measure the friction between adsorbed layers and crystalline substrates. The QCM is usually used to determine the mass of an adsorbed layer from the decrease in resonance frequency of the quartz crystal. Krim and collaborators [2,4,130] have shown that the increase in the width of the resonance can be used to determine the amount of dissipation due to sliding. The crystal is cut so that the applied voltage drives a shear mode, and the dissipation is studied as a function of drive amplitude.

Measurements have been made on a wide variety of molecules adsorbed on Au, Ag, or Pb surfaces [3,4,131,132]. The phase of the adsorbed layer changes from fluid to crystal as the density is increased. As expected, motion of fluid layers produces viscous dissipation; that is, the friction vanishes linearly with the sliding velocity. The only surprise is that the ratio between friction and velocity, called the *drag coefficient*, is orders of magnitude smaller than would be implied by the conventional no-slip boundary condition. When the layer enters an incommensurate phase, the friction retains the viscous form. Not only does the incommensurate crystal slide without measurable static friction, the drag coefficient is as much as an order of magnitude smaller than for the liquid phase!

The viscous drag observed for these adsorbed layers represents a linear response and can thus be described by the fluctuation–dissipation theorem. This has been used to measure the drag coefficient in MD simulations of Xe on Ag [133]. A simple analytic expression for the drag coefficient can be derived in terms of the phonon damping rate and static response of the film, and it agrees well with MD simulations of fluid and incommensurate layers [134,135]. The decrease in drag on going from liquid to incommensurate states is found to reflect the decreased deformability of the solid film, and simulations quantitatively reproduce the changes observed in experiments on Kr on Au [134,135] and Xe on Ag [133] (Fig. 12).

Persson [48,136] has noted that energy can also be dissipated through coupling to low-lying electronic excitations of the metallic substrate. The relative importance of this mechanism is expected to increase with decreasing temperature (since the phonon lifetime increases [132]) and should be larger for molecules with a quadrupole moment like N_2 [137]. Studies of N_2 on Pb [131,138] show a sharp drop in dissipation at the onset of superconductivity which could be attributed to vanishing of the electronic loss mechanism [139,140]. Another group saw neither a drop nor any evidence of sliding [141], perhaps due to increased disorder on the surface [138,142]. Indeed, one might expect from the arguments of the preceding section that sufficiently large islands of adsorbate would always be pinned by disorder and exhibit a static friction. Simulations suggest that a common form of defect, step edges, should

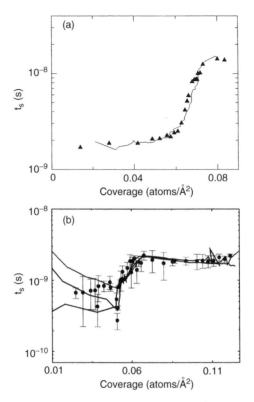

Figure 12. Inverse drag coefficients or slip times t_s versus coverage for (a) Kr on Au [4,134] and (b) Xe on Ag [133]. Calculated values are indicated by symbols, and experimental results are indicated by solid lines. Experimental data for three different runs are shown in (b). With permission from Ref. 133. *Science* **265**, 1209 (1994); Ref. 134. *Science* **265**, 1209 (1994).

pin films [143]; however, the measured friction sometimes *decreases* with increasing disorder [144]. The most recent studies [138] indicate that films may be thermally depinned at high enough temperatures. Another possibility is that the layer deforms plastically rather than elastically and flows around disorder. This type of motion has been directly observed in type II superconductors where elastic pinning only occurs in finite systems [145].

IV. DRY, NONELASTIC CONTACTS

A. Commensurate Interfaces

Sufficiently large normal or lateral stresses can lead to plastic deformation and wear. Both processes contribute to the friction force. The simple geometry

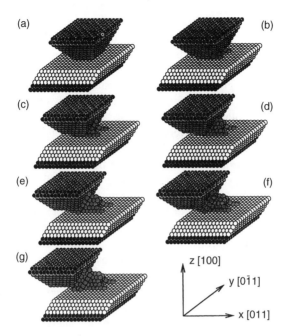

Figure 13. Snapshots showing the evolution of a Cu(100) tip on a Cu(100) substrate during sliding to the left. With permission from Ref. 63. *Phys. Rev. B* **53**, 2101 (1996).

illustrated in Fig. 13 has been the subject of many studies [12]. Here a single crystalline asperity with a clean flat tip is slid over a flat substrate. If the surfaces are commensurate, there will be a finite energy barrier for lateral motion. While thermal activation will allow exponentially slow sliding to occur at any finite stress [24], more rapid motion requires lateral stresses large enough to overcome these energy barriers. If this stress is large enough, it may lead to plastic deformation within the tip rather than interfacial sliding.

Sørensen et al. [63] simulated contacts between clean tips and surfaces using realistic potentials for copper atoms. In the case of commensurate (111) surfaces, sliding occurred without wear. The tip followed a zigzag trajectory, jumping between close-packed configurations. Detailed analysis of the jumps showed that they occurred via a dislocation mechanism. Dislocations were nucleated at the corner of the tip, and then moved rapidly along the interface through the contact region. Analysis of the local stress tensor showed that the stress in these nucleating regions was much larger than the mean values in the contact. When two infinite periodic surfaces were used, there was no region of stress concentration that could nucleate dislocations, and the static friction increased substantially.

The stress concentration near corners follows naturally from the equations of continuum elasticity, which predict diverging stresses at corners between sufficiently dissimilar materials like a tip and the surrounding vacuum [146–148]. The best-known example is the case of a crack tip, where the stress at a distance r from the tip grows as $r^{-1/2}$. In the general case the stress diverges as r^{-x} with $x \leq 1/2$. Vafek and Robbins [149] have recently compared continuum predictions to molecular simulations of the stress near sharp corners. The microscopic stress follows the divergence predicted by continuum theory until it is cut off at some small scale l by the discreteness of the lattice, by anharmonicity, or by plastic yield. Of particular relevance is that the maximum stress diverges as $(L/l)^x$, where L is the system size. This implies that the ratio between the stress at the edge of a finite contact and the mean stress will grow as a power of contact size. Nucleation of dislocations will be possible at lower and lower stresses, leading to decreased static friction. Only in the artificial case of periodic boundary conditions can the edge be removed and nucleation suppressed.

This conclusion is closely related to work by Hurtado and Kim [150,151], who used continuum theories of plastic flow to analyze the variation in friction with contact area. They identify three different regimes of behavior. When the contact diameter is less than or comparable to the core size of a dislocation, the entire contact will slip together. In this regime the lateral stress is close to the large values obtained for periodic boundary conditions. As the contact becomes bigger than the core size, nucleation of dislocations at the edges leads to a steady decrease in frictional stress with contact size. At the largest sizes there are many dislocations in the contact that move smoothly along the interface. The lateral stress approaches a constant lower value in this regime. Hurtado and Kim argue that AFM tips and SFA experiments are in the small- and large-scale limits, respectively, leading to different frictional forces. However, there are many other differences between these two types of experiments, including the chemical constituents of the surfaces.

The above mechanism for sliding involves dislocations, but no wear. Sørensen et al. found no wear between (111) tips and surfaces up to the largest loads studied. Adhesion led to a large static friction at zero load ($\tau_0 \sim 3$ GPa), and the friction increased very slowly with load. The slope $\alpha = 0.03$ is comparable to the values found for the adsorbed layers described in Section V. Because of this small slope, friction was observed down to very large *negative* loads corresponding to tensile stresses of ~ 6 GPa. Friction at negative loads is not uncommon for experiments between flat, adhesive surfaces. Although it contradicts Amontons's laws [Eq. (1)], it is consistent with the more general relation based on a local shear stress [Eq. (2)]. One can think of τ_0/α as an adhesive stress that acts like an extra external pressure.

The (111) plane is the easy slip plane for fcc crystals, and Sørensen et al. [63] observed very different behavior for (100) tips and surfaces. Sliding in the (011) direction led to interplane sliding between (111) planes inside the tip rather than at the contact. As shown in Fig. 13, this plastic deformation led to wear of the tip, which left a trail of atoms in its wake. This adhesive wear led to a steady rise in the contact area, total energy, and friction with sliding distance. The friction showed cyclical fluctuations as stress built up and was then released by nucleation of slip at internal (111) planes.

Nieminen et al. [152] observed a different mechanism of plastic deformation in essentially the same geometry, but at higher velocities (100 m/s versus 5 m/s) and with a different model for the potential between Cu atoms. Sliding took place between (100) layers inside the tip. This led to a reduction of the tip by two layers that was described as the climb of two successive edge dislocations under the action of the compressive load. Although wear covered more of the surface with material from the tip, the friction remained constant at constant normal load. The reason was that the portion of the surface where the tip advanced had a constant area. As in Sørensen et al.'s work [63], dislocations nucleated at the corners of the contact and then propagated through it.

Both the above simulations considered identical tips and substrates. Failure moved away from the interface for geometric reasons, and the orientation of the interface relative to easy slip planes was important. In the more general case of two different materials, the interfacial interactions may be stronger than those within one of the materials. If the tip is the weaker material, it will be likely to yield internally regardless of the crystallographic orientation. This behavior has been observed in experiments between clean metal surfaces where a thin tip is scraped across a flat substrate [31]. When the thin tip is softer than the substrate, failure is localized in the tip, and it leaves material behind as it advances. However, the simulations considered in this section treated the artificial case of a commensurate interface. It is not obvious that the shear strength of an interface between two incommensurate surfaces should be sufficient to cause such yield, nor is it obvious how the dislocation model of Hurtado and Kim applies to such surfaces.

B. Incommensurate Interfaces

Sørensen et al. [63] also examined the effect of incommensurability. The tip was made incommensurate by rotating it about the axis perpendicular to the substrate by an angle θ. The amount of friction and wear depended sensitively on the size of the contact, the load, and θ. The friction between large slabs exhibited the behavior expected for incommensurate surfaces: There was no wear, and the kinetic friction was zero within computational accuracy. The friction on small tips was also zero until a threshold load was exceeded. Then elastic instabilities were observed leading to a finite friction. Even larger loads lead to wear like that found for commensurate surfaces.

The transition from zero to finite friction with increasing load for small tips can be understood from the Prandtl–Tomlinson model. The control parameter $\tilde{k}$ decreases with load because the interaction between surfaces is increased and the internal stiffness of the solid and tip is relatively unchanged. The pinning potential is an edge effect that grows more slowly than the area (Section II). Thus the transition to finite friction occurs at larger loads as the area increases. Tips that were only 5 atoms in diameter could exhibit friction at very small loads. However, for some starting positions of the tip, no friction was observed even at 7.3 GPa. When the diameter was increased to 19 atoms, no friction was observed for any position or load considered.

The above studies indicate that while finite contacts are never truly incommensurate, contact areas as small as a few hundred atoms are large enough to exhibit incommensurate behavior. They also confirm Hirano and Shinjo's conclusion [22,23] that even bare metal surfaces of the same material will not exhibit static friction if the surfaces are incommensurate. Other studies indicate that the interactions between the two surfaces must be many times larger than the internal interactions before elastic multistability leads to pinning. Müser and Robbins considered a simple Prandtl–Tomlinson-like model [24] and concluded that the intersurface interactions ϵ_1 needed to be about six times larger than the intrasurface interactions ϵ_0. Müser [25] later studied contact between two and six layers of incommensurate surfaces and found no elastic metastability. Instead, spontaneous interdiffusion welded the surfaces together at ϵ_1/ϵ_0 between 6 and 8. Thus elastic metastability is very unlikely to be the origin of static friction between incommensurate surfaces.

Many experiments between clean metals do exhibit static friction even though there is no effort to make them commensurate, so some other mechanism must operate. One possibility is that diffusion allows rearrangements of atoms in the contact into locally commensurate regions on experimental time scales. For example, a high-angle grain boundary could roughen to produce a greater surface area, but with lower surface energy. This would naturally lock the surfaces together, leading to static friction. Large-scale rearrangements are too slow to access directly with simulations of bulk surfaces using realistic interactions. However, diffusion and plastic deformation can be accelerated by magnifying the driving energy [25]. Note that in Müser's study [25], interdiffusion was thermodynamically favored as soon as ϵ_1/ϵ_0 exceeded unity. However, it was not observed on simulation time scales until the driving force became strong enough to overcome thermal barriers for diffusion. Barriers to diffusion are also minimized by considering very small tips, as we now discuss.

C. Indentation

A large number of simulations have considered plastic deformation caused by applying a large normal load to a small tip. An extensive review has been given by Harrison et al. [12]. Typically the tip is only 2 or 3 atoms across, in order to mimic AFM tips used for imaging. This small radius of curvature maximizes the interfacial energy driving atomic rearrangements. The barriers for diffusion, along with the distances the atoms must diffuse, are also minimized.

Among the indentation studies of metals are simulations of an Ni tip indenting Au(100) [153], an Ni tip coated with an epitaxial gold monolayer indenting Au(100) [154], an Au tip indenting Ni(001) [155,156], an Ir tip indenting a soft Pb substrate [157], an Au tip indenting Pb(110) [158], and a Cu tip indenting Cu [63]. In general, plastic deformation occurs mainly in the softer of the two materials, typically Au or Pb in the cases above. The force between the tip and solid remains nearly zero until a separation of order 1 Å. Then there is a rapid jump to contact that produces a large contact area and significant plastic deformation. In the case of Au indenting Ni [156], the Au atoms in the tip displace by 2 Å within ∼1 ps. The strong adhesive stresses produce reconstruction of the Au layers through the fifth layer of the tip. There is additional reconstruction as the load increases, allowing the contact area to grow much more rapidly than in a purely elastic contact.

Further plastic deformation is generally observed when the tip is withdrawn [12,63,159]. In many cases a crystalline neck of the softer material forms between the tip and substrate. The pulling force oscillates as stress gradually builds and then is suddenly released. During each drop in stress, slip along planes within the neck leads to an increase in the number of atomic layers along its length. The neck gradually thins and may expand many times in length before it ultimately breaks. Sørensen et al. [63] examined the friction due to such necks. The lateral force also shows a cyclical building and release of stress, with a large average friction force. The drops in stress result from slip along (111) planes within the tip. The tip gradually thins and then breaks when the top is displaced by about the contact diameter.

Similar cold-welding and yield may occur in the contacts between asperities on two clean metal surfaces; and many textbooks, following Bowden and Tabor [31], advance this as the molecular-scale origin of friction. However, any metal exposed to air will acquire a layer of oxide and other contamination. Contact between the underlying metal will only occur at high stresses. Moreover, this mechanism implies extraordinarily high wear: Damage occurs over a depth of several atomic layers during sliding by a similar lateral distance. Many experimental systems slide meters before an atomic layer is worn away. Thus this mechanism is likely to be applicable to a relatively restricted set of circumstances. An important example is the case of machining discussed in the next subsection.

Harrison et al. have considered indentation of nonmetallic systems including hydrogen terminated diamond [12,160]. Unlike the metal–metal systems, diamond–diamond systems did not show a pronounced jump to contact. This is because the adhesion between diamond (111) surfaces is quite small if at least one is hydrogen-terminated [161]. For effective normal loads up to 200 nN (i.e., small indentations), the diamond tip and surface deformed elastically and the force–distance curve was reversible. A slight increase to 250 nN led to plastic deformation that produced hysteresis. If the tip was removed after plastic deformation occurred, there was some transfer of material between tip and substrate. However, because diamond is not ductile, pronounced necking was not observed.

Müser [25] examined yield of much larger tips modeled as incommensurate Lennard-Jones solids. The tips deformed elastically until the normal stress became comparable to the ideal yield stress and then deformed plastically. No static friction was observed between elastically deformed surfaces, while plastic deformation always led to pinning. Sliding led to mixing of the two materials like that found in larger two-dimensional simulations of copper discussed in Section IV.E.

D. Plowing and Machining

One of the commonly described mechanisms for producing friction is plowing (Fig. 1e) [31]. In this case a hard tip is indented into a softer material and "plows" a permanent groove into the material as it slides. The work needed to produce this plastic deformation of the substrate has to be provided by the frictional force. This mechanism clearly occurs whenever a substrate is scratched during sliding. This naturally leads to rapid wear, which may be desirable in the context of machining.

Large-scale, two- and three-dimensional molecular dynamics simulations of the indentation and scraping of metal surfaces were carried out by Belak and Stowers [162]. A blunted carbon tip was first indented into a copper surface and then pulled over the surface. In the two-dimensional simulation, the contact followed Hertzian behavior up to a load $L \approx 2.7$ nN and an indentation of about 3.5 Cu layers. The surface then yielded through creation of a series of single dislocation edges along the easy slip planes. After indenting by about 6 Cu layers, the carbon tip was slid parallel to the original Cu surface. The work per unit volume of removed material was found to scale as $(D)^{-0.6}$, where D is the depth of cut. The same power law was observed in machining of Cu with a 25 nm tip [163], but macroscopic experiments typically give an exponent of -0.2. Belak and Stowers note that the change in exponent occurs in experiments when the depth of cut is comparable to the grain size, leading to new deformation mechanisms.

In the three-dimensional (3D) simulations, the substrate contained as many as 36 layers or 72,576 atoms. Hence long-range elastic deformations were included. The surface yielded plastically after an indentation of only 1.5 layers, through the creation of a small dislocation loop. The accompanying release of load was much bigger than in two-dimensional (2D) simulations. Further indentation to about 6.5 layers produced several of these loading-unloading events. When the tip was pulled out of the substrate, both elastic and plastic recovery was observed. Surprisingly, the plastic deformation in the 3D studies was confined to a region within a few lattice spacings of the tip, while dislocations spread several hundred lattice spacings in the 2D simulations. Belak and Stowers concluded that dislocations were not a very efficient mechanism for accommodating strain at the nanometer length scale in 3D.

When the tip was slid laterally at $v = 100$ m/s during indentation, the friction or "cutting" force fluctuated around zero as long as the substrate did not yield (Fig. 14). This nearly frictionless sliding can be attributed to the fact that the

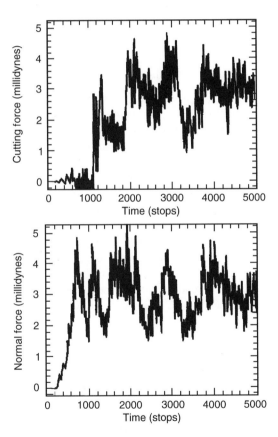

Figure 14. Normal (*bottom*) and lateral (*top*) force on a three-dimensional, pyramidal Si tip on a copper surface as a function of time. No plastic flow was reported up to 1000 time steps. The indentation stopped at about 5 layers after 2000 time steps. With permission from Ref. 162. *Fundamentals of Friction: Macroscopic and Microscopic Processes*, Kluwer Academic Publishers, 1992.

surfaces were incommensurate, and the adhesive force was too small to induce locking. Once plastic deformation occurred, the cutting force increased dramatically. Figure 14 shows that the lateral and normal forces are comparable, implying a friction coefficient of about one. This large value was expected for cutting by a conical asperity with small adhesive forces [164].

More recently, Zhang and Tanaka have performed similar simulations of 2D [165] and 3D [166] machining. The latter work considered both a spherical tip translating rigidly over a silicon surface and a spherical "wear" particle that rotated as it advanced. At low loads the substrate deforms elastically and there is no friction or wear. At higher loads the top layer of silicon is plastically deformed into an amorphous state, and some atoms may adhere to the tip. Still higher loads lead to the plowing of a permanent groove in the material, and finally to cutting, where a chip forms ahead of the tip and then flakes off. Cutting is very unlikely to occur when the tip is replaced by a wear particle that is free to rotate.

E. Plastic Deformation and Material Mixing

All of the above simulations were done in relatively small systems where there are no preexisting dislocations and the yield stress is near its theoretical maximum. At the larger scales relevant to many experiments, the yield stresses and the oscillations in stress during yield events may be much smaller. On these scales, and in ductile materials, plastic deformation is likely to occur throughout a region of some characteristic width about the nominal sliding interface [167]. Sliding-induced mixing of material from the two surfaces and sliding-induced grain boundaries are two processes that have been observed in experiments and in recent simulations of large systems.

Hammerberg et al. [168] performed large-scale simulations of a two-dimensional model for copper. The simulation cell contained 256×256 Cu atoms that were subject to a constant normal pressure $P_\perp$. Two reservoir regions at the upper and lower boundaries of the cell were constrained to move at opposite lateral velocities $\pm u_p$. The initial interface was midway between the two reservoirs.

The friction was measured at $P_\perp = 30 \, \text{GPa}$ as a function of the relative sliding velocity v. Different behavior was seen at velocities above and below about 10% of the speed of transverse sound. At low velocities, the interface welded together and the system formed a single work-hardened object. Sliding took place at the artificial boundary with one of the reservoirs. At higher velocities the friction was smaller, and it decreased steadily with increasing velocity. In this regime, intense plastic deformation occurred at the interface. Hammerberg et al. [168] found that the early time dynamics of the interfacial structure could be reproduced with a Frenkel–Kontorova model. As time increased, the interface was unstable to the formation of a fine-grained polycrystalline microstructure, which coarsened with distance away from the interface as a function of time. Associated with this microstructure was the mixing of material across the interface.

V. EMBEDDED SYSTEMS AND LUBRICANTS

A. Structural Changes at a Solid–Fluid Interface

There have been many analytic and numerical studies of the structure that solids induce in an adjacent fluid. Early studies focussed on layering in planes parallel to a flat solid surface. The sharp cutoff in fluid density at the wall induces density modulations with a period set by oscillations in the pair correlation function for the bulk fluid [169–173]. An initial fluid layer forms at the preferred wall–fluid spacing. Additional fluid molecules tend to lie in a second layer, at the preferred fluid–fluid spacing. This layer induces a third, and so on. The pair correlation function usually decays over a few molecular diameters, except near a critical point or in other special cases. Simulations of simple spherical fluids show on the order of 5 clear layers [174–176], while the number is typically reduced to 3 or less for chain or branched molecules that have several competing length scales [177–180].

Crystalline walls also induce commensurate density modulations within the plane of the fluid layers [174,176,181–184]. In extreme cases, one or more layers may actually crystallize, forming a solid wetting layer [155,174,176]. These changes in structure near the wall modify the local viscosity. Somewhat surprisingly, the strength of layering has almost no direct effect on viscosity [174]. Indeed, layering is strongest for ideal flat surfaces that cannot exert any lateral stress. In contrast, there is a one-to-one correspondence between the degree of commensurate in-plane order and the viscous coupling between the first layer and the substrate [174]. This can be understood from simple perturbation arguments [185] that are directly related to an analytic theory for the friction between an adsorbed monolayer and a substrate described above [134,135].

The viscosity change near a solid interface can be quantified by a slip length S that represents the change in the effective hydrodynamic width of the film. If v is the lateral velocity, then the shear rate in the center of a film confined by two identical walls separated by distance h is changed from v/h to $v/(h + 2S)$. Positive values of S imply a decrease in viscosity or "slip" at the wall, so that the effective width of the channel is larger. Negative values imply that fluid within a distance $|S|$ is "stuck" and moves with the same velocity as the wall. Simulations show that S can be of either sign, and its magnitude is usually comparable to a molecular diameter [174,175,186]. It can grow to much larger scales if the coupling between the wall and fluid is weak or the lateral density modulations are frustrated by incompatible spatial periods [174,187]. The slip length can also increase when the stress exceeds the maximum value the interface can support [188].

B. Phase Transitions in Thin Films

When the thickness of a fluid film is reduced below about 10 molecular diameters, the ordering influences of the two confining walls begin to interfere. As a result, the normal pressure P, viscosity, and other properties of the film show periodic oscillations as a function of the thickness h. One dramatic result is that the system becomes linearly unstable in regions where $dP/dh > 0$, leading to discontinuous changes in the number of layers of atoms between the two surfaces [189].

Pioneering grand canonical Monte Carlo simulations by Schoen et al. [176,183,184] examined the equilibrium behavior of a film of spherical molecules between two commensurate walls. As h and the lateral translation $\vec{x}$ of the top wall were varied, the film underwent transitions between fluid and crystalline states. The crystal was stable when the thickness was near an integral multiple of the spacing between crystalline layers and when the lateral translation was also consistent with the crystalline repeat. In this case, the ordering influences of the two walls interfered constructively and provided a strong ordering influence. Other values of h and $\vec{x}$ led to destructive interference and stabilized the fluid phase.

The resulting variations in free energy with $\vec{x}$ imply a strong static friction mediated by the film. Müser and Robbins later argued that the in-plane order induced by commensurate walls should always lead to a finite static friction—even if the film is in the fluid state [24]. The reason is that the linear response to one wall produces a commensurate modulation in the density of the fluid that propagates to the second wall. The energy of the second wall must depend on its phase with respect to this commensurate disturbance. Although the perturbation may be exponentially weak at large thicknesses, it is proportional to the contact area and will eventually pin the walls. Müser and Robbins verified this conclusion using simulation studies of the lateral diffusion of the walls in the absence of any lateral force [24]. The walls remained pinned by a periodic potential even when the intervening fluid layers were in a gaseous state! Curry et al. [190] also observed pinning by a film with rapid diffusion. As for bulk crystals, diffusion is not incompatible with the long-range structural order that separates a fluid and solid state.

Thompson et al. [178,191] studied phase transitions of spherical and chain molecules between commensurate walls. As the film thickness h decreased or pressure P increased, films of spherical molecules underwent a first-order transition to a commensurate crystalline state. However, short-chain molecules underwent a continuous glass transition. The divergence of the characteristic relaxation time, viscosity, and inverse of the diffusion constant could be fit to a free volume theory [178] like that used to describe the glass transition in many bulk systems [192]. The amount of commensurate structure induced in solid glassy films [178] was too small to explain the large static friction. The transition is analogous instead to "jamming" transitions in many other equilibrium and nonequilibrium systems, such as glass-forming materials, granular media, and foams [193]. Jamming of spherical and chain molecules has also been observed between amorphous [61,181] and incommensurate walls [30,61,194,195] where the effect of any induced lateral modulations averages to zero. A finite yield stress that corresponds to a static friction is required to unjam the system and produce shear.

Gao et al. [179,180] have examined changes in the equilibrium structure of films with film thickness. Their work uses more realistic potentials and allows molecules to be exchanged between the film and a surrounding reservoir. However, both these simulations and experiments may have difficulties in achieving equilibrium due to the sluggish dynamics of films as they approach a glassy state. Gao et al. contrast the behavior of simple linear and more complex branched molecules. Simple linear alkanes show much more pronounced layering. There is also more in-plane order within the layers, although at the intrinsic spacing of an alkane crystal rather than at the period of the wall. Branched molecules show less order in both directions. Gao et al. also calculate the normal load, or solvation force, needed to maintain a given film thickness. This force oscillates as the layering influences from the two walls change from constructive to destructive interference. The oscillations are more pronounced for the linear molecules that show sharper layers. The more ordered layers also take more time to squeeze out, indicating that they are more solid-like; however, the viscosity of the films was not studied. Gao et al. [196] also investigated the effects of (sub)nanoscale roughness and found that even a small amount can strongly reduce the solvation force and lead to better stick boundary conditions for the confined fluid.

C. Connection to Experimental Studies

The above studies have benefited from a fertile exchange with experimental groups using the Surface Force Apparatus (SFA). The SFA allows the mechanical properties of fluid films to be studied as a function of thickness over a range from hundreds of nanometers down to contact. The fluid is confined between two atomically flat surfaces. The most commonly used surfaces are mica, but silica, polymers, and mica coated with amorphous carbon, sapphire, or aluminum oxide have also been used. The surfaces are pressed together with a constant normal load, and the separation between them is measured using optical interferometry. The fluid can then be sheared by translating one surface at a constant tangential velocity [7,39] or by oscillating it at a controlled amplitude and frequency in a tangential [197–200] or normal [201,202] direction. The steady sliding mode mimics a typical macroscopic friction measurement, while the oscillatory mode is more typical of bulk rheological measurements (see Section VI.C). Both modes reveal the same sequence of transitions in the behavior of thin films.

When the film thickness h is sufficiently large, one observes the rheological behavior typical of bulk fluids [201,202]. Flow can be described by the bulk viscosity μ_B and a slip length S at each wall. As in simulations, typical values of S are comparable to molecular dimensions and would be irrelevant at the macroscopic scale. However, a few systems show extremely large slip lengths, particularly at high shear rates [203,204].

When the film thickness decreases below about 10 molecular diameters, the deviations from bulk behavior become more dramatic [39,199,205]. There are pronounced oscillations in the solvation force. As in simulations [179,180], these oscillations are most pronounced for simple spherical molecules, less pronounced for short-chain molecules, and smallest for branched molecules. There is also a dramatic increase in the stress needed to shear the films as h decreases. The shear stress τ is far too large to interpret in terms of a bulk viscosity and slip length, because the width of the fluid region $h + 2S$ would have to be much smaller than an angstrom. Experimental results are often expressed in terms of an effective viscosity $\mu_{eff} \equiv \tau h / v$ and effective shear rate $\dot{\gamma}_{eff} \equiv v/h$. For almost all fluids studied, the low-frequency limit of μ_{eff} rises by many orders of magnitude as the film thickness drops to a few molecular diameters [39,197,205]. At the same time, the shear rate at which the low-frequency limit is reached decreases by more than 10–12 orders of magnitude. Both observations imply a divergence in the relaxation time for structural rearrangements in the film. By the time h reaches two or three molecular diameters, most films behave like solids over the accessible range of time scales. The main exception is water, which seems to retain its fluid character [206–208].

In some cases the transition looks like a continuous glass transition, but at temperatures and pressures that are far from the glass region in the bulk phase diagram. In fact, the molecules may not readily form glasses in the bulk. In other cases the transition from liquid to solid behavior of the film appears to occur discontinuously, as if the film underwent a first-order freezing transition [200,205]. Simulations suggest that this is most likely to happen when the crystalline phase of the film is commensurate with the solid walls and when the molecules have a relatively simple structure that facilitates order [178–180,191,209]. Unfortunately, no direct experimental determination of the structure of confined films has been made to determine whether they are crystalline or amorphous, and different behavior has been reported for the same fluid. Part of the variation in experimental results is attributable to different mechanical properties and modes of operation (Section VI.C) of the experimental setups and perhaps to variations in the purity of the fluids; however, there are also differences in definition. The discontinuous jumps in shear response reported by Klein and Kumacheva [200,205] occurred during a discontinuous decrease in the number of layers confined between the surfaces [200,205]. A mechanical instability leading to a change in the number of layers will always produce discontinuous changes in mechanical properties. One can only expect a continuous glass transition if the density, or free volume, is varied continuously. Because the thickness is not a unique function of load, the density can be varied by decreasing the load with a fixed number of layers. Klein and Kumacheva do report a fairly continuous decrease of the yield stress when the load is decreased in this way. If the fluid were crystalline, one would expect a discontinuous transition during this decrease in load. Another observation that suggests that films form glassy states is that a crystal would in general be incommensurate with at least one of the two unaligned surfaces, leading to much smaller friction forces than observed.

Clearly the film thickness, fluid density, and perhaps the orientation of the surfaces [207] are additional thermodynamic variables that may shift or alter phase boundaries. Cases where a single interface stabilizes a different phase than the bulk are central to the field of wetting. The presence of two interfaces separated by only a few nanometers leads to more pervasive phase changes.

The rheological response of bulk materials frequently obeys time–temperature scaling near the glass transition [210,211]. The real and imaginary parts of the elastic moduli describe the stiffness and viscosity as a function of frequency. Results from different temperatures can be collapsed onto a universal curve if the frequency is multiplied by a temperature-dependent relaxation time, and the viscosity is normalized by its low frequency limit. Demirel and Granick [197] achieved a similar collapse by scaling their data by a relaxation time that depended on film thickness rather than temperature. This suggested that their film underwent a glass transition much like that of a bulk material.

Figure 15 shows that simulation results for the viscosity of thin films can also be collapsed using Demirel and Granick's approach [212]. Data for different thicknesses, normal pressures, and interaction parameters [178,191] were scaled by the low shear rate viscosity μ_0. The shear rate was then scaled by the rate $\dot{\gamma}_c$ at which the viscosity dropped to $\mu_0/2$. Also shown on the plot (circles) are data for different temperatures that were obtained for longer chains in films that are thick enough to exhibit bulk behavior [212–214]. The data fit well on to the same curve, providing a strong indication that a similar glass transition occurs whether thickness, normal pressure, or temperature is varied.

As in the case of bulk glasses, the issue of whether there is a true glass transition at finite temperature is controversial because observation times are

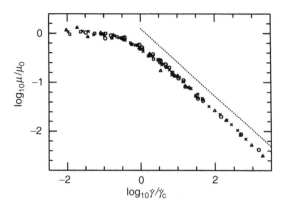

Figure 15. Collapse of simulation data for the effective viscosity versus shear rate as a glass transition is approached by decreasing temperature (circles), increasing normal pressure at fixed number of fluid layers (triangles), or decreasing film thickness at fixed pressure with two different sets of interaction potentials (squares and crosses). The dashed line has a slope of −0.69. With permission from Ref. 212. *Microstructure and Microtribiology of Polymer Surfaces,* American Chemical Society, 2000.

limited, particularly for simulations. However, thin films of most molecules act like solids over experimentally accessible times [39]. No motion occurs until a yield stress is exceeded, and the calculated yield stress is comparable to values for bulk solid phases of the molecules at higher pressure or lower temperature. The response of the walls to a lateral force becomes typical of static friction, and Gee et al. analyze their results in terms of Eq. (2) [39]. Their values of τ_0 are of the order of a few to 20 MPa, and values of α range from 0.3 to 1.5. However, the latter are influenced by the change in film thickness at the relatively low pressures used (< 40 MPa) and would probably not be representative of the behavior in a typical mechanical device.

D.   Jammed Monolayers and Amontons's Laws

Physisorbed molecules, such as the short hydrocarbons in the SFA experiments described above, can be expected to sit on any surface exposed to atmospheric conditions. Even in ultrahigh vacuum, special surface treatments are needed to remove strongly physisorbed species from surfaces. Since such films jam between randomly oriented mica surfaces, Robbins and Smith [215] suggested they could be responsible for the finite static friction between many macroscopic objects. Recent simulations [24,25,30,61,194] and analytic work [61] have supported this speculation.

The static friction between bare incommensurate or amorphous surfaces vanishes in the thermodynamic limit because the density modulations on two bare surfaces cannot lock into phase with each other unless the surfaces are unrealistically compliant [61]. However, a submonolayer of molecules that form no strong covalent bonds with the walls can simultaneously lock to the density modulations of both walls (Fig. 1d). Once they adapt to the existing configuration of the walls, they will resist any translation that would change their energy minima. This gives rise to a finite static friction for all surface symmetries: commensurate, incommensurate, and amorphous [61]. In essense, the layer plays the role of weak springs in the Prandtl–Tomlinson model (Section III.A), although the detailed picture may be different for kinetic friction [216]. In order to also induce finite F_k, sudden jumps of the confined atoms must occur between symmetrically inequivalent positions when one wall is slid with respect to the other. Such jumps are frequently absent for commensurate surfaces, which implies zero kinetic but finite static friction.

The variation of the friction with wall geometry, interaction potentials, temperature, velocity, and other parameters has been determined with simulations using simple spherical or short-chain molecules to model the monolayer [24,25,30,61,194]. In all cases, the shear stress shows a linear dependence on pressure that is consistent with Eq. (2) up to the gigapascal pressures expected in real contacts. Moreover, the friction is relatively insensitive to parameters that are not controlled in typical experiments, such as the orientation of crystalline walls, the direction of sliding, the density of the layer, the length of the hydrocarbon chains, and so on. Variations are of order 20%, which is comparable to variations in results from different laboratories [44,217]. Larger variations are only seen for commensurate walls, which may also exhibit a nonlinear pressure dependence [218].

The linear variation of τ with P [Eq. (2)] can be understood from a simple hard-sphere picture. Once the adsorbed molecules have conformed to the surface, they produce an interlocking configuration like that envisioned by early researchers (Fig. 1a), but at an atomic scale. The value of α is the effective slope that the surface must be lifted up over this interlocking layer. The main factor that changed τ_s in the above simulations was the ratio of the characteristic length for wall–adsorbate interactions, σ_{wa}, to the nearest-neighbor spacing on the walls, d. As the ratio σ_{wa}/d decreased, adsorbed atoms could penetrate more deeply into the corrugation on the wall and both the effective slope and α increased.

Physisorbed molecules also provide a natural explanation for the logarithmic increase in kinetic friction with sliding velocity that is observed for many materials and represented by the coefficient A in the rate-state model of Eq. (5). Figure 16 shows calculated values of τ_0 and α as a function of $\log v$ for a submonolayer of chain molecules between incommensurate surfaces [195]. The value of τ_0 becomes independent of v at low velocities. The value of α, which

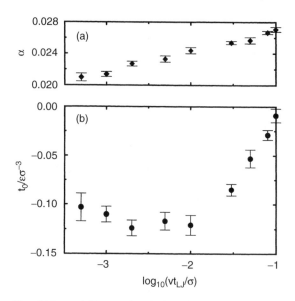

Figure 16. Plot of (a) α and (b) τ_0 against the logarithm of the velocity for incommensurate walls separated by a monolayer of chain molecules containing six monomers each. Velocities are in Lennard-Jones units of σ/t_{LJ}, which corresponds to roughly 100 m/s. With permission from Ref. 165. *Wear* **211**, 44 (1997).

dominates the friction at large loads, rises linearly with $\log_{10} v$ over more than two decades in velocity. He and Robbins were able to correlate this scaling with thermally activated hops of the chain molecules. At low velocities, almost all the molecules are stuck in a local potential minimum at any instant [61,195]. At zero temperature, they can only move when the applied stress makes their local minimum become mechanically unstable. At finite temperature, thermal activation allows motion at a lower stress. As the velocity decreases, there is more time for activated hops, and the friction decreases. If the shift in barrier height drops linearly with stress, τ varies as $A \log v$ with A proportional to temperature. He and Robbins found that A did rise roughly linearly with temperature. Calculated values of A/μ_0 at the relatively large temperatures they used were between 0.025 and 0.05. Experimental values for rocks are in roughly the same range, about 0.008 to 0.025.

The above results suggest that adsorbed molecules and other "third bodies" may prove key to understanding macroscopic friction measurements. It will be interesting to extend these studies to more realistic molecular potentials and to rough surfaces. Similar arguments apply to other so-called "third bodies" that are trapped between two surfaces, such as wear debris, grit, dust, and so on.

These particles are too large to lock in the valleys between individual atoms, but can lock in places where the two opposing rough surfaces both have larger valleys.

VI. UNSTEADY SLIDING

A. Stick–Slip Dynamics

The dynamics of sliding systems can be very complex and depend on many factors, including the types of metastable states in the system, the times needed to transform between states, and the mechanical properties of the device that imposes the stress. At high rates or stresses, systems usually slide smoothly. At low rates the motion often becomes intermittent, with the system alternately sticking and slipping forward [31,44]. Everyday examples of such stick–slip motion include the squeak of hinges and the music of violins.

The alternation between stuck and sliding states of the system reflects changes in the way energy is stored. While the system is stuck, elastic energy is pumped into the system by the driving device. When the system slips, this elastic energy is released into kinetic energy and eventually dissipated as heat. The system then sticks once more and begins to store elastic energy, and the process continues. Both elastic and kinetic energy can be stored in all the mechanical elements that drive the system. The whole coupled assembly must be included in any analysis of the dynamics, see e.g., Fig. 18.

The simplest type of intermittent motion is the atomic-scale stick–slip that occurs in the multistable regime ($\tilde{k} < 1$) of the Prandtl–Tomlinson model. Energy is stored in the springs while atoms are trapped in a metastable state, and it is converted to kinetic energy as the atoms pop to the next metastable state. This phenomenon is quite general and has been observed in many simulations of wearless friction [12,13] as well as in the motion of atomic force microscope tips [6,9]. In these cases, motion involves a simple ratcheting over the surface potential through a regular series of hops between neighboring metastable states. The slip distance is determined entirely by the periodicity of the surface potential. Confined films, adsorbed layers, and other nonelastic systems have a much richer potential energy landscape due to their many internal degrees of freedom. One consequence is that stick–slip motion between neighboring metastable states can involve microslips by distances much less than a lattice constant. [181,212,213] An example is seen at $t/t_{LJ} = 620$ in Fig. 17b. Such microslips involve atomic-scale rearrangements within a small fraction of the system. Closely related microslips have been studied in granular media and foams [219–221].

Many examples of stick–slip involve a rather different type of motion that can lead to intermittency and chaos [17a,39,51,53,181]. Instead of jumping between neighboring metastable states, the system slips for very long distances before sticking. For example, Gee et al. [39] and Yoshizawa and Israelachvili [222] observed slip distances of many microns in their studies of confined films. This distance is much larger than any characteristic periodicity in the potential, and it varies with velocity, load, and the mass and stiffness of the SFA. The fact that the SFA does not stick after moving by a lattice constant indicates that sliding has changed the state of the system in some manner, so that it can continue sliding even at forces less than the yield stress.

One simple property that depends on past history is the amount of stored kinetic energy. This can provide enough inertia to carry a system over potential energy barriers even when the stress is below the yield stress. The simplest example is the underdamped Prandtl–Tomlinson model, which has been thoroughly studied in the mathematically equivalent case of an underdamped Josephson junction [223]. If one pulls with a constant force, one finds a hysteretic response function where static and moving steady states coexist over a range of forces between F_{min} and the static friction F_s. There is a minimum stable steady-state velocity v_{min} corresponding to F_{min}. At lower velocities, the only steady state is linearly unstable because $\partial v / \partial F < 0$; pulling harder slows the system. It is well-established that this type of instability can lead to stick–slip motion [31,44]. If the top wall of the Tomlinson model is pulled at an average velocity less than v_{min} by a sufficiently compliant system, it will exhibit large-scale stick–slip motion.

Experimental systems typically have many other internal degrees of freedom that may depend on past history. Most descriptions of stick–slip motion have assumed that a single unspecified state variable is relevant and used the phenomenological rate-state models described in Section I.D. This approach has captured experimental behavior from a wide range of systems including surface force apparatus experiments [48,224], sliding paper [53], and earthquakes [52]. However, in most cases there is no microscopic model for the state variable and its time dependence. This makes it difficult to understand how small changes in materials, such as the percentage of recycled aluminum cans, can effect stick–slip in a process such as aluminum extrusion.

One set of systems where microscopic pictures for the state variable have been considered are confined films in SFA experiments. Confined films have structural degrees of freedom that can change during sliding, and this provides an alternative mechanism for stick–slip motion [181]. Some of these structural changes are illustrated in Fig. 17, which shows stick–slip motion of a two-layer film of simple spherical molecules. The bounding walls were held together by a constant normal load. A lateral force was applied to the top wall through a spring k attached to a stage that moved with fixed velocity v in the x direction. The equilibrium configuration of the film at $v = 0$ is a commensurate crystal that resists shear. Thus at small times, the top wall remains pinned at $x_W = 0$.

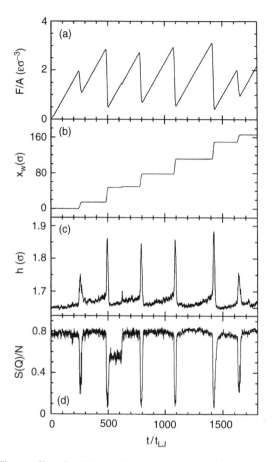

Figure 17. Time profiles of (a) frictional force per unit area F/A, (b) displacement of the top wall x_w, (c) wall spacing h, and (d) Debye–Waller factor $S(Q)/N$ during stick–slip motion of spherical molecules that form two crystalline layers in the static state. Note that the system dilates (c) during each slip event. The coinciding drop in Debye–Waller factor shows a dramatic decrease from a typical crystalline value to a characteristic value for a fluid. (Courtesy of Peter A. Thompson.)

The force grows linearly with time, $F = kv$, as the stage advances ahead of the wall. When F exceeds F_s, the wall slips forward. The force drops rapidly because the slip velocity $\dot{x}_W$ is much greater than v. When the force drops sufficiently, the film recrystallizes, the wall stops, and the force begins to rise once more.

One structural change that occurs during each slip event is dilation by about 10% (Fig. 17c). Dhinojwala and Granick [225] have recently confirmed that dilation occurs during slip in SFA experiments. The increased volume makes it easier for atoms to slide past each other and is part of the reason that the sliding friction is lower than F_s. The system may be able to keep sliding in this dilated state as long as it takes more time for the volume to contract than for the wall to advance by a lattice constant. Dilation of this type plays a crucial role in the yield, flow, and stick–slip dynamics of granular media [219,226,227].

The degree of crystallinity also changes during sliding. Deviations from an ideal crystalline structure can be quantified by the Debye–Waller factor $S(Q)/N$ (Fig. 17d), where Q is one of the shortest reciprocal lattice vectors and N is the total number of atoms in the film. When the system is stuck, $S(Q)/N$ has a large value that is characteristic of a 3D crystal. During each slip event, $S(Q)/N$ drops dramatically. The minimum values are characteristic of simple fluids that would show a no-slip boundary condition (Section V.A). The atoms also exhibit rapid diffusion that is characteristic of a fluid. The periodic melting and freezing transitions that occur during stick–slip are induced by shear and not by the negligible changes in temperature. Shear-melting transitions at constant temperature have been observed in both theoretical and experimental studies of bulk colloidal systems [228,229]. While the above simulations of confined films used a fixed number of particles, Ref. 230 found equivalent results at fixed chemical potential.

Very similar behavior has been observed in simulations of sand [226], chain molecules [212], and incommensurate or amorphous confining walls [181]. These systems transform between glassy and fluid states during stick–slip motion. As in equilibrium, the structural differences between glass and fluid states are small. However, there are strong changes in the self-diffusion and other dynamic properties when the film goes from the static glassy to the sliding fluid state.

In the cases just described, the entire film transforms to a new state, and shear occurs throughout the film. Another type of behavior is also observed. In some systems shear is confined to a single plane—either a wall–film interface or a plane within the film [212,213]. There is always some dilation at the shear plane to facilitate sliding. In some cases there is also in-plane ordering of the film to enable it to slide more easily over the wall. This ordering remains after sliding stops, and it provides a mechanism for the long-term memory seen in some experiments [39,222,231]. Buldum and Ciraci [232] found stick–slip motion due to periodic structural transformations in the bottom layers of a pyramidal Ni(111) tip sliding on an incommensurate Cu(110) surface.

The dynamics of the transitions between stuck and sliding states are crucial in determining the range of velocities where stick–slip motion is observed, the shape of the stick–slip events, and whether stick–slip disappears in a continuous or discontinuous manner. Current models are limited to energy balance arguments [233,234] or phenomenological models of the nucleation and growth of "frozen" regions [48,53,222,224]. Microscopic models and detailed experimental data on the sticking and unsticking process are still lacking. However, the two simple models with reduced degrees of freedom discussed next provide useful insight into the origins of the complex dynamics seen in experiments.

B.   Relation to Prandtl–Tomlinson and Rate-State Models

As noted in the introduction, there has been an ongoing quest to identify simple one degree of freedom models, like the Prandtl–Tomlinson (PT) model, that capture the sliding dynamics of macroscopic solids. This concerns not only dry friction but also lubricated friction. Baumberger and Caroli [84] proposed a phenomenological model of boundary lubricated junctions in terms of the PT model. Their contact consists of two plane-parallel walls of area a^2 which are separated by a distance h. The gap between the plates is filled with a fluid of viscosity η. The equation of motion for the top plate is identical to Eq. (19); however, the damping is related to the internal viscous response, suggesting the following substitution for the damping coefficient:

$$m\gamma \rightarrow \eta a^2/h \qquad (38)$$

Caroli and Baumberger argue that such a description is legitimate if there is a time-scale separation between the motion of the top plate and that of the lubricant atoms. However, if there is such a separation, one can object that in almost all cases there should be no significant, corrugation potential. For symmetry reasons the corrugation potential must be periodic in the upper wall's lattice constant b_u and that of the lower wall b_l. Unless identical and perfectly aligned crystals are employed, b_u and b_l can be assumed to be effectively incommensurate, in which case the common periodic becomes extremely large and the corrugation potential becomes arbitrarily small.

Nevertheless, Baumberger and Caroli's comparison of the PT model to boundary lubrication remains appealing. The PT model is known to show two stable modes of motion under certain circumstances: If the externally applied force is in between static friction force and kinetic friction force, then the running solution and the pinned solution are both stable against a small perturbation. This mimics the behavior of shear melting of confined lubricants. As in Fig. 17, the shear force necessary to induce shear melting are larger than the shear force where a liquefied lubricant freezes back into a glassy or a crystalline state.

As discussed in Section I.D, the dependence of friction on past history is often modeled by the evolution of a "state" variable (Eq. 6) in a rate-state model [50,51]. Heslot et al. [53] have compared one such model, where the state variable changes the height of the potential in a finite-temperature PT model, to their detailed experimental studies of stick-slip motion. They slid two pieces of a special type of paper called Bristol board and varied the slider mass M, pulling

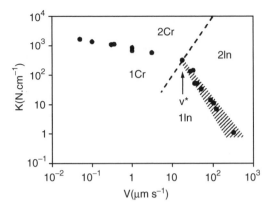

Figure 18. Phase diagram for stick-slip motion with $M = 1.2$ kg. Stick-slip occurs in region 1 and smooth sliding in region 2. The transition between the two is continuous in the creep regime (CR) and discontinuous in the inertial regime (In). With permission from Ref. 53. *Phys. Rev. E* **49**, 4973 (1994).

spring stiffness K and average velocity V. As shown in Fig. 18, they found stick-slip motion at low velocities and stiffnesses. Increasing the ratio K/M produced a continuous transition to smooth sliding, while increasing V produced a discontinuous transition with hysteresis (hatched region). Their data and model show that the continuous transition occurs when the time for creep within the many contacts between the surfaces (Fig. 2) is shorter than the inertial time given by $\sqrt{M/K}$. The discontinous transition appears to occur when the inertial time becomes shorter than the time to traverse a characteristic distance, but they were unable to verify that the ratio M/K was the only control parameter. Note that a continuous transition can only occur when the kinetic friction decreases with increasing velocity. Stick-slip motion may still occur when the kinetic friction rises with velocity as long as it remains smaller than the static friction [181], but the transition to steady sliding must then be first order.

C. Rheological Versus Tribological Response

Two approaches are usually used to investigate shear forces in confined liquids: the tribological one, where a constant driving velocity is applied, and the rheological one, which applies an oscillatory drive. In the latter case the coordinate of the laterally driven stage changes as $X_d = X_0 \sin(\omega t)$; X_0 and ω are the amplitude and the frequency of the drive. One of the goals of rheological measurements is to obtain information on the energy loss in the system. This is usually done in terms of response theory. Practically, linear response is assumed in the analysis of rheological results, leading to a description of the properties of the system in terms of the complex dynamic shear modulus $G(\omega) = G'(\omega) + iG''(\omega)$. The shear modulus is defined as a linear coefficient in the relationship between Fourier components of a shear force acting on the embedded system and the lateral displacement, Y, of the driven plate [210,235]. The storage modulus G' and loss modulus G'' are obtained from measurements and determine the elastic and dissipative properties of the confined system.

Since in rheological measurements the drive is oscillatory, one can fix either the driving amplitude or frequency and vary the other quantity. For small driving amplitudes X_0 the displacement of the plate is smaller than a characteristic microscopic length, and the response of the system is linear. After exceeding a critical driving amplitude X_0^c, the top plate executes oscillations with amplitude larger than the characteristic distance. If the rheological behavior of a confined fluid is mimicked in terms of a simple model such as the Prandtl–Tomlinson (PT) model, then this characteristic length can be estimated to be $X_0^c \approx F_s/K$, where F_s is the static friction force obtained from tribological experiments and K is the external spring constant [236].

For values of the driving amplitude above X_0^c, the PT model exhibits stick–slip motion before it crosses over to smooth sliding at amplitudes larger than a second threshold V_c/ω [236]. The stick–slip motion in this regime $X_0^c < X_0 < V_c/\omega$ is similar to stick–slip motion in tribological experiments; however, there are intermittent regimes during rheological driving, where the top wall remains pinned for a certain amount of time. Such rheological dynamics is shown in Fig. 19. Similar behavior is found for a small chain embedded between two corrugated, one-dimensional plates [237], and the calculations are also consistent with available experimental data of confined fluids under rheological shear [238–243].

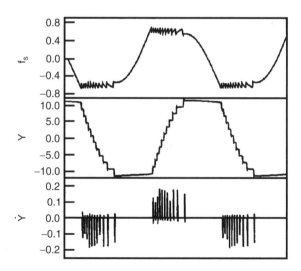

Figure 19. Time dependence of spring force f_s, top plate displacement Y, and velocity $\dot{Y}$ as a function of time. With permission from Ref. 237. *Phys. Rev. Lett.* **81**, 1227 (1998).

For very high driving amplitudes, $X_0 > V_c/\omega$, the PT model and also the embedded model apparently follows linear response again. In this latter regime, the corrugation barriers are small compared to the (average) kinetic energy of the involved atoms, and friction is dominated by the damping coefficient γ [see Eqs. (12) and (19) and the discussion of Fig. 4]. Thus $X_0 > V_c/\omega$ roughly corresponds to the smooth sliding regime in tribological experiments.

One feature of the rheological response is that the microscopic degrees of freedom can be in three different regimes for a given combination of X_0 and ω. (i) At times where $|\dot{Y}|$ is small, the atoms can oscillate around their potential energy minima and the observed loss will be small. (ii) At intermediate $|\dot{Y}|$, the microscopic degrees of freedom exhibit instabilities, akin to the elastic instabilities in the PT model or the phase transition instabilities of shear molten/frozen lubricants. (iii) Only if $|\dot{Y}|$ is sufficiently large does the system go into the smooth sliding or shear molten regime. The precise choice of X_0 and ω will thus determine how much time the system spends in what regime. However, it is not possible to associate the friction force measured in a rheological experiment with amplitude X_0 and frequency ω with a tribological experiment done at sliding velocity $v = X_0\omega$. This is shown in Fig. 20. If the driving amplitude is large, the system is more likely to be forced to exhibit instabilities and thus it can be expected to show larger friction. This is confirmed by a detailed analysis of the two above-mentioned model systems [236,237].

From the above analysis it follows that the increase in η_{eff} at small v cannot be interpreted as shear thickening. Tribological driving (at positive tempera- tures) would result in a curve similar to that shown in Fig. 15, in which the linear-response viscosity is obtained at the smallest shear rates. However, as the two model systems mentioned above are athermal, their tribologically determined η_{eff} would tend to infinity at zero shear rate [216].

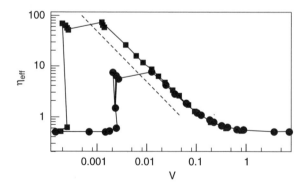

Figure 20. Effective viscosity as a function of the product $v = X_0\omega$ at fixed driving frequencies ω. Circles represent a frequency 10 times larger than squares. Consequently, the driving amplitude X_0 is 10 times larger for the squares than for the circles at a given v. With permission from Ref. 237. *Phys. Rev. Lett.* **81**, 1227 (1998).

D. Lateral-Normal Coupling: Reducing Static Friction Spikes

While most studies of frictional forces have focused on the lateral response to a lateral driving force, there have been a few theoretical [81,181,226,244,245] and experimental [238,241,246,247] observations of response in the normal direction. A feature characteristic of the response in the normal direction is the dilatancy during slippage and sliding, observed in experiments and molecular dynamics calculations. The interest in coupling between lateral and normal motions has recently gained additional attention due to the prediction that one should be able to modify the lateral response by imposing a controlling normal drive [81,244,245,248–251]. Main features of the lateral–normal coupling can be demonstrated by considering a generalization of the one-dimensional Prandtl–Tomlinson model to include also the normal motion [245]. Introducing lateral–normal coupling leads to the following coupled equations of motion which describe the motion of the top plate of mass M:

$$M\ddot{X} = -\eta_{\parallel}\dot{X} - \frac{\partial U(X,Z)}{\partial X} + K_d(Vt - X) \tag{39}$$

$$M\ddot{Z} = -\eta_{\perp}\dot{Z} - \frac{\partial U(X,Z)}{\partial Z} + K_n[Z_p(t) - Z] \tag{40}$$

where

$$\eta_{\parallel,\perp} = \eta^0_{\parallel,\perp}\exp(1 - Z/\Lambda) \tag{41}$$

and

$$U(X,Z) = U_0 \exp(1 - Z/\Lambda)[1 - \sigma^2 \cos(2\pi X/b)] \tag{42}$$

The top plate is connected to a laterally driven spring, of spring constant K_d, and to a spring K_n that is used to control the motion in the normal direction. $U(X,Z)$ is the effective potential experienced by the plate due to the presence of the embedded system, b is its periodicity, and σ characterizes the corrugation of the potential in the lateral direction. The parameters $\eta_{\parallel}$ and $\eta_{\perp}$ are responsible for the dissipation of the plate kinetic energy due to the motion in the lateral and normal directions. In contrast to the traditional Prandtl–Tomlinson model, here the dependence of U and $\eta_{\parallel,\perp}$ on the distance Z between plates is taken into account. The detailed distance dependence is determined by the nature of the interaction between the plate and embedded system. As an example, we assume an exponential decrease of U and $\eta_{\parallel,\perp}$ with a rate Λ^{-1} as Z increases. The possibility of an external modulation of the normal load $L_n(t) = K_n[Z_p(t) - Z]$ is taken into account by introducing a time dependence into the position of the normal stage, $Z_p(t)$.

A relevant property of the coupled equations is the existence of two stable surface separations Z_0 and Z_s, which correspond to the plate being either at rest or in fast motion. At rest the plate feels the lateral corrugation of the potential, $U(X, Z)$, and sits at the minima of the potential, while during fast motion there is a decoupling from this potential corrugation. From Eqs. (39) and (40), one obtains the following self-consistent equation for the maximal dilatancy:

$$\Delta Z_D = Z_s - Z_0 = \frac{L_n}{K_n}[(1 - \sigma^2)^{-1} \exp(-\Delta Z_D/\Lambda) - 1] \qquad (43)$$

where

$$L_n = (1 - \sigma^2)(U_0/\Lambda) \exp(1 - Z_0/\Lambda) \qquad (44)$$

is the normal load at rest. Equation (43) demonstrates that the dilatancy ΔZ_D explicitly depends on the potential corrugation and on the normal spring constant. It is quite evident that measuring the dilatancy helps in determining the amplitude of the potential corrugation.

Figures 17b and 17c show the response in the lateral and normal directions to a lateral constant velocity drive for the stick–slip regime that occurs at low driving velocities. This behavior is similar for the presently discussed model. The separation between the plates, which is initially Z_0 at equilibrium, starts growing before slippage occurs and stabilizes at a larger interplate distance as long as the motion continues. Since the static friction is determined by the amplitude of the potential corrugation $\sigma^2 U_0 \exp(1 - Z/\Lambda)$, it is obvious that the dilatancy leads to a decrease of the static friction compared to the case of a constant distance between plates.

Over a wide range of system parameters the dilatancy is smaller than the characteristic length Λ. Under this condition the generalized Prandtl–Tomlinson model predicts a linear increase of the static friction with the normal load, which is in agreement with Amontons's law. It should be noted that, in contrast to the multi-asperity surfaces discussed in Section VII, here the contact area is independent of the load. The fulfillment of Amontons's law in the present model results from the enhancement of the potential corrugation, $\sigma^2 U_0 \exp(1 - Z/\Lambda)$, experienced by the driven plate with an increase of the normal load.

While the present discussion is concerned with the interaction of two commensurate walls only, similar dilatancy effects occur under more general circumstances. Indeed, dilatancy is expected in a broad range of systems where coupling between lateral and normal directions exists. Recently this phenomenon has been intensively studied in the context of sheared granular media [226,246,252] and seismic faults [253,254]. The observation of dilatancy is limited by the choice of the normal spring, which should be weaker than the potential elasticity, $K_n \Lambda^2 \ll U_0$. From the dependence of the dilatancy ΔZ_D on the normal spring constant K_n it is clear that the weaker the K_n, the larger the ΔZ_D, and therefore the frictional force tends to decrease.

From Eq. (43) one can see that as the spring constant K_n decreases, the dilatancy grows. This leads to smoothing of the stick–slip motion present for stiffer springs. Weakening the normal spring constant results in lowering the critical velocity at which stick–slip turns to sliding. The dependence of the critical velocity on K_n becomes most pronounced for underdamped conditions, when $\eta_{\parallel,\perp} \ll MK_d$. This consideration emphasizes again that the experimental observations in SFA strongly depend on the choice of the mechanical parameters.

Dilatancy has been shown above to occur naturally during slippage events. It has also been demonstrated that the measured frictional forces are sensitive to the interplate distance. One can therefore expect to be able to control frictional behavior by monitoring the motion in the normal direction. From a practical point of view, one wishes to be able to control frictional forces so that the overall friction is reduced or enhanced, the stick–slip regime is eliminated, and instead, smooth sliding is achieved. Such control can be of high technological importance for micromechanical devices—for instance in computer disk drives, where the early stages of motion and the stopping processes, which exhibit chaotic stick–slip, pose a real problem [255].

Controlling frictional forces has been traditionally approached by chemical means, namely, using lubricating liquids. A different approach, which has attracted considerable interest recently [81,244,245,248–251], is by controlling the system mechanically. The idea is not to change physical properties of interfaces but to stabilize desirable modes of motion which are unstable in the absence of the control. The goal of this approach is twofold: (a) to achieve smooth sliding at low driving velocities, which otherwise correspond to the stick–slip regime, and (b) to decrease the frictional forces. Two different methods of control have been discussed: The first one uses a feedback control, and the second one relies on a "brute-force" modification of the system dynamics without feedback.

We start from the feedback mechanism of control. The analysis of the mechanism has been done with a one-dimensional model, which includes two rigid plates and noninteracting particles embedded between them [244]. In order to mimic the effect of normal load, the dependence of the amplitude of the periodic potential on the load has been introduced:

$$U_0(L_n) = U_0[1 + \chi(L_n - L_n^0)] \qquad (45)$$

Here, U_0 is the value of the potential for some nominal value of the normal load L_n^0, and χ is a dimensional constant. The normal load has been used as the control parameter to modify friction. Equation (45) assumes small load variations around L_n^0, which, as shown in Ref. 244, are sufficient to achieve control. The control method is characterized by two independent steps: (a) reaching the vicinity of an unstable sliding mode of motion and (b) stabilizing it. The control has been realized by small variations of the normal load, which has been externally adjusted by employing a proportional feedback mechanism [244]. A delayed-feedback mechanism has also been used for stabilizing the sliding states [248].

The aim of this approach is to stabilize a sliding state for driving velocities $V < V_c$, where one would expect chaotic stick–slip motion. Within the model discussed here, sliding states correspond to periodic orbits of the system with two periods: (a) period $T = b/V$, which corresponds to a motion of the particles while trapped by one of the plates, and (b) period $T = 2b/V$, which corresponds to the particles moving with the drift velocity $V/2$. In the chaotic stick–slip region both orbits still exist, but are unstable. The approach is therefore to drive the system into a sliding state by controlling these unstable periodic orbits. This makes it possible to extend the smooth sliding to lower velocities. The control of such orbits in dynamical systems has been proposed [256] and experimentally applied to a wide variety of physical systems including mechanical systems, lasers, semiconductor circuits, chemical reactions, biological systems, and so on (see Ref. 257 for references).

Figure 21 demonstrates the effect of the mechanical control on the time dependence of the spring force. The results correspond to the control of the trapped–sliding state [244], in which the particles cling to one of the plates and move either with velocity zero (sticking at the bottom plate) or velocity V (sticking at the top plate). The control is switched on at time t_1 and is shut down

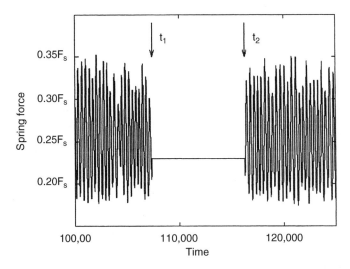

Figure 21. Friction force F as a function of time. Stick–slip motion and static friction spikes are suppressed through small variations in the normal load at times $t_1 < t < t_2$. With permission from Ref. 244. *Phys. Rev. E* **57**, 7340 (1998).

at time t_2. We clearly see that as a result of the control the chaotic motion of the top plate is replaced by smooth sliding that corresponds to a trapped state with constant frictional force.

In Refs. 244 and 248 the possibility to control friction has been discussed in model systems described by differential equations. Usually, in realistic systems, time series of dynamical variables—rather than governing equations—are experimentally available. In this case the time-delay embedding method [258] can be applied in order to transform a scalar time series into a trajectory in phase space. This procedure allows one to find the desired unstable periodic orbits and to calculate variations of parameters required to control friction.

The brute-force method of control based on a harmonic modulation of the normal load $L_n(t)$ has been studied within various approaches that include the generalized Tomlinson model [245], one-dimensional rate-state models [247,249,251], and grand-canonical MD simulations [249]. All calculations demonstrated that oscillations of the normal load could lead to a transition from a high-friction stick–slip dynamics to a low-friction sliding state. This effect can be controlled through the selection of the oscillation frequency. Mechanism of this phenomenon in lubricated junctions has been clarified by MD simulations [249] which show that oscillations of the normal load frustrate ordering in the lubricated film, maintaining it in a nonequilibrium sliding state with low friction. Theoretical predictions have been supported by recent experiments [250,251] which indicate that normal vibrations generally stabilize the system against stick–slip oscillations, at least for a modulation frequency much larger than the stick–slip one.

VII.   MULTI-ASPERITY CONTACTS

Most mechanical junctions discussed in the previous sections are single nanoscale or microscale contacts where the two opposing solids are in intimate, atomic contact and where the geometry of the walls can be approximated as either flat or uniformly curved. On larger length scales, however, surfaces are almost always rough or self-affine. A measure for the roughness is the so-called roughness exponent H, which states how fast the mean corrugation σ_z increases with the area $A = \pi R^2$, in which σ_z is determined, namely,

$$\sigma_z(R) = \langle z^2 \rangle_R - \langle z \rangle_R^2 \propto R^H \tag{46}$$

where $\langle z^n \rangle_R$ denotes the average nth moment of the surface height z measured in an area of radius R. In practical applications, H is frequently assessed in a range $1\,\text{nm} \le R \le 1\,\text{mm}$. The observed roughness depends strongly on the instruments used to determine σ_z, and we refer to Ref. 259 for further details. Since for most surfaces $H > 0$, the actual area of molecular contact between two surfaces, A_{real}, is generally a small fraction of the apparent macroscopic area [31,33].

The motion of the individual contacts is neither strictly correlated nor strictly uncorrelated, because different microjunctions interact with each other through elastic deformations within the bulk. The interplay of local shear forces in individual contacts and/or surface roughness and long-range elasticity will be discussed in this section. One of the pioneering studies of this competition was made by Burridge and Knopoff [260]. They suggested a model similar to the FKT model shown in Fig. 7 and discussed in Section III.C.6. The main difference is that one degree of freedom in the Burridge–Knopoff (BK) rather than an individual atom model represents a whole mechanical block. The underlying assumption is that within one BK block all microcontacts move completely correlated, while motion of neighboring blocks is much less correlated. One nonmoving junction can counterbalance an intrabulk elastic force up to a static friction F_s. Once a junction slides, its lateral motion is counteracted by a kinetic friction force $F_k < F_s$. The BK model is mainly employed in the context of seismology, and we refer to a review by Carlson et al. [261] for more details.

In the remainder of this section, we will first analyze the competition between local shear and long-range elastic interaction in the case of hard solids. One of the central questions is how one can determine the size of one BK block. Finally, in Section VII.B, the sliding motion of a soft elastic body with large internal friction on a corrugated substrate will be discussed. In the latter case, friction arises not only as a consequence of finite shear forces in the individual microcontacts but also because of the internal friction of the elastic medium.

A. Elastic Coherence Length

Consider a 3-dimensional, corrugated solid placed on a smooth substrate as a simplified model for a mechanical contact, which is a subtle case (Section III.C.5). The macroscopic contact will then consist of individual junctions where asperities from the corrugated solid touch the substrate. A microscopic point of contact p then carries a normal load l_p, and a shear force $\vec{f}_p$ will be exerted from the substrate to the asperity and vice versa. These random forces $\vec{f}_p$ will try to deform both solids. For the sake of simplicity, let us only consider elastic deformations in the top solid. Asperities in intimate contact with the substrate will be subject to a competition between the (elastic) coupling to the top solid and the interaction with the substrate. If the elastic stress $\sigma_{e,p}$ exceeds the local critical shear stress $\sigma_{c,p}$ of junction p, the contact will break and asperity p will find a new mechanical equilibrium position. In order for $\sigma_{e,p}$ to exceed $\sigma_{c,p}$, the area $A = \pi L^2$ over which the random forces accumulate must be sufficiently large. The value of L where this condition is satisfied is called the elastic coherence length $\mathscr{L}$. It is believed to play an important role in various aspects of friction.

The concept of an elastic coherence length was first introduced in the context of charge density waves [28,62] and flux line lattices in superconducters [262]. The relevance of this concept for friction was suggested by Baumberger and Caroli [54]: The contact can be decomposed into pinning centers, also called Larkin domains, with a typical linear size $\mathscr{L}$, which constitute the BK blocks introduced in the beginning of this section.

We are now concerned with the calculation of $\mathscr{L}$. We roughly follow the derivation given by Persson and Tosatti [27]. The condition for the elastic coherence length can be formulated as follows. Over what area $A_{\mathscr{L}} = \pi \mathscr{L}^2$ does one need to accumulate the elastic displacement field $u(\mathbf{x})$ in order to exceed the critical value a_c (corresponding to σ_c) at which an individual asperity breaks? Up to that point, linear response theory applies; beyond that point, it fails. Hence $\mathscr{L}$ can be obtained from the condition

$$a_c^2 = \frac{2}{A_{\mathscr{L}}} \int_0^{\mathscr{L}} d^2x' [\langle u(\mathbf{x})^2 \rangle - \langle u(\mathbf{x}) u(\mathbf{x} + \mathbf{x}') \rangle] \tag{47}$$

The displacement field $u(\mathbf{x})$ is related to the externally imposed stress $\sigma(\mathbf{x})$ through

$$u(\mathbf{x}) = \int dx'' G(\mathbf{x} - \mathbf{x}'') \sigma(\mathbf{x}'') \tag{48}$$

where $G(\mathbf{x})$ is a Green's function. Although $u(\mathbf{x})$, $G(\mathbf{x})$, and $\sigma(\mathbf{x})$ correspond to vectors and matrices in principle, indices are skipped for the sake of simplicity. $G(\mathbf{x})$ can be related to Young's modulus E through a relation of the form $G(\mathbf{x}) = 1/E|\mathbf{x}|$, modulo geometric factors incorporating Poisson's ratio. One can now write the last term in the integrand of Eq. (47) as

$$\int d^2x'' \int d^2x''' G(\mathbf{x} - \mathbf{x}'') G(\mathbf{x} + \mathbf{x}' - \mathbf{x}''') \langle \sigma(\mathbf{x}'') \sigma(\mathbf{x}''') \rangle \tag{49}$$

It is commonly assumed [26,27] that the shear is highly correlated within one contact and completely uncorrelated between different microjunctions. This introduces a lower cutoff at the (average) radius R_c of a microcontact in the integrals above, and it allows one to replace $\langle \sigma(\mathbf{x}'') \sigma(\mathbf{x}''') \rangle$ with δ-correlated terms of the form $\pi R_c^2 \langle \sigma_c^2 \rangle \delta(\mathbf{x}'' - \mathbf{x}''')$. Thus the expression in Eq. (49) can be reformulated to

$$\pi R_c^2 \langle \sigma_c^2 \rangle \int_{R_c}^{\mathscr{L}} d^2x'' G(\mathbf{x} - \mathbf{x}'') G(\mathbf{x} + \mathbf{x}' - \mathbf{x}'') \tag{50}$$

The condition for the breakdown of linear response theory, Eq. (47), now reads

$$a_c^2 = \frac{2}{A_{\mathscr{L}}} \pi R_c^2 \langle \sigma_c^2 \rangle \int_{R_c}^{\mathscr{L}} d^2x' \, d^2x'' \, [G(\mathbf{x} - \mathbf{x}'')^2 - G(\mathbf{x} - \mathbf{x}'')G(\mathbf{x} + \mathbf{x}' - \mathbf{x}'')] \quad (51)$$

$$\approx 4\pi^2 R_c^2 \frac{\langle \sigma_c^2 \rangle}{E^2} \ln \frac{\mathscr{L}}{R_c} \qquad \text{for } \mathscr{L} \gg R_c \quad (52)$$

The elastic coherence length $\mathscr{L}$ thus depends exponentially on the square of the contact radius R_c and the square of the elastic displacement a_c of a junction before failure. Both R_c and a_c will depend on material properties, surface geometry, and interface properties. Thus R_c and a_c are both statistical quantities whose average value is extremely hard to assess, and any calculation of $\mathscr{L}$ can only be interpreted as an attempt to estimate its order of magnitude. This may be the origin of the large discrepancies in the calculated values for $\mathscr{L}$. Persson and Tosatti obtain an astronomically large value for steel on steel, namely, $\mathcal{O}(\mathscr{L}) \approx 10^{100\,000}$ m, while Sokoloff [26] concludes that $\mathcal{O}(\mathscr{L}) \approx 1\,\mu\text{m}$ is a typical value for two generic materials in contact.

It might yet be possible to predict trends from Eq. (52). Persson and Tosatti [27] argue that $\mathscr{L}$ decreases quickly with increasing normal pressure. This has important consequences for tectonic motion; for instance, small earthquakes typically do not occur close to the earth's surface. Sokoloff [122] concluded recently that even for his small value of $\mathscr{L}$, the contribution to kinetic friction due to elastic instabilities that result from a competition between surface roughness and elastic interactions would lead to rather small friction coefficients on the order of 10^{-5}, provided that no contamination layer or other local friction mechanisms were present in the contacts.

B. Rubber Friction

The friction force between a rubber and a rough surface is often modelled by an interfacial adhesive and a bulk hysteretic component [263]. While the adhesive component is only important for very clean surfaces, the pioneering studies of Grosch [264] have shown that sliding rubber friction is in many cases dominated by the *internal* friction of rubber. In particular, the friction coefficient of rubber sliding on silicon carbide paper and glass surfaces have the same temperature dependence as that of the complex elastic modulus $E(\omega)$ of bulk rubber. This behavior is in contrast to most other frictional systems discussed here, for which the resistance to sliding originates from elastic or plastic instabilities within or near the interface between the two solid bodies. The large internal friction within rubber at small frequencies is, however, due to thermal configurational changes rather than to "simple" phonon damping. In this sense, the relatively large rubber friction at small velocities also originates from instabilities.

In a series of recent papers, Persson and coworkers [265–267] as well as Klüppel and Gläser [268] have developed a theory that connects the friction force with the energy loss due to the deformations that rubber experiences while sliding over a rough substrate. The rubber is able to accommodate the surface roughness profile over a range of length scales that depends on slider velocity. In this theory, the friction coefficient can be calculated as a function of surface roughness and the frequency-dependent loss and storage modulus. Strictly speaking, the static friction force is zero in such a treatment because for infinitely low sliding speeds, $\omega = 0$, would be the main relevant frequency in $E(\omega)$ and hence there would be no loss. However, owing to the extremely broad distribution of loss frequencies and roughness scales, one may argue that also the quasi-static friction forces that make our tires operate on the street are related to the above-mentioned dissipation processes.

Persson and co-workers [265–267] consider a rough, rigid surface with a height profile $h(\mathbf{x})$, where $\mathbf{x}$ is a two-dimensional vector in the x-y plane. In reaction to $h(\mathbf{x})$ and its externally imposed motion, the rubber will experience a (time-dependent) normal deformation $\delta z(\mathbf{x}, t)$. If one assumes the rubber to be an elastic medium, then it is possible to relate $\delta\tilde{z}(\mathbf{q}, \omega)$, which is the Fourier transform (F.T.) of $\delta z(\mathbf{x}, t)$, to the F.T. of the stress $\tilde{\sigma}(\mathbf{q}, \omega)$. Within linear-response theory, one can express this in the rubber-fixed frame (indicated by a prime) via

$$\delta\tilde{z}'(\mathbf{q}, \omega) = G(\mathbf{q}, \omega)\tilde{\sigma}'(\mathbf{q}, \omega) \tag{53}$$

In the present context it will be helpful to remember that differentiating a function with respect to $\mathbf{x}$ and/or with respect to t corresponds to multiplying the F.T. of the function with $i\mathbf{q}$ and/or $-i\omega$, respectively. Bearing this is mind, $i\mathbf{q}\delta\tilde{z}'(\mathbf{q}, \omega)$ can be associated with a strain and thus $-iG(\mathbf{q}, \omega)/q$ can be interpreted in the continuum limit as the inverse of a generalized frequency-dependent elastic constant $E(\omega)$. Strictly speaking, $G(\mathbf{q}, \omega)$ is a tensor, for which in general $(G_{\alpha,\beta})^{-1} \neq (G^{-1})_{\alpha,\beta}$; but for the sake of a clear presentation, we omit indices here. Instead we refer to Appendix A in Ref. 266 for a more detailed discussion of the precise form of $G(\mathbf{q}, \omega)$ and $E(\omega)$ and their dependence on the Poisson ratio.

The crucial assumption is to relate the macroscopic rate of dissipation (which appears on the left-hand side of the following equation) with the microscopic rate of dissipation (appearing on the right-hand side):

$$\sigma_f A_0 \dot{x} = \int_{A_0} d^2x \, \delta\dot{z}'(\mathbf{x}, t) \, \sigma'(\mathbf{x}, t) \tag{54}$$

Here σ_f denotes the frictional shear stress occurring at the interface of area A_0. It is first assumed that during sliding, the whole rubber interfacial area moves forward according to $\mathbf{r}(t)$ with $\dot{\mathbf{r}}(t) = \dot{x}(t)\mathbf{e}_x$ and it is also assumed that the rubber is able to deform and completely follow the substrate surface profile $h(\mathbf{x})$. The second assumption will later be replaced by assuming a velocity-dependence for the extent to which the rubber can accommodate the substrate's surface profile. For now, however, the transformation between rubber-fixed and substrate-fixed frame reads

$$\delta z'(\mathbf{x}, t) = \delta z(\mathbf{x} - \mathbf{r}(t)) \tag{55}$$

which, with the above-made remarks, is equivalent to

$$\delta z'(\vec{q}, \omega) = \tilde{h}(\vec{q}) f(\vec{q}\,\omega) \tag{56}$$

with the definition of

$$f(\vec{q}, \omega) = \frac{1}{2\pi} \int dt\, e^{-i[\mathbf{q}\mathbf{x}(t) - \omega t]} \tag{57}$$

Substituting Eqs. (53) and (55)–(57) into Eq. (54), yields

$$\sigma_f(t) = \int d^2q\, d\omega\, d\omega'\, iE(-\mathbf{q}, \omega')\, \tilde{h}(\mathbf{q})\, \tilde{h}(-\mathbf{q}) f(\mathbf{q}, \omega) f(-\mathbf{q}, \omega') e^{i(w+w')t} \tag{58}$$

where we have made use of the comments following Eq. (53) and of the identity $\delta \dot{z}' = -\dot{x}\nabla\delta z$, which follows from Eq. (55). This equation relates the macroscopic dissipative response $\sigma_f(t)$ to the surface corrugation $h(\vec{q})$, the dynamic elastic module $E(\mathbf{q}, \omega)$, and the trajectory of the rubber, which is reflected in $f(\mathbf{q}, \omega)$. The special case of constant sliding velocity $\mathbf{r}(t) = vt\mathbf{e}_x$ corresponds to $f(\mathbf{q}, \omega) = \delta(q_x v - \omega)$ and thus for steady-state sliding we obtain

$$\sigma_f(v) = \int d^2q\, iE(-\mathbf{q}, -q_x v))\, |\, h(\mathbf{q})\, |^2 \tag{59}$$

This result was obtained by Persson [266] before the more general equation for nonsteady state was obtained in Ref. 267. An example for the connection between the loss part of the elastic modulus and the friction coefficient is shown in Fig. 22. We also want to reemphasize that $E(\mathbf{q}, \omega)$ should not be taken literally as the bulk modulus or a coefficient of the tensor of elastic constants but as a more generalized expression, which is discussed in detail in Ref. 266. For a more detailed presentation that also includes contact mechanics and that allows one to calculate the friction coefficient, we refer to the original literature [267].

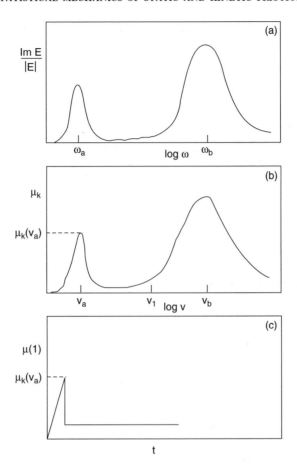

Figure 22. (a) Loss spectrum $\mathrm{Im}[E(\omega)]/|E(\omega)|$ as a function of frequency ω. (b) Kinetic friction coefficient at constant sliding velocity v. (c) Transient behavior of the friction coefficient as a function of time t. At the smallest t, the velocity is abruptly changed from zero velocity to finite velocity. With permission from Ref. 267. *Phys. Rev. B* **65**, 134106 (2002).

While Eq. (59) gives the shear stress for an externally imposed constant sliding velocity, Eq. (58) provides a way to incorporate the history of the past motion into the calculation of the friction force. Persson and Volokitin suggest that rubber friction may also lead to a so-called stiction peak [267]; that is, some short time after sliding is initiated, the lateral force takes a large value that it will not reach again in the following motion. This phenomenon is often observed during stick–slip motion and can easily be explained by the fact that static friction increases with the age of the contact. Here, however, the stiction peak would have to be associated with relaxation processes within the rubber. Figure 22 demonstrates the striction peak.

VIII. CONCLUDING REMARKS

The theoretical studies undertaken over the last decade have led to significant breakthroughs in our understanding of tribology. However, much work remains to be done. In concluding, we would like to emphasize some of the important lessons learned, unresolved problems, and directions for future studies.

The theoretical approaches that have been used to investigate frictional forces fall into three classes: large scale MD simulations, "minimalistic" models based on simple microscopic interactions, and phenomenological rate-state models for macroscopic variables. Each approach has limitations and strengths, and emphasizes different aspects of the problem. The phenomenological and minimalistic approaches have been able to reproduce many experimental results, but it is unclear how to relate the parameters in these models to microscopic interactions or how to extrapolate them from one system to another. MD simulations allow direct correlations between microscopic interactions and sliding behavior to be established, but are unable to access the large length and long time scales of macroscopic experiments. Future theoretical studies that connect "exact" molecular level calculations to macroscopic models could allow these limitations to be overcome, extending simulations from the nm/fs to the μm/μs scales, and offering a practical approach for investigating molecular systems under shear for a wide range of parameters.

One key message of the theories presented above is that shear-induced instabilities on the microscopic scale control macroscopic frictional behavior. The embedded molecular systems at the interface must exhibit mechanical, and perhaps electronic, instabilities in order to produce both static friction and a kinetic friction that remains finite in the limit of small sliding velocities. The presence of these instabilities implies that certain degrees of freedom move rapidly even when the center of mass of the slider moves arbitrarily slowly. The important questions to be asked about a given sheared system are (a) what are the relevant types of microscopic instabilities, and (b) how does each couple to the macroscopic tribological behavior. The relative contributions of different types of instability depend not only on the properties of the embedded system but also on the stiffness, size and mass of the experimental system, and, for instance, may be different in SFA and AFM experiments. A better understanding of molecular processes and their coupling to normal and lateral displacements of the slider may allow one to optimize frictional forces and stabilize desirable phases of motion, such as steady sliding.

A key characteristic of microscopic instabilities is their time scale. Other important intrinsic time scales that need to be determined when analyzing a system are those for energy redistribution, thermal activation, plastic deformation, and the inertial response of the mechanical system that applies stress to the slider. There are also important extrinsic time scales related to the rate of change in the applied external force or displacement. The ratio of the charateristic intrinsic and extrinsic times controls the types of response that will occur and approximations that can be made. A thorough understanding of the time scales will help to formulate more accurate, and molecularly motivated, rate-state models of friction.

Two important areas of investigation have been relatively overlooked. One is the influence of roughness at nm and larger scales on frictional forces. Most studies have considered single asperities with flat or uniformly curved surfaces. The second is strongly irreversible tribological processes, which include material mixing, cold welding, tribochemical and triboelectrical effects. A common feature of all these processes is the absence of a steady state. So far only a few MD simulations have been used to study strongly irreversible phenomena. Simpler theoretical models are needed to provide deeper insight into these complex processes, which are of great technological importance.

Acknowledgments

We are grateful to K. Binder and J. Klafter for critical comments on the manuscript. MHM and MU acknowledge support from the Israeli-German D.I.P.-Project No 352-101. MOR acknowledges support from National Science Foundation Grant DMR-DMR-0083286.

References

1. D. Dowson, *History of Tribology*, Longman, New York, 1979.

2. J. Krim and A. Widom, *Phys. Rev. B* **38**, 12184 (1988).

3. J. Krim, E. T. Watts, and J. Digel, *J. Vac. Sci. Technol. A* **8**, 3417 (1990).

4. J. Krim, D. H. Solina, and R. Chiarello, *Phys. Rev. Lett.* **66**, 181 (1991).

5. G. Binning, C. F. Quate, and C. Gerber, *Phys. Rev. Lett.* **56**, 930 (1986).

6. C. M. Mate, G. M. McClelland, R. Erlandsson, and S. Chiang, *Phys. Rev. Lett.* **59**, 1942 (1987).

7. J. N. Israelachvili, P. M. McGuiggan, and A. M. Homola, *Science* **240**, 189 (1988).

7a. J. N. Israelachvili, *Surf. Sci. Rep.* **14**, 109 (1992).

8. S. Granick, *Science* **253**, 1374 (1992).

9. R. W. Carpick and M. Salmeron, *Chem. Rev.* **97**, 1163 (1997).

10. P. M. McGuiggan, J. Zhang, and S. M. Hsu, *Tribol. Lett.* **10**, 217 (2001).

11. F. Tiberg G. Bogdanovic, and M. W. Rutland, *Langmuir* **17**, 5911 (2001).

12. J. A. Harrison, S. J. Stuart, and D. W. Brenner, in B. Bhushan, ed., *Handbook of Micro/Nanotribology*, CRC Press, Boca Raton, FL, 1999, p. 525.

13. M. O. Robbins and M. H. Müser, in B. Bhushan, ed., *Modern Tribology Handbook*, CRC Press, Boca Raton, FL, 2001, p. 717.

14. Amontons, *Histoire de l'Académie Royale des Sciences avec les Mémoires de Mathématique et de Physique*, 1699, p. 206.

15. J. Leslie, *An Experimental Inquiry into The Nature and Propagation of Heat*, printed for J. Newman, No. 22 (1804).

16. M. Brillouin, *Notice sur les Traveaux Scientifique*, Gauthiers-Vilars, Paris, 1990.

17. G. A. Tomlinson, *Philos. Mag. Series* **7**, 905 (1929).

17a. M. G. Rozman, M. Urbakh, and J. Klafter, *Phys. Rev. Lett.* **77**, 683 (1996).

17b. M. G. Rozman, M. Urbakh, and J. Klafter, *Phys. Rev. E* **54**, 6485 (1996).

18. L. Prandtl, *Z. Angew. Math. Mech.* **8**, 85 (1928).

19. E. Gerde and M. Marder, *Science* **413**, 285 (2001).

20. T. Baumberger, C. Caroli, and O. Ronsin, *Phys. Rev. Lett.* **88**, 075509 (2002).

21. C. Caroli and P. Nozières, *Eur. Phys. J. B* **4**, 233 (1998).

22. M. Hirano and K. Shinjo, *Phys. Rev. B*, **41**, 11837 (1990).

23. M. Hirano and K. Shinjo, *Wear* **168**, 121 (1993).

24. M. H. Müser and M. O. Robbins, *Phys. Rev. B* **61**, 2335 (2000).

25. M. H. Müser, *Tribol. Lett.* **10**, 15 (2001).

26. J. B. Sokoloff, *Phys. Rev. Lett.* **86**, 3312 (2000).

27. B. N. J. Persson and E. Tosatti, *Solid State Commun.* **109**, 739 (1999).

28. H. Fukuyama and P. A. Lee, *Phys. Rev. B* **17**, 535 (1978).

29. D. S. Fisher, *Phys. Rev. B* **31**, 1396 (1985).

30. G. He, M. H. Müser, and M. O. Robbins, *Science* **284**, 1650 (1999).

31. F. P. Bowden and D. Tabor, *The Friction and Lubrication of Solids*, Clarendon Press, Oxford, 1986.

32. P. Berthoud and T. Baumberger, *Proc. R. Soc. A* **454**, 1615 (1998).

33. J. H. Dieterich and B. D. Kilgore, *Pure Appl. Geophysics* **143**, 238 (1994).

34. J. H. Dieterich and B. D. Kilgore, *Tectonophysics* **256**, 219 (1996).

35. I. L. Singer, Solid lubrication processes, in *Fundamentals of Friction: Macroscopic and Microscopic Processes*, Elsevier, Amsterdam, 1992, pp. 237–261.

36. B. J. Briscoe and D. Tabor, *J. Adhesion* **9**, 145 (1978).

37. B. J. Briscoe, *Philos. Mag. A* **43**, 511 (1981).

38. B. J. Briscoe and D. C. B. Evans, *Proc. R. Soc. A* **380**, 389 (1982).

39. M. L. Gee, P. M. McGuiggan, J. N. Israelachvili, and A. M. Homola, *J. Chem. Phys.* **93**, 1895 (1990).

40. A. Berman, C. Drummons, and J. N. Israelachvili, *Tribol. Lett.* **4**, 95 (1998).

41. J. A. Greenwood and J. B. P. Williamson. *Proc. R. Soc. A* **295**, 300 (1966).

42. A. Volmer and T. Natterman, *Z. Phys. B* **104**, 363 (1997).

43. B. N. J. Persson, *Phys. Rev. Lett.* **87**, 116101 (2001).

44. E. Rabinowicz, *Friction and Wear of Materials*, Wiley, New York, 1965.

45. Sir W. B. Hardy, *Collected Works*, Cambridge University Press, Cambridge, 1936.

46. D. M. Tolstoi, *Wear* **10**, 199 (1967).

47. J. M. Carlson and A. A. Batista, *Phys. Rev. E* **53**, 4153 (1996).

48. B. N. J. Persson, *Sliding Friction: Physical Principles and Applications*, Springer, Berlin, 1998.

49. E. Rabonowicz, *Proc. Phys. Soc. London* **71**, 668 (1958).

50. J. H. Dieterich, *J. Geophys. Res.* **84**, 2161 (1979).

51. A. Ruina, *J. Geophys. Res.* **88**, 10359 (1983).

52. J. H. Dieterich and B. D. Kilgore, *Proc. Natl. Acad. Sci. USA* **93**, 3787 (1996).

53. F. Heslot, T. Baumberger, B. Perrin, B. Caroli, and C. Caroli, *Phys. Rev. E* **49**, 4973 (1994).

54. T. Baumberger and C. Caroli, *Comments Cond. Mater. Phys.* **17**, 306 (1995).

55. P. Hänggi, P. Talkner, and M. Borkovec. *Rev. Mod. Phys.* **62**, 251 (1990).

56. P. Berthoud and T. Baumberger, *Europhys. Lett.* **41**, 617 (1998).

57. Y. Estrin and Y. Bréchet, *Pageoph* **147**, 745 (1996).

58. P. Le Doussal, P. Chauve, and T. Giamarchi, *Europhys. Lett.* **44**, 110 (1998).

59. P. Le Doussal, P. Chauve, and T. Giamarchi, *Phys. Rev. B* **62**, 6241 (2000).

60. S. Panyukov and Y. Rabin, *Phys. Rev. E* **56**, 7053 (1997).

61. M. H. Müser, L. Wenning, and M. O. Robbins. *Phys. Rev. Lett.* **86**, 1295 (2001).

62. P. A. Lee and T. M. Rice, *Phys. Rev. B* **19**, 3970 (1979).

63. M. R. Sørensen, K. W. Jacobsen, and P. Stoltze, *Phys. Rev. B* **53**, 2101 (1996).

64. H. Risken, *The Fokker Planck Equation*, Springer, Heidelberg, 1984.

65. G. E. Uhlenbeck and L. S. Ornstein, *Phys. Rev.* **36**, 823 (1930).

66. H. Risken and H. D. Vollmer, *Z. Phys. B*, **33**, 297 (1979).

67. M. R. Falvo et al., *Nature* **397**, 236 (1999).

68. M. R. Falvo, J. Steele, R. M. Taylor II, and R. Superfine, *Phys. Rev. B* **62**, 10665 (2000).

69. M. F. Yu, B. I. Yakobson, and R. S. Ruoff, *J. Phys. Chem. B* **104**, 8764 (2000).

70. J. Liu et al., *Science* **280**, 1253 (1998).

71. A. Buldum and S. Ciraci, *Phys. Rev. Lett.* **83**, 5050 (1999).

72. M. Damnjanovic, I. Milosevic, T. Vukovic, and R. R. Sredanovic, *J. Phys. A* **32**, 4097 (1999).

73. A. N. Kolmogorov and V. H. Crespi, *Phys. Rev. Lett.* **85**, 4727 (2000).

74. L. Wenning and M. H. Müser, *Europhys. Lett.* **54**, 693 (2001).

74a. K. Miura and S. Kamiya, *Europhys. Lett.* **58**, 601 (2002).

75. C. A. J. Putman, M. Igarashi, and R. Kaneko, *Appl. Phys. Lett.* **66**, 3221 (1995).

76. P. Schwarz, O. Zwörner, P. Köster, and R. Wiesendanger, *Phys. Rev. B* **56**, 6987 (1997).

77. H. Klein, D. Pailharey, and Y. Mathey, *Surf. Sci.* **387**, 227 (1997).

78. M. Weiss and F.-J. Elmer, *Phys. Rev. B* **53**, 7539 (1996).

79. T. Gyalog and H. Thomas, *Z. Phys. B* **104**, 669 (1997).

80. E. Gnecco, R. Bennewitz, T. Gyalog, C. Loppacher, M. Bammerlin, E. Meyer, and H. J. Güntherodt, *Phys. Rev. Lett.* **84**, 1172 (2000).

81. M. Porto, V. Zaloj, M. Urbakh, and J. Klafter, *Tribol. Lett.* **9**, 45 (2000).

82. E. Gnecco, R. Bennewitz, T. Gyalog, and E. Meyer, *J. Phys. Condens. Mater.* **13**, R619 (2001).

83. D. S. Fisher, *Phys. Rev. Lett.* **50**, 1486 (1983).

84. T. Baumberger and C. Caroli, *Eur. Phys. J. B* **4**, 13 (1998).

85. A. A. Middleton, *Phys. Rev. Lett.* **68**, 670 (1992).

86. E. Evans, *Annu. Rev. Biophys. Biomol. Struct.* **30**, 105 (2001).

87. Y. Sang, M. Dubé, and M. Grant, *Phys. Rev. Lett.* **87**, 174301 (2001).

88. O. K. Dudko, A. E. Filippov, J. Klafter, and M. Urbakh, *Chem. Phys. Lett.* **352**, 499 (2002).

89. E. Evans and K. Ritchie, *Biophys. J.* **72**, 1541 (1997).

90. J. N. Glosli and G. McClelland, *Phys. Rev. Lett.* **70**, 1960 (1993).

91. T. Ohzono, J. N. Glosli, and M. Fujihara, *Jpn. J. Appl. Physics*, **37**, 6535 (1998).

91a. J. A. Harrison, C. T. White, R. J. Colton, and D. W. Brenner, *Phys. Rev. B* **46**, 9700 (1992).

91b. J. A. Harrison, C. T. White, R. J. Colton, and D. W. Brenner, *J. Phys. Chem.* **97**, 6573 (1993).

92. T. Bouhacina, S. Gauthier, J. P. Aimé, D. Michel, and V. Heroguez, *Phys. Rev. B* **56**, 7694 (1997).

93. E. L. Florin, V. T. Moy, and H. E. Gaub, *Science* **264**, 415 (1994).

94. M. Rief, F. Oesterhelt, B. Heymann, and H. E. Gaub, *Science* **275**, 1295 (1997).

95. A. F. Oberhauser, P. E. Marszalek, H. P. Erickson, and J. M. Fernandez, *Nature* **393**, 181 (1998).

96. G. U. Lee, L. A. Chrisey, and R. J. Colton, *Science* **266**, 771 (1994).

97. T. Charitat and J.-F. Joanny, *Eur. Phys. J. E* **3**, 369 (2000).

98. Y. I. Frenkel and T. Kontorova, *Zh. Eksp. Teor. Fiz.* **8**, 1340 (1938).

99. U. Dehlinger, *Ann. Phys.* **5**, 749 (1929).

100. O. M. Braun and Yu. S. Kivshar, *Phys. Rep.* **306**, 1 (1998).

101. D. S. Fisher, *Phys. Rep.* **301**, 113 (1998).

102. S. Aubry, *J. Phys. (Paris)* **44**, 147 (1983).

103. M. Peyrard and S. Aubry, *J. Phys. C* **16**, 1593 (1983).

104. R. S. MacKay, *Physica D* **50**, 71 (1991).

105. P. M. Chaikin and T. C. Lubensky, Cambridge University Press, Cambridge, 1995, Chapter 10.

106. F. C. Frank and J. H. Van der Merwe, *Proc. R. Soc.* **198**, 205 (1949).

107. A. Kochendörfer and A. Seeger, *Z. Physik* **127**, 533 (1950).

108. M. J. Rice, A. R. Bishop, J. A. Krumhansl, and S. E. Trullinger, *Phys. Rev. Lett.* **36**, 432 (1976).

109. D. J. Bergman, E. Ben-Jacob, Y. Imry, and K. Maki, *Phys. Rev. A* **27**, 3345 (1983).

110. R. Peierls, *Proc. R. Soc. London* **52**, 34 (1940).

111. F. R. N. Nabarro, *Proc. R. Soc. London* **59**, 256 (1947).

112. V. G. Bar'yakhtar, B. A. Ivanov, A. L. Sukstanskii, and E. V. Tartakovskaya, *Teor. Mat. Fiz.* **74**, 46 (1988).

113. P. Bak, *Rep. Prog. Phys.* **45**, 587 (1982).

114. P. Bak and Fukuyama, *Phys. Rev. B* **21**, 3287 (1980).

115. B. Hu, B. Li, and H. Zhao, *Europhys. Lett.* **53**, 342 (2001).

116. H. Miyake and H. Matsukawa, *J. Phys. Soc. Jpn.* **67**, 3891 (1998).

117. V. L. Popov, *Phys. Rev. Lett.* **83**, 1632 (1999).

118. C.-I. Chou, C.-I. Ho, and B. Hu, *Phys. Rev. E* **55**, 5092 (1997).

119. S. F. Mingaleev, Y. B. Gaididei, E. Majernikova, and S. Shpyrko, *Phys. Rev. E* **61**, 4454 (2000).

120. A. Vanossi, J. Röder, A. R. Bishop, and V. Bortolani, *Phys. Rev. E* **63**, 017203 (2001).

121. T. Kawaguchi and H. Matsukawa, *Phys. Rev. B* **56**, 13932 (1997).

122. J. B. Sokoloff, *Phys. Rev. B* **65**, 115415 (2002).

123. M. Weiss and F.-J. Elmer, *Z. Phys. B* **104**, 55 (1997).

124. F. Lançon, J. M. Penisson, and U. Dahmen, *Europhys. Lett.* **49**, 603 (2000).

125. F. Lançon, *Europhys. Lett.* **57**, 74 (2002).

126. M. Hirano, K. Shinjo, R. Kaneko, and Y. Murata, *Phys. Rev. Lett.* **67**, 2642 (1991).

127. J. M. Martin, C. Donnet, and T. Le Mogne, *Phys. Rev. B* **48**, 10583 (1993).

128. I. L. Singer, private communication.

129. M. Hirano, K. Shinjo, R. Kaneko, and Y. Murata, *Phys. Rev. Lett.* **78**, 1448 (1997).

130. J. Krim, *Sci. Am.* **275**(4), 74 (1996).

131. A. Dayo, W. Alnasrallay, and J. Krim, *Phys. Rev. Lett.* **80**, 1690 (1998).

132. M. O. Robbins and J. Krim, *MRS Bull.* **23**(6), 23 (1998).

133. M. S. Tomassone, J. B. Sokoloff, A. Widom, and J. Krim, *Phys. Rev. Lett.* **79**, 4798 (1997).

134. M. Cieplak, E. D. Smith, and M. O. Robbins, *Science* **265**, 1209 (1994).

135. E. D. Smith, M. Cieplak, and M. O. Robbins, *Phys. Rev. B.* **54**, 8252 (1996).

136. B. N. J. Persson, *Phys. Rev. B* **44**, 3277 (1991).

137. L. W. Bruch, *Phys. Rev. B* **61**, 16201 (2000).

138. B. L. Mason, S. M. Winder, and J. Krim, *Tribol. Lett.* **10**, 59 (2001).

139. T. Novotný and B. Velický, *Phys. Rev. Lett.* **83**, 4112 (1999).

140. V. L. Popov, *JETP Lett.* **69**, 558 (1999).

141. R. L. Renner, J. E. Rutledge, and P. Taborek, *Phys. Rev. Lett.* **83**, 1261 (1999).

142. J. Krim, *Phys. Rev. Lett.* **83**, 1262 (1999).

143. M. S. Tomassone and J. B. Sokoloff, *Phys. Rev. B* **60**, 4005 (1999).

144. C. Mak and J. Krim, *Faraday Discuss.* **107**, 389 (1997).

145. A. Brass, H. J. Jensen, and A. J. Berlinsky, *Phys. Rev. B* **39**, 102 (1989).

146. M. J. Williams, *J. Appl. Mech.* **19**, 526 (2001).

147. M. C. Kuo and D. B. Bogy, *J. Appl. Mech.* **41**, 197 (1974).

148. E. D. Reedy, Jr., *Eng. Fract. Mech.* **36**, 575 (1990).

149. O. Vafek and M. O. Robbins, *Phys. Rev. B* **60**, 12002 (1999).

150. J. A. Hurtado and K. S. Kim, *Proc. R. Soc. Ser. A* **455**, 3363 (1999).

151. J. A. Hurtado and K. S. Kim, *Proc. R. Soc. Ser. A* **455**, 3385 (1999).

152. J. A. Nieminen, A. P. Sutton, and J. B. Pethica, *Acta Metall. Mater.* **40**, 2503 (1992).

153. U. Landman, W. D. Luedtke, N. A. Burnham, and R. J. Colton, *Science* **248**, 454 (1990).

154. U. Landman, W. D. Luedtke, and E. M. Ringer, *Wear* **153**, 3 (1992).

155. U. Landman and W. D. Luedtke, *Surf. Sci. Lett.* **210**, L117 (1989).

156. U. Landman and W. D. Luedtke, *J. Vac. Sci. Technol. B* **9**, 414 (1991).

157. H. Raffi-Tabar, J. B. Pethica, and A. P. Sutton, *Mater. Res. Soc. Symp. Proc.* **239**, 313 (1992).

158. O. Tomagnini, F. Ercolessi, and E. Tosatti, *Surf. Sci.* **287/288**, 1041 (1993).

159. U. Landman, W. D. Luedtke, and J. Gao, *Langmuir* **12**, 4514 (1996).

160. J. A. Harrison, C. T. White, R. J. Colton, and D. W. Brenner, *Surf. Sci.* **271**, 57 (1992).

161. J. A. Harrison, D. W. Brenner, C. T. White, and R. J. Colton, *Thin Solid Films* **206**, 213 (1991).

162. J. Belak and I. F. Stowers, in I. L. Singer and H. M. Pollock, eds., *Fundamentals of Friction: Macroscopic and Microscopic Processes*, Kluwer Academic Publishers, Dordrecht (1992), p. 511.

163. T. Moriwaki and K. Okuda, *Ann. CIRP* **38**, 115 (1989).

164. N. P. Suh, *Tribophysics*, Prentice-Hall, Englewood Cliffs, NJ, 1986.

165. L. Zhang and H. Tanaka, *Wear* **211**, 44 (1997).

166. L. Zhang and H. Tanaka, *Tribol. Int.* **31**, 425 (1998).

167. D. A. Rigney and J. E. Hammerberg, *MRS Bull.* **23**(6), 32 (1998).

168. J. E. Hammerberg, B. L. Holian, J. Röder, A. R. Bishop, and J. J. Zhou, *Physica D*, **123**, 330 (1998).

169. F. F. Abraham, *J. Chem. Phys.* **68**, 3713 (1978).

170. S. Toxvaerd, *J. Chem. Phys.* **74**, 1998 (1981).

171. I. K. Snook and W. van Megen, *J. Chem. Phys.* **72**, 2907 (1980).

172. S. Nordholm and A. D. J. Haymet, *Aust. J. Chem.* **33**, 2013 (1980).

173. M. Plischke and D. Henderson, *J. Chem. Phys.* **84**, 2846 (1986).

174. P. A. Thompson and M. O. Robbins, *Phys. Rev. A* **41**, 6830 (1990).

175. J. Magda, M. Tirrell, and H. T. Davis, *J. Chem. Phys.* **83**, 1888 (1985).

176. M. Schoen, C. L. Rhykerd, D. J. Diestler, and J. H. Cushman, *J. Chem. Phys.* **87**, 5464 (1987).

177. I. Bitsanis and G. Hadziioannou, *J. Chem. Phys.* **92**, 3827 (1990).

178. P. A. Thompson, M. O. Robbins, and G. S. Grest, *Isr. J. Chem.* **35**, 93 (1995).

179. J. Gao, W. D. Luedtke, and U. Landman, *J. Phys. Chem. B* **101**, 4013 (1997).

180. J. Gao, W. D. Luedtke, and U. Landman, *J. Chem. Phys.* **106**, 4309 (1997).

181. P. A. Thompson and M. O. Robbins, *Science* **250**, 792 (1990).

182. U. Landman, W. D. Luedtke, and M. W. Ribarsky, *J. Vac. Sci. Technol. A* **7**, 2829 (1989).

183. M. Schoen, J. H. Cushman, D. J. Diestler, and C. L. Rhykerd, *J. Chem. Phys.* **88**, 1394 (1988).

184. M. Schoen, C. L. Rhykerd, D. J. Diestler, and J. H. Cushman, *Science* **245**, 1223 (1989).

185. J.-L. Barrat and L. Bocquet, *Faraday Discuss.* **112**, 1 (1999).

186. L. Bocquet and J.-L. Barrat, *Phys. Rev. E* **49**, 3079 (1994).

187. J.-L. Barrat and L. Bocquet, *Phys. Rev. Lett.* **82**, 4671 (1999).

188. P. A. Thompson and S. M. Troian, *Nature* **389**, 360 (1997).

189. R. G. Horn and J. N. Israelachvili, *J. Chem. Phys.* **75**, 1400 (1981).

190. J. E. Curry, F. Zhang, J. H. Cushman, M. Schoen, and D. J. Diestler, *J. Chem. Phys.* **101**, 10824 (1994).

191. P. A. Thompson, G. S. Grest, and M. O. Robbins, *Phys. Rev. Lett.* **68**, 3448 (1992).

192. G. S. Grest and M. H. Cohen, in *Advances in Chemical Physics*, Wiley, New York, 1981, p. 455.

193. A. J. Liu and S. R. Nagel, *Jamming and Rheology: Constrained Dynamics on Microscopic and Macroscopic Scales*, Taylor and Francis, London, 2000.

194. G. He and M. O. Robbins, *Phys. Rev. B* **64**, 035413 (2001).

195. G. He and M. O. Robbins, *Tribol. Lett.* **10**, 7 (2001).

196. J. P. Gao, W. D. Luedtke, and U. Landman, *Tribol. Lett.* **9**, 3 (2000).

197. A. L. Demirel and S. Granick, *Phys. Rev. Lett.* **77**, 2261 (1996).

198. J. Van Alsten and S. Granick, *Phys. Rev. Lett.* **61**, 2570 (1988).

199. S. Granick, *MRS Bull.* **16**, 33 (1991).

200. J. Klein and E. Kumacheva, *Science* **269**, 816 (1995).

201. D. Y. C. Chan and R. G. Horn, *J. Chem. Phys.* **83**, 5311 (1985).

202. J. N. Israelachvili, *J. Colloid Interface Sci.* **110**, 263 (1986).

203. J. Baudry, E. Charlaix, A. Tonck, and D. Mazuyer, *Langmuir* **17**, 5232 (2001).

204. Y. Zhu and S. Granick, *Phys. Rev. Lett.* **87**, 096105 (2001).

205. J. Klein and E. Kumacheva, *J. Chem. Phys.* **108**, 6996 (1998).

206. J. N. Israelachvili, P. M. McGuiggan, M. Gee, A. M. Homola, M. O. Robbins, and P. A. Thompson, *J. Phys. Condens. Matter* **2**, SA89 (1990).

207. Y. Zhu and S. Granick, *Phys. Rev. Lett.* **87**, 096104 (2001).

208. U. Raviv, P. Laurat, and J. Klein, *Nature* **413**, 51 (2001).

209. M. J. Stevens, M. Mondello, G. S. Grest, S. T. Cui, H. D. Cochran, and P. T. Cummings, *J. Chem. Phys.* **106**, 7303 (1997).

210. J. D. Ferry, *Viscoelastic Properties of Polymers*, 3rd ed., Wiley, New York, 1980.

211. W. Götze and L. Sjögren, *Rep. Prog. Phys.* **55**, 241 (1992).

212. M. O. Robbins and A. R. C. Baljon, in V. V. Tsukruk and K. J. Wahl, eds., *Microstructure and Microtribology of Polymer Surfaces*, American Chemical Society, Washington D.C., 2000, p. 91.

213. A. R. C. Baljon and M. O. Robbins, in B. Bhushan, ed., *Micro/Nanotribology and Its Applications*, Kluwer, Dordrecht, 1997, p. 533.

214. A. R. C. Baljon and M. O. Robbins, *Science* **271**, 482 (1996).

215. M. O. Robbins and E. D. Smith, *Langmuir* **12**, 4543 (1996).

216. M. H. Müser, *Phys. Rev. Lett.* **89**, 224301(2002); cond-mat/0204395.

217. H. Czichos, S. Becker, and J. Lexow, *Wear* **135**, 171 (1989).

218. G. T. Gao, X. C. Zeng, and D. J. Diestler, *J. Chem. Phys.* **113**, 11293 (2000).

219. S. Nasuno, A. Kudrolli, and J. P. Gollub. *Phys. Rev. Lett.* **79**, 949 (1997).

220. C. T. Veje, D. W. Howell, and R. P. Behringer, *Phys. Rev. E* **59**, 739 (1999).

221. A. D. Gopal and D. J. Durian. *Phys. Rev. Lett.* **75**, 2610 (1995).

222. H. Yoshizawa and J. N. Israelachvili, *J. Phys. Chem.* **97**, 11300 (1993).

223. D. E. McCumber, *J. App. Phys.* **39**, 3113 (1968).

224. A. A. Batista and J. M. Carlson, *Phys. Rev. E* **57**, 4986 (1998).

225. A. Dhinojwala, S. C. Bae, and S. Granick, *Tribol. Lett.* **9**, 55 (2000).

226. P. A. Thompson and G. S. Grest, *Phys. Rev. Lett.* **67**, 1751 (1991).

227. H. M. Jaeger, S. R. Nagel, and R. P. Behringer, *Rev. Mod. Phys.* **68**, 1259 (1996).

228. M. J. Stevens and M. O. Robbins, *Phys. Rev. E* **48**, 3778 (1993).

229. B. J. Ackerson, J. B. Hayter, N. A. Clark, and L. Cotter, *J. Chem. Phys.* **84**, 2344 (1986).

230. M. Lupowski and F. van Swol, *J. Chem. Phys.* **95**, 1995 (1991).

231. A. L. Demirel and S. Granick, *Phys. Rev. Lett.* **77**, 4330 (1996).

232. A. Buldum and S. Ciraci, *Phys. Rev. B* **55**, 12892 (1997).

233. M. O. Robbins and P. A. Thompson, *Science* **253**, 916 (1991).

234. P. A. Thompson, M. O. Robbins, and G. S. Grest, in D. Dowson, C. M. Taylor, T. H. C. Childs, M. Godet, and G. Dalmaz, eds., *Thin Films in Tribology*, Elsevier, Amsterdam, 1993, p. 347.

235. J. Peachey, J. V. Alste, and S. Granick, *Rev. Sci. Instrum.* **62**, 462 (1991).

236. V. Zaloj, M. Urbakh, and J. Klafter, *J. Chem. Phys.* **110**, 1263 (1999).

237. V. Zaloj, M. Urbakh, and J. Klafter, *Phys. Rev. Lett.* **81**, 1227 (1998).

238. A. L. Demirel and S. Granick, *J. Chem. Phys.* **109**, 6889 (1998).

239. H.-W. Hu, G. A. Carson, and S. Granick, *Phys. Rev. Lett.* **66**, 2758 (1991).

240. Y.-K. Cho and S. Granick, *Wear* **200**, 346 (1996).

241. G. Luengo, F. J. Schmitt, R. Hill, and Israelachvili, *J. Macromolecules* **30**, 2482 (1997).

242. G. Luengo, J. Israelachvili, and S. Granick, *Wear* **200**, 1263 (1999).

243. A. L. Demirel and S. Granick, *J. Chem. Phys.* **115**, 1498 (2001).

244. M. G. Rozman, M. Urbakh, and J. Klafter, *Phys. Rev. E* **57**, 7340 (1998).

245. V. Zaloj, M. Urbakh, and J. Klafter, *Phys. Rev. Lett.* **82**, 4823 (1999).

246. J.-C. Géminard, W. Losert, and J. P. Gollub, *Phys. Rev. E* **59**, 5881 (1999).

247. L. Bureau, T. Baumberger, and C. Caroli, *Phys. Rev. E* **62**, 6810 (2000).

248. F.-J. Elmer, *Phys. Rev. E* **57**, R4903 (1998).

249. J. P. Gao, W. D. Luedtke, and U. Landman, *J. Phys. Chem. B* **102**, 5033 (1998).

250. M. Heuberger, C. Drummond, and J. N. Israelachvili, *J. Phys. Chem. B* **102**, 5038 (1998).

251. A. Cochard, L. Bureau, and T. Baumberger. cond-mat/0111369.

252. F. Lacombe, S. Zapperoi, and H. J. Herrman, *Eur. Phys. J. E* **2**, 181 (2000).

253. M. Linker and J. H. Dieterich, *J. Geophys. Res.* **97**, 4923 (1992).

254. E. Richardson and C. Marone, *J. Geophys. Res.* **104**, 28859 (1999).

255. C. M. Mate and A. M. Homola, in B. Bhushan, ed., *Micro/Nanotribology and Its Applications, Vol. 330 of NATO Advanced Sciences Institutes Series E: Applied Sciences*, Kluwer Academic Publishers, Dordrecht, 1997, p. 647.

256. E. Ott, C. Grebogi, and J. A. York, *Phys. Rev. Lett.* **64**, 1196 (1990).

257. E. Barreto, E. J. Kostelich, C. Grebogi, E. Ott, and J. A. York, *Phys. Rev. E* **51**, 4169 (1995).

258. H. D. I. Abarbanel, R. Brown, J. L. Sidorowich, and L. S. Tsimring, *Rev. Mod. Phys.* **65**, 1331 (1993).

259. B. Bhushan, in B. Bhushan, ed., *Modern Tribology Handbook*, CRC Press, Boca Raton, FL, 2001, p. 49.

260. R. Burridge and L. Knopoff, *Bull. Seismol. Soc. Am.* **57**, 341 (1967).

261. J. M. Carlson, J. S. Langer, and B. E. Shaw, *Rev. Mod. Phys.* **66**, 657 (1994).

262. A. I. Larkin and Y. N. Ovchinnikov, *J. Low Temp. Phys.* **34**, 409 (1979).

263. A. D. Roberts, *Rubber Chem. Technol.* **65**, 3 (1992).

264. K. A. Grosch, *Proc. R. Soc. London* **A274**, 21 (1963).

265. B. N. J. Persson and E. Tosatti, *J. Chem. Phys.* **112**, 2021 (2000).

266. B. N. J. Persson, *J. Chem. Phys.* **115**, 3840 (2001).

267. B. N. J. Persson and A. I. Volokitin, *Phys. Rev. B* **65**, 134106 (2002).

268. M. Klüppel and G. Heinrich, *Rubber Chem. Technol.* **73**, 578 (2000).

AUTHOR INDEX

Numbers in parentheses are reference numbers and indicate that the author's work is referred to although his name is not mentioned in the text. Numbers in *italic* show the pages on which the complete references are listed.

SUBJECT INDEX